Current-Driving of Loudspeakers

CURRENT-DRIVING OF LOUDSPEAKERS

Eliminating Major Distortion
and Interference Effects by the
Physically Correct Operation Method

Esa Meriläinen, M.Sc.

First Edition
2010

Copyright © 2010 Esa T. Meriläinen

Homepage: www.current-drive.info

ISBN: 1450544002
EAN-13: 9781450544009

Published in the U.S.A. through a print-on-demand service

Preface

The subtitle of the book may at first sound too much promising or inflated, which it, however, is not at all; for as will show up especially in chapter 4, the issue is indeed about a fundamental fallacy in sound reproduction technology.

We have namely been deceived – not so much with intent but by negligence or ignorance anyway – by providing to our use solely audio power equipment that disregards the realities of physics, and by establishing, as the backing of the practice, yet odd myths that do not stand up closer examination. Most regrettable in the present practice is that the manifoldly indefinite electromotive forces of the loudspeaker are freely allowed to mingle with the reproduced signal.

The book you have opened is the result of many years of investigation and work and probably the first volume ever to delve into loudspeaker current-drive. The start has been made with a clean slate, and the reader is not required to necessarily have prior knowledge in the field. The symbology used is as simple as possible, not referring to the general, difficultly configured symbology of the art.

The book is intended as an eye opener and guide into a new kind of technology as well as a source of new innovations and inspiration and clearer of common illusions for all who have any interest in sound reproduction implements or music. The content and way of representation are suitable for both the academic community and hobbyists as well as all interested; but above all for the equipment industry, the principles and ideas outlined afford a wealth of new opportunities.

A problem in entering the market may be, however, the baggage of history, that is, the justification of former products and the fact that a current-drive speaker is not suited to be used with a conventional voltage amplifier and a conventional speaker is not well suited for current-drive. Such cross-usage should be prevented by some means, and as the wayshowers would perhaps best be suited new actors that are capable of manufacturing both amplifiers and speakers.

The presented design and construction guidelines provide all DIY-minded with the opportunity to enter into experiencing in practice the dramatic sound quality improvement brought by the current-principle, without stepping into those pitfalls in which deficient knowledge easily leads and without needing to await the awakening of industrial manufacturers. At the same time, also the general pursuing of and knowledge on analogue electronics may gain new substance.

Besides actual current-drive information, novel ideas and methods e.g. for filter design, computer modelling, and measurements are also presented. The operation equations and modelling of loudspeakers are expounded in an understandable manner, and the fundamentals of linear systems are taught to those unskilled in them.

Existing faults are brought out at times even quite vigorously. However, the issues treated are universal, and nobody has reason to react to them personally although recognizing one's own misconceptions may sometimes produce inconvenience.

When it comes to the sound quality achievable by current-drive, it is probably pointless to describe it very much by words since all adjectives have already worn out in the assessment of conventional hifi gear down the ages. From my own experience, I can state, however, that although I have been able to listen to, e.g. in exhibitions, both expensive and extremely expensive voltage drive systems, I have never been satisfied with what I have heard, especially with any electro-dynamic speakers. Instead, the current-drive equipment I have used is like from a different world, and finally I have been able to be fully satisfied. The difference is so decisive that I rather listen to a current-drive system even in mono than a voltage drive system in stereo, and there is no turning back.

If you also agree that the design of loudspeaker operations should conform to the known laws of electrodynamics and other relevant sciences rather than to some old tradition or popular imagery, don't fail to do something about it. You may also consider posting a review on the store page where you shopped or elsewhere.

E. M.

Contents

Preface   V

Chapter 1   Some Parallels   1
  1.1   The Era of Direct Current   1
  1.2   Modulation Methods   2
  1.3   Deflection Coils   2

Chapter 2   Second-Order Systems   4
  2.1   Characteristic Frequency and Q Factor   5
  2.2   Response Types   9
  2.3   Resonance Networks   13

Chapter 3   Operation of the Electro-Dynamic Transducer   19
  3.1   Magnetic Force Effects   19
  3.2   Motional Equations under Current-Drive   23
  3.3   Effect of Motional EMF   27
  3.4   Inductive EMF   32
  3.5   Total Impedance   35
  3.6   Motional Equations under Voltage Drive   37
  3.7   The Origin of Sound   40
  3.8   Directivity and Horn Effect   46
  3.9   Power Consumption   49
  3.10 Microphone EMF   52

Chapter 4   The Consequences of Voltage Drive   55
  4.1   Circulation of the Electromotive Forces   56
  4.2   Microphone Feedback   57
  4.3   Outward Microphone Coupling   63
  4.4   Feedback of Mechanical Disturbances   67
  4.5   Phase Modulation   72
  4.6   Inductance Modulation   74
  4.7   Inductance Nonlinearity   78
  4.8   Resistance Variations   82
  4.9   Circuit Element Nonlinearities   84
  4.10 Drive at Low Frequencies   86

VIII

Chapter 5   The Principles of Current-Drive   89
  5.1   Thévenin and Norton Equivalents   89
  5.2   Evaluation of the Driving Mode   93
  5.3   Effect of Sensitivity Parameters   95
  5.4   Multiway Systems   97
  5.5   Suitability of Drive Units   102
  5.6   Microphones and Phono Cartridges   107
  5.7   The Secret of Tube Amplifiers   111

Chapter 6   Compensation of the Rising Response   113
  6.1   RCL Equalization   113
  6.2   Dual Coil Equalization   119
  6.3   Equalization Using Two Drivers   125

Chapter 7   Modelling and Simulation   127
  7.1   Low-Frequency Modelling   128
  7.2   Generalization of the Analogy   132
  7.3   Modelling of Voice Coil Inductance   135
  7.4   Dual Coil Drivers   138
  7.5   Two Drivers in the Same Space   139
  7.6   Filter Coils   141
  7.7   Operational Amplifiers   144

Chapter 8   Resonance Compensation   146
  8.1   Passive Equalization   146
  8.2   Voltage Divider Equalizer   148
  8.3   Pole Shifting   153
  8.4   The Linkwitz Equalizer   155
  8.5   Equalizer with Active Feedback   157
  8.6   Equalizer with RCL Feedback   161
  8.7   Double-Integrator Method   166
  8.8   Adjustable Equalizer with Fixed Poles   169
  8.9   Adjustable Equalizer with Tracking Poles   172

Chapter 9   Resonance Variations   175
  9.1   Current-Dependence   176
  9.2   Temperature Dependence   178
  9.3   "Burn-in"   179
  9.4   UV Sensitivity   181

Chapter 10   Amplifier Realizations   183
  10.1 The Series Resistor Method   183
  10.2 Current-Feedback   184
  10.3 Stability   189

10.4 Ground-Connecting the Load   190
10.5 Using One-Sided Power Supply   191
10.6 Bridge Connection   193
10.7 The Modified Howland Transconductor   194
10.8 A Do-It-Yourself Project   197
10.9 Using Headphones   206

Chapter 11   Speaker Realizations   **208**
11.1 Principles of Enclosing   208
11.2 Damping Materials   212
11.3 Usage of Electrolytic Capacitors   214
11.4 Project CS-12   217
11.5 Project CS-8   230
11.6 A Minimum-Phase Active Filter   237

Chapter 12   Protections   **242**
12.1 Amplifier Protection   242
12.2 Speaker Protection   243
12.3 Voice Coil Temperature Indication   248
12.4 Overexcursion Indication   249

Chapter 13   Measurement Techniques   **252**
13.1 Resonance Parameter Extraction   252
13.2 EMF Extractor   254
13.3 Capacitance Measurement   260
13.4 Inductance Measurement   262
13.5 Lossless Current Meter   264
13.6 Coil Loss Measurement   266

Chapter 14   Myths and Attitudes   **268**
14.1 The Whole Picture of Electrical Damping   268
14.2 Slew Rate Distortion   271
14.3 The Plague of Compression   273
14.4 What RMS Power?   277
14.5 The Real Meaning of Group Delay   278
14.6 Music Counts   280

Appendix A   Introduction to Complex Numbers   **285**

Appendix B   Properties of Linear Systems   **290**
B1   Definitions   291
B2   Differentiators and Integrators   294
B3   Time Domain Representation   298
B4   Frequency Response   304

B5   Transfer Functions   308
B6   Poles and Zeroes   311
B7   Minimum-Phase Systems   315
B8   Phase Linearity   318

Appendix C   Frequency Content of Signals   **320**
C1   Periodic Signals   320
C2   One-Time Signals   323
C3   Continuous-Nature Signals   326
C4   The Relationship between Time and Frequency Domains   330
C5   About Test Signals   333

Appendix D   Solving HDEs   338

Appendix E   Decibel Scale   339

Appendix F   Amplifier Artwork   340

Appendix G   Power Supply Artwork   341

Appendix H   EMF Extractor Artwork   342

1
SOME PARALLELS

1.1 *The Era of Direct Current*

The issue whether loudspeakers should be excited by a voltage or current signal is quite well comparable to a dispute that took place over a century ago, concerning whether the production and distribution of electricity should operate on direct or alternating current.

Thomas A. Edison had opened, in New York, the world's first power generating plant, that supplied a DC voltage of 110 volts for an area of a few square kilometres in Manhattan. Another pioneer of electrical technology, Croatian-born Nikola Tesla, instead, believed strongly in the superiority of the three-phase AC system he had developed. In 1886, George Westinghouse founded an electric company to utilize the inventions and patents of Tesla and to compete with Edison.

Edison was not at all pleased seeing a rivaling system threatening his dominating stature in power production. The conflict caused a breach between Edison and Tesla, and a public struggle about which system would become prevalent. Edison even resorted to a trick campaign in his attempt to defame AC power, that he thought was dangerous.

The technical superiority of AC became, however, soon apparent to the public, and the power plant built at Niagara Falls in 1895 denoted a breakthrough for AC technology although DC systems were still used in cities for a couple of decades.

This story teaches how even top-talented persons may, blinded by their own human limitations, strive to advocate technical solutions that are ineffective and anything but optimal in fulfilling their purpose.

How would things have turned out without the innovations brought by Nikola Tesla? Would somebody else have filled his place and turned the unfavorable direction of development? Or, would it be that even this day our wall sockets would supply direct current and conversion of one voltage for another would only be possible by chopper techniques?

Reason thus prevailed in power technology, and this is also possible to happen in audio technology, for moving to current-drive does not even require new investments from the industry, only a little reformative thinking.

1.2 Modulation Methods

The difference between loudspeaker driving methods is, by its quality and significance, also comparable to the difference of the methods used for modulating radio waves.

In amplitude modulation (AM), that has been in use since the '20s, the amplitude of the carrier wave is modified in accordance with the transmitted signal. Unfortunately though, many interfering factors along the way also tend to modulate the same amplitude, so the sound quality of the received transmission is quite poorly controlled.

Instead, in frequency modulation (FM), developed in 1928 by Edwin H. Armstrong, the transmitted signal is used to modify the carrier frequency, which is much less affected by climatic and technical disturbances than is the amplitude. The message is thus recovered with better quality since, in the transmission chain, there is not used such quantities, or conversions between quantities, that are subject to uncontrolled factors.

Correspondingly, by operating a speaker driver directly by a current signal, one can avoid the interference mechanisms pertaining to the relationship between voltage and current. After experience, it is not so overstated to assert that the difference in sound quality between current- and voltage-driven speakers is of the same order than the quality difference between AM and FM broadcasts in normal receiving conditions (ignoring the stereophony of FM broadcasts and the present, very lamentable practice to compress transmissions into unnatural growling).

1.3 Deflection Coils

A tangible example of the improvement brought by current-drive is found in TV technology.

In the past, it was customary to drive the vertical deflection coils of picture tubes by voltage, in a like manner that loudspeakers are still driven today. As a consequence, when the coils warmed up and their resistance increased, the amplitude of the deflection current altered, causing changes in picture size. Even thermistors had to be employed when trying to compensate these temperature effects.

Later on, these coils were learned to be operated directly by current, so that the changes in load impedance were no more able to affect the strength of the deflection field, and the picture kept more stable.

Thermal compression is also a familiar phenomenon in loudspeaker technology. In voltage-driven speakers, the variations in sound level and frequency response, caused by voice coil heating, are a significant problem, especially in high-power systems. One might expect this alone to be a sufficient reason to try out the possibilities offered by current-drive; but, for some reason, the required change in the mindset has not yet at all reached the designers of the audio field.

2

SECOND-ORDER SYSTEMS

Before we can truly set about examining the behavior of the electro-dynamic speaker driver in various situations, we need certain theoretical background and insight into 2nd-order linear systems. A driver in a closed cabinet can namely be modelled, in the region of the lower corner frequency, quite accurately by 2nd-order transfer functions. This also applies to the acoustic as well as electrical and mechanical behavior of the speaker. Therefore, it is natural to also implement bass response correction networks in a corresponding manner. In practice, second-order systems form a foundation that can also be used in the implementation of higher-order filters, when necessary.

Perfect comprehension of the equations presented is not a requisite for the assimilation of the factual content itself, so nobody needs to give up acquaintance with current-drive for the reason that the mathematics used seems unfamiliar. For readers to whom transfer functions are not familiar, but who are interested to internalize the using of these practical representations, appendix B provides down-to-earth introduction to the subject. As a foundation, one then mostly needs a general understanding of differentiation and integration operations. Studying the appendices is also otherwise very recommendable since the necessary basic concepts and the background information related to them is presented there.

Concerning the behavior of first-order (one-pole) filter functions, there is sufficiently information available from other sources, so these systems are not discussed here separately.

2.1 Characteristic Frequency and Q Factor

A transfer function describing a second-order system (so-called biquad function) can be written as a ratio of real-coefficient polynomials; generally:

$$H(s) = \frac{a_2 s^2 + a_1 s + a_0}{s^2 + b_1 s + b_0} \tag{2.1}$$

The system is stable when b_1 and b_0 are positive. The numerator coefficients a_2, a_1, and a_0 may instead be also negative, and often only one of them differs from zero. The degree of the denominator and the number of poles is consequently always 2, but the degree of the numerator is either 0, 1, or 2.

When the numerator is a constant (degree 0), the system is of low-pass type, and the frequency response is found by substituting $s = j\omega$:

$$H(\omega) = \frac{a_0}{(j\omega)^2 + b_1 j\omega + b_0} \tag{2.2}$$

When frequency ω is low, $H(\omega) \approx a_0/b_0$, the gain being then a frequency-independent constant. At high frequencies, in turn, $H(\omega) \approx a_0/(j\omega)^2 = -a_0/\omega^2$, so the gain is inversely proportional to the square of frequency. These two asymptotes of gain intersect when $|a_0/b_0| = |{-a_0/\omega^2}|$, that is, when $\omega = \sqrt{b_0}$. At this frequency, the denominator terms $(j\omega)^2$ and b_0 cancel each other establishing a resonance, whose strength is determined by coefficient b_1.

The gain at this characteristic frequency (designated ω_0) is therefore

$$|H(\omega_0)| = \frac{|a_0|}{b_1 \sqrt{b_0}} \tag{2.3}$$

which is, depending on b_1, greater or smaller than the value represented by the mentioned asymptotes, $|a_0|/b_0$.

The ratio of these two gains is termed *Q factor*, or simply Q, i.e.:

$$Q = \frac{|a_0|}{b_1 \sqrt{b_0}} \bigg/ \frac{|a_0|}{b_0} = \frac{\sqrt{b_0}}{b_1} \tag{2.4}$$

In the low-pass case, the Q thus indicates the gain at the characteristic frequency with respect to frequency 0.

The concept Q denotes so-called quality factor, that originally has been used to characterize the ideality of coils, that is, the ratio of their inductive impedance (reactance) to resistance, $\omega L/R$, in tuned circuits. When relating to coils, the Q value should thus be as high as possible. However, the concept of quality factor is also used in a more general sense to indicate the strength or sharpness of resonance in which case a high Q does not necessarily signify any "quality", the target value being instead always case-dependent.

A second-order low-pass function can now be expressed using the characteristic frequency ω_0 and Q factor as follows:

$$H(s) = \frac{a_0}{s^2 + \dfrac{\omega_0}{Q} s + \omega_0^2} \tag{2.5}$$

The shape of the frequency response is therefore fully determined by two parameters, ω_0 and Q, while a_0 only affects the scaling of amplitude. Equation (2.5) represents a general and practical way to describe 2nd-order transfer functions.

In literature, so-called damping ratio, ζ, is often used instead of Q. ζ is inversely proportional to Q so that $\zeta = 1/(2Q)$. ω_0 is sometimes replaced by its reciprocal, that is termed time constant.

Figure 2.1 shows the frequency response (Bode plot) of a 2nd-order low-pass system with varying Q. The Figure has been scaled for general purpose by choosing $\omega_0 = 1$ and $a_0 = \omega_0^2$. Thereby, the gain approaches unity (0 dB) at low frequencies, and at $\omega = 1$ the gain takes the value of Q. (on a dB scale, $20 \cdot \log_{10} Q$). As frequency increases further, the gain curves approach an asymptote indicated by the dashed line, that has a slope of -40 dB per decade (2nd-degree slope).

When Q is greater than $1/\sqrt{2}$ (≈ 0.707), the gain curve exhibits a peak that reaches higher than the low-frequency gain. This peak is not exactly at $\omega = \omega_0$ but at somewhat lower frequency, which is

$$\omega_{peak} = \omega_0 \sqrt{1 - \frac{1}{2Q^2}} \quad , \quad Q > \frac{1}{\sqrt{2}} \tag{2.6}$$

At high values of Q, the peak becomes sharp, ω_{peak} and ω_0 being then virtually equal.

When Q is $1/\sqrt{2}$, a condition is reached where the gain at low frequencies is as flat as possible and peaking no more occurs. In this case the filter is of Butterworth type and has its cut-off frequency (3 dB atte-

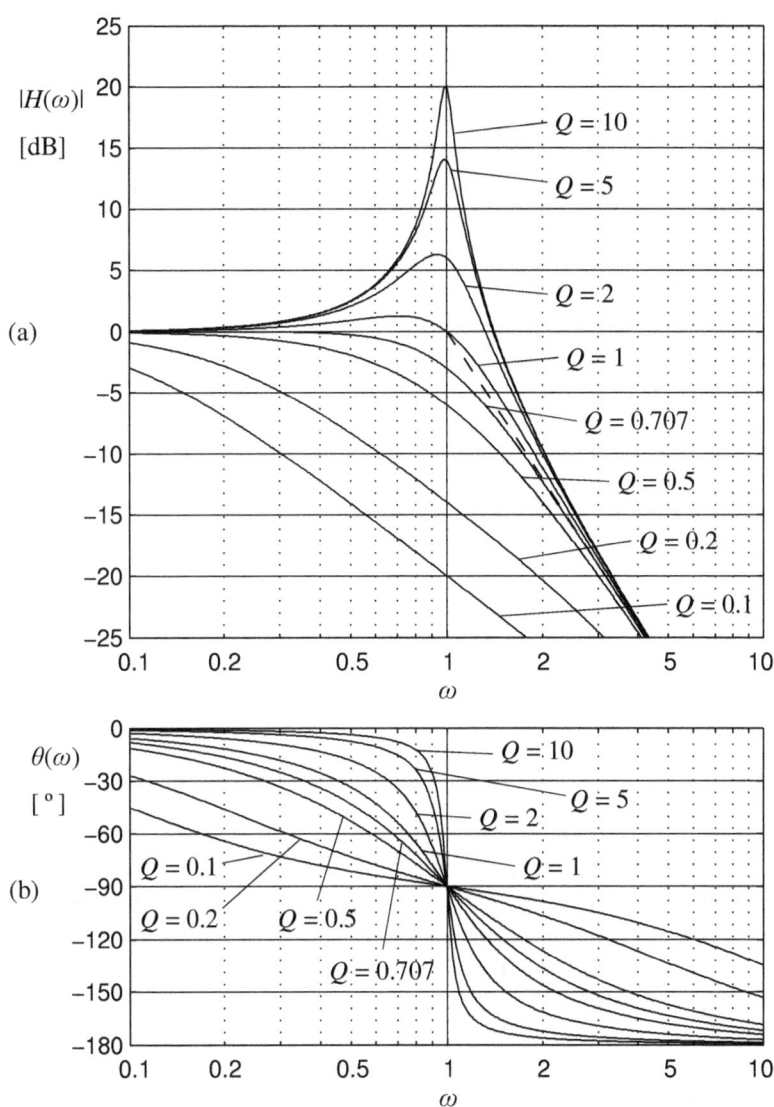

Figure 2.1. a) Amplitude response of second-order low-pass system, with Q as a parameter. Frequency axis has been normalized so that $\omega_0 = 1$. $|H(\omega)|$ axis has been normalized by letting $a_0 = \omega_0^2$ in expression (2.5). Dashed line indicates the asymptote that $|H(\omega)|$ approaches at high frequencies. b) Phase shift curves corresponding to the amplitude curves of Fig. 'a'.

nuation) at $\omega = \omega_0$. Butterworth responses are preferred in many audio applications just because of their flatness.

When Q decreases enough, the curve straightens around ω_0 into 1st-degree slope, achieving 2nd-degree slope only near the aforementioned asymptote. Actual resonance cannot then be recognized any more.

The phase shift $\theta(\omega)$ starts from zero at low frequencies, reaches $-90°$ at the characteristic frequency, and ends up to $-180°$ at high frequencies, regardless of the Q value. When $|H(\omega)|$ reaches the 2nd-degree slope of the asymptote, the output has thus turned antiphase with respect to the input (assuming $a_0 > 0$). The rate of change of phase near the characteristic frequency is proportional to the Q value. The curves are symmetric, so that $\theta(\omega)$ differs from zero as much as $\theta(1/\omega)$ differs from $-180°$.

Transfer function (2.5) may be written in the factored form:

$$H(s) = \frac{a_0}{(s - p_1)(s - p_2)} \qquad (2.7)$$

where p_1 and p_2 are the system poles. By opening the brackets we have:

$$H(s) = \frac{a_0}{s^2 - (p_1 + p_2)s + p_1 p_2} \qquad (2.8)$$

Comparing this with expression (2.5), we can reason that $p_1 p_2 = \omega_0^2$, so:

$$\omega_0 = \sqrt{p_1 p_2} \qquad (2.9)$$

In practice, the poles p_1 and p_2 can be either negative real numbers or a complex conjugate pair with negative real part. In the case of complex conjugates, we obtain from equation (2.9) $\omega_0 = |p_1| = |p_2|$, so the distance of the poles from the origin equals their characteristic frequency.

The position of the poles can be found by equating the denominator of function (2.5) to zero. By applying the general solution formula of second-order equations, we obtain as the result:

$$p_1, p_2 = \omega_0 \left(-\frac{1}{2Q} \pm \sqrt{\frac{1}{4Q^2} - 1} \right) \qquad (2.10)$$

The poles are real-valued when the expression underneath the root sign is non-negative, that is, when $Q \leq 1/2$. When Q is greater than $1/2$, the poles are located in the complex plane on the perimeter of an ω_0-radius circle, at

$$p_1, p_2 = \omega_0 \left(-\frac{1}{2Q} \pm j\sqrt{1 - \frac{1}{4Q^2}} \right), \quad Q > \frac{1}{2} \qquad (2.11)$$

This case is depicted in Fig. 2.2a. As Q increases, the poles move closer to the imaginary axis. In the case of $Q = 1/2$ (Fig. 2.2b), the poles are equal, corresponding to a condition in which two identically operating 1st-order systems have been connected in cascade. When Q is smaller than 1/2 (Fig. 2.2c), the poles have different frequencies, manifesting in the amplitude response as two distinct corner points when the poles are far enough from each other.

A transfer function containing only real poles and zeroes can always be partitioned into a product of 1st-order transfer functions, but with complex roots this cannot be done.

2.2 Response Types

Above, we considered the low-pass system, for which it is characteristic that of the numerator coefficients, in the transfer function (2.1), only a_0 differs from zero. Within all other response types, also the numerator is frequency-dependent, and besides the poles, there are also zeroes involved in the operation.

If, in the expression (2.1), $a_2 = a_0 = 0$, the system is of band-pass

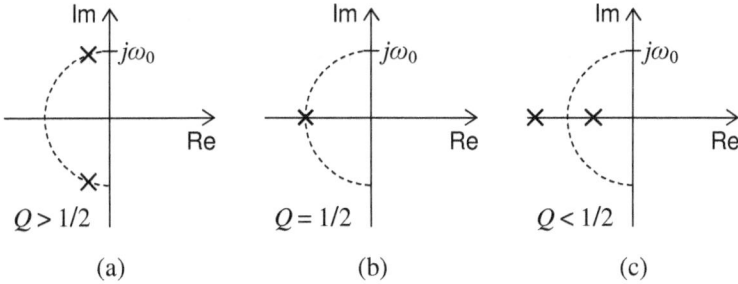

Figure 2.2. Pole-zero diagram of second-order low-pass system at three different Q values. In Fig. 'a', $Q > 1/2$, the poles being complex conjugates at distance ω_0 from the origin. In Fig. b, $Q = 1/2$, the poles being coincident at $-\omega_0$. In Fig. c, $Q < 1/2$, the poles lying on the negative real axis on both sides of $-\omega_0$.

type, that can be thought to develop so that in series with the above-described low-pass system is connected an ideal differentiator, whose transfer function is s (multiplied by some constant).

The band-pass transfer function can now be expressed in the manner of equation (2.5):

$$H(s) = \frac{a_1 s}{s^2 + \dfrac{\omega_0}{Q} s + \omega_0^2} \tag{2.12}$$

The shape of the response is still determined according to the characteristic frequency and Q, but the differentiation operation (multiplication by $j\omega$) increases the slope of $|H(\omega)|$ by 20 dB per decade, for all frequencies.

Curve B, in Fig. 2.3a, depicts the amplitude response of a band-pass system, while curve A represents the corresponding low-pass response, having the same characteristic frequency and Q. The former has been scaled so that, in the function (2.12), $a_1 = 1$, thus making the gain at frequency 1 ($= \omega_0$) equal to Q, like in the low-pass case. In the band-pass case, the amplitude peak occurs exactly at ω_0. At low and high frequencies, the gain approaches the asymptotes, indicated by dashed lines, that exhibit 1st-degree slope. The peak value is always Q-fold relative to the intersection point of the asymptotes.

Corresponding phase responses are shown in Fig. 2.3b. The curves are identical in shape, but, due to the differentiation operation, the band-pass phase (B) is, for all frequencies, 90° ahead of the corresponding low-pass phase (A), and reaches zero at ω_0.

A pivotal property relating to band-pass systems is the bandwidth, which usually refers to the span of the frequency range within which the gain is at least $1/\sqrt{2}$ times its maximum value. This bandwidth is called the half-power or 3-dB bandwidth (more precisely -3.01 dB).

Equation (2.12) gives as the amplitude response

$$|H(\omega)| = \frac{|a_1|\omega}{\sqrt{(\omega_0^2 - \omega^2)^2 + (\omega_0\omega/Q)^2}} \tag{2.13}$$

The 3-dB frequencies (designated ω_1 and ω_2) can best be solved by choosing $a_1 = \omega_0/Q$ which yields $|H(\omega_0)| = 1$, and then finding the frequencies at which $|H(\omega)|^2 = 1/2$. One can see that this condition is satisfied when $(\omega_0^2 - \omega^2)^2 = (\omega_0\omega/Q)^2$. At the same time, the real and imaginary parts of the frequency response function denominator become equal in magnitude, corresponding to a 45° phase change with respect to

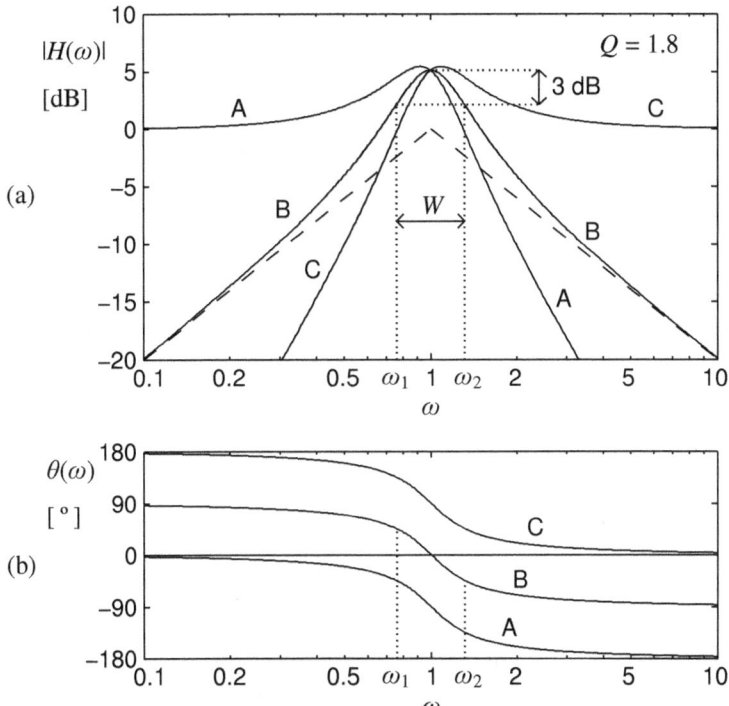

Figure 2.3. a) Amplitude responses of second-order low-pass (A), band-pass (B), and high-pass (C) systems, ω_0 and Q being same for all. The graphs have been scaled so that $\omega_0 = 1$ and $|H(\omega_0)| = Q$. The band-pass response has been supplemented with the 3-dB boundary frequencies ω_1 and ω_2 and bandwidth W. b) Phase shift curves corresponding to Fig. 'a'. At frequencies ω_1 and ω_2, the band-pass phase (B) equals ±45°.

the center frequency ω_0.

By solving the mentioned equation, one obtains as the boundary frequencies (considering only positive solutions)

$$\omega_1, \omega_2 = \omega_0 \left(\sqrt{\frac{1}{4Q^2} + 1} \pm \frac{1}{2Q} \right) \tag{2.14}$$

From this, one can easily find that $\omega_1 \omega_2 = \omega_0^2$, so ω_0 is the geometric mean of the 3-dB boundary frequencies, as can be expected due to the symmetry of the graph.

The bandwidth becomes now

$$W = \omega_2 - \omega_1 = \frac{\omega_0}{Q} \tag{2.15}$$

So, the 3-dB bandwidth and Q are inversely proportional to each other. Result (2.15) can be used e.g. in the evaluation of the Q value.

Graphs C, in Fig. 2.3, depict the response of a 2nd-order high-pass system, ω_0 and Q being still the same as before. A high-pass system can be thought to develop so that in series with a band-pass system is added an ideal differentiator, that increases the slope of the amplitude response by 20 dB per decade, for all frequencies. Correspondingly, one can get from the high-pass system back to the band-pass system, and further to the low-pass system, through an ideal integrator.

The denominator of the transfer function also remains the same in the high-pass case, but the numerator consists of a 2nd-degree term. The high-pass transfer function is thus generally

$$H(s) = \frac{a_2 s^2}{s^2 + \dfrac{\omega_0}{Q} + \omega_0^2} \tag{2.16}$$

By setting $a_2 = 1$, the high-frequency gain equals 0 dB $((j\omega)^2/(j\omega)^2)$, as in Fig. 2.3a. The low-pass response (A) and high-pass response (B) are then mirror images of each other about the line $\omega = \omega_0$. By virtue of this symmetry, the characteristic curves of e.g. Fig. 2.1a may also be used to evaluate high-pass systems, by changing ω into $1/\omega$.

The phase of a high-pass system is 90° ahead of the corresponding band-pass phase (Fig. 2.3b), which in turn is 90° ahead of the low-pass phase. In regions where the amplitude response is constant, the phase response approaches zero, as is natural to minimum-phase systems.

From the above, we can see that

The shape of the frequency response of 2nd-order low-pass, band-pass, and high-pass systems is completely determined by two parameters: ω_0 and Q.

The same also holds for time domain responses since time behavior is always directly related to frequency behavior.

Many kinds of other response types can also be described by the general biquad transfer function (2.1), by letting the zeroes of the system lie also elsewhere than at the origin. By zeroes, one can establish upward-leading turning points in the amplitude response, similarly as the poles produce downward-leading turning points. The numerator of the transfer function then incorporates two or all three terms, instead of one. Stability does not impose restrictions with respect to the zeroes, so all the coefficients of the numerator polynomial can be of either sign.

In systems dealt with in practice, the 2nd-degree term and constant term in the numerator are usually of the same sign, meaning that the zeroes are located on the same side of the imaginary axis. Then, the numerator can be expressed in terms of a characteristic frequency and Q, just the same way as the denominator, and the transfer function may be written as

$$H(s) = K\frac{s^2 + \dfrac{\omega_{0Z}}{Q_Z}s + \omega_{0Z}^2}{s^2 + \dfrac{\omega_{0P}}{Q_P}s + \omega_{0P}^2} \qquad (2.17)$$

where subscript Z refers to the zeroes, or the numerator; and subscript P to the poles, or the denominator, K being the gain constant. Q_P must always be positive, but Q_Z may also be negative, in which case the system is non-minimum-phase.

By altering the numerator's and denominator's parameters and their relative proportions in function (2.17), one can establish different types of frequency responses, the most important of which are presented in Table 1, together with the basic types discussed earlier. The Table also includes the locations of the poles and zeroes for each case and the necessary conditions to realize the response.

For our purposes, of particular interest of the examples described by equation (2.17) is the "low-boost", which can be used to straighten out the fundamental resonance of a speaker driver and to shape the bass response.

2.3 Resonance Networks

The concept of resonance relates closely to 2nd-order systems since in these transfer functions both the numerator and denominator are able to produce a resonance phenomenon, as the 2nd- and zero-degree terms cancel out each other at a certain frequency. Mechanical resonances are thus possible to be modelled by electrical equivalent circuits since both are based on similar mathematical description. Electrical resonance networks can be used, among other things, to model loudspeaker impedance and the effects of enclosing and, physically, as parts of crossover fil-

Table 1. Second-order filter types

Function	Bode plot	Transfer function	Conditions	Pole-zero d.
Low-pass	$\lvert H\rvert$ $(Q>1)$	$\dfrac{K}{s^2+\dfrac{\omega_0}{Q}s+\omega_0^2}$	–	
Band-pass	$\lvert H\rvert$ $(Q>1)$	$K\dfrac{s}{s^2+\dfrac{\omega_0}{Q}s+\omega_0^2}$	–	
High-pass	$\lvert H\rvert$ $(Q>1)$	$K\dfrac{s^2}{s^2+\dfrac{\omega_0}{Q}s+\omega_0^2}$	–	
All-pass	$\lvert H\rvert$	$K\dfrac{s^2+\dfrac{\omega_{0Z}}{Q_Z}s+\omega_{0Z}^2}{s^2+\dfrac{\omega_{0P}}{Q_P}s+\omega_{0P}^2}$	$\omega_{0P}=\omega_{0Z}$ $Q_Z=-Q_P$	
Band-stop	$\lvert H\rvert$	$-,,-$	$\omega_{0P}=\omega_{0Z}$ $\lvert Q_Z\rvert\to\infty$	
Band-boost	$\lvert H\rvert$	$-,,-$	$\omega_{0P}=\omega_{0Z}$ $Q_P>\lvert Q_Z\rvert$	
Low-boost	$\lvert H\rvert$ $(Q_P>1)$ $(Q_Z>1)$	$-,,-$	$\omega_{0P}<\omega_{0Z}$	
High-boost	$\lvert H\rvert$ $(Q_Z>1)$ $(Q_P>1)$	$-,,-$	$\omega_{0P}>\omega_{0Z}$	

ters.

The impedance of a passive resonance network made up of a coil, a capacitor, and a resistor is expressible as an impedance function similar to 2nd-order transfer functions. These impedance functions differ, however, from general transfer functions in the regard that the magnitude of the direction angle cannot exceed 90°, that is, the real part (resistance) cannot become negative. (With active circuits, it is possible to produce negative resistances too, but even with them, the real part of the impedance can never change its sign.) Correspondingly, the frequency dependence of the impedance magnitude outside the resonance region can be at most of 1st degree.

The schemes of passive resonance networks are shown in Fig. 2.4.

The symmetric parallel resonance network of Fig. 'a' has also been considered in appendix B, where expression (b25) has been derived as the transfer function from current to voltage. Thus, the impedance of network 'a' is

$$Z(\omega) = \frac{(1/C)j\omega}{(j\omega)^2 + (1/RC)j\omega + 1/LC} \qquad (2.18)$$

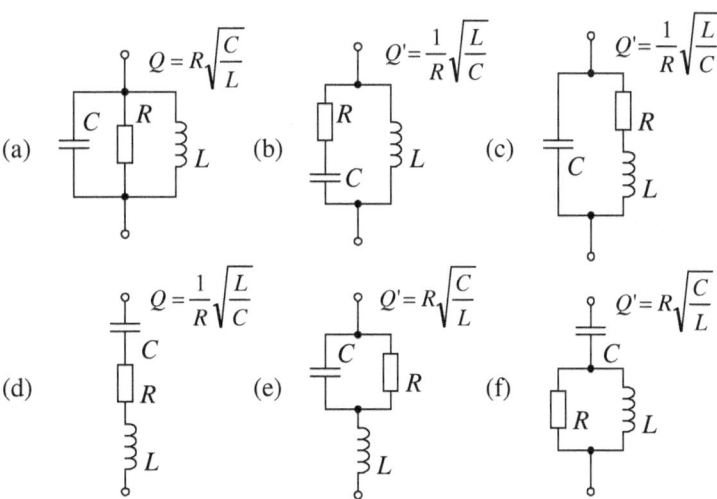

Figure 2.4. Passive resonance networks. a) Symmetric parallel resonator. b) & c) Asymmetric parallel resonators. d) Symmetric series resonator. e) & f) Asymmetric series resonators. Figures b, c, e, and f provide a computational quality factor Q', that is applicable when the resonance is sufficiently sharp.

This is a band-pass type function, from which one can recognize, in a familiar way, the characteristic frequency (resonant frequency):

$$\omega_0 = 1/\sqrt{LC}$$

or

$$f_0 = \frac{1}{2\pi\sqrt{LC}} \tag{2.19}$$

The same result can also be achieved by equating the absolute values of the coil and capacitor impedances.

According to equation (2.4), we obtain as the Q factor of network 'a'

$$Q = R\sqrt{\frac{C}{L}} \tag{2.20}$$

The Q is thus directly proportional to resistance R, which is also the maximum value of the impedance, $Z(\omega_0)$, as can be concluded from equation (2.18).

The essential behavior of the magnitude of impedance (2.18) is depicted in Fig. 2.5 (solid line). At low frequencies, $|Z|$ is almost solely made up of the coil impedance, and at high frequencies, correspondingly, of the capacitor impedance. At f_0, the currents in the coil and capacitor are equal but in opposite phase, cancelling out each other, whereupon we are left only with the resistance's effect.

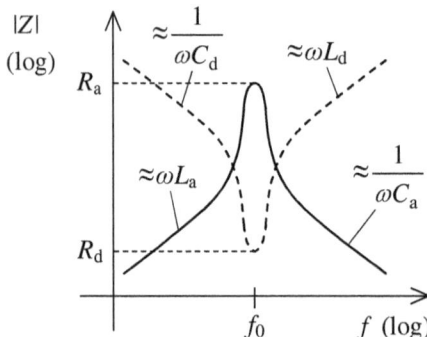

Figure 2.5.   Impedance magnitude of the parallel resonance network of Fig. 2.4a as a function of frequency (solid line), and the same for the series resonance network of Fig. 2.4b (dashed line). Subscript 'a' refers to the elements of the parallel network and 'd' to those of the serial network. Both axes are logarithmic.

By moving the resistor into series with the capacitor or coil, one can bring about asymmetric parallel resonators (Fig. 2.4b-c) whose impedance shape is otherwise similar to that of case 'a' but approaches value R at high or low frequencies. The expression for the resonant frequency, (2.19), also applies to these networks, but the impedance peak is not exactly at f_0. Because of the asymmetry, the normal definition of Q can be applied to these only approximately and only when the resonance clearly exists (R sufficiently low).

The impedance of the series resonance network of Fig. 2.4d is the sum of the element impedances, so in this case

$$Z(\omega) = j\omega L + R + \frac{1}{j\omega C}$$

$$= \frac{(j\omega)^2 + (R/L)j\omega + 1/LC}{(1/L)j\omega} \tag{2.21}$$

(In impedance functions, the degree difference between the numerator and denominator can be either 1, 0, or −1.) The numerator is now in a form from which one can see the resonant frequency being still expressed by formula (2.19). The Q of the zeroes is now the reciprocal of expression (2.20), as attached to Fig. 2.4d.

The impedance of the series connection, (2.21), is inverse in shape with respect to the parallel connection impedance, (2.18), so there develops a minimum at the resonant frequency, as sketched in Fig. 2.5 (dashed line). At low frequencies, the dominating impedance is now due to the capacitor, and at high frequencies, correspondingly, due to the coil. At f_0, these impedances again cancel out each other, and merely the resistance is left.

Figures 2.4e-f represent asymmetric series resonators, whose impedance otherwise resembles case d but approaches at low or high frequencies value R. For f_0 and Q, the same remarks hold as in cases b and c, but in networks e and f the resonance sharpens with increasing R.

Figure 2.6a shows the locus of the parallel connection impedance (2.18) in the complex plane as a function of frequency (Nyquist plot). The locus forms a circle as is explained in appendix section B4. The Figure also depicts the resonance impedance $Z(\omega_0)$, which is real, and the impedances at the 3-dB boundary frequencies (2.14), marked $Z(\omega_1)$ and $Z(\omega_2)$ ($\omega_2 > \omega_1$). These impedances have an angle of ±45°, relative to the real axis, and their absolute value is $Z(\omega_0)/\sqrt{2}$. If the coil or the capacitor were removed from the circuit, the impedance locus would form a semicircle.

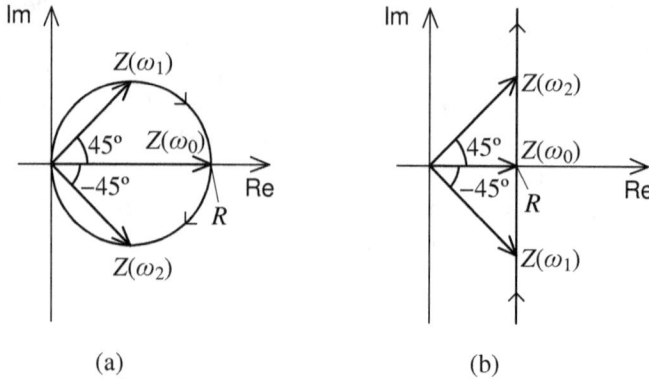

Figure 2.6.   a) Nyquist plot of the impedance of the resonance network of Fig. 2.4a. Points $Z(\omega_1)$ and $Z(\omega_2)$, representing the 3-dB boundary frequencies, lie at the highest and lowest points of the circular locus. b) Nyquist plot of the impedance of the resonance network of Fig. 2.4d. The 3-dB boundary frequencies ω_1 and ω_2 have been defined correspondingly as for the band-pass function of Fig. 'a'.

Figure 2.6b shows the Nyquist plot of the series connection imped-ance (2.21). The locus is a vertical line, along which the point moves upward as ω increases. The equations relating to bandwidth, (2.14) and (2.15), are also applicable to this case, but now $Z(\omega_1)$ and $Z(\omega_2)$ have an absolute value of $\sqrt{2} \cdot Z(\omega_0)$. If the coil or the capacitor were short-cir-cuited, the impedance locus would be a half-line, starting from point R.

3

OPERATION OF THE ELECTRO-DYNAMIC TRANSDUCER

Overwhelmingly the most popular means of converting electrical signals to sound is still the electro-dynamic principle, wherein movement of the vibrating diaphragm is accomplished by interaction between current and a magnetic field. Taking into account the enormous amount of this kind of speaker drivers, all over the world, and their significance in our daily life as producers and modifiers of sound, there is a need to have a little deeper than usual overview into the physics of operation of these basic necessities. Thereby we acquire requisites for understanding the interference mechanisms acting in a driver and are able to dispel some erroneous notions, that one often sees presented.

Here, we consider moving-coil drive units whose diaphragm is structurally rigid. The electro-dynamic principle is also used in less common, mostly for high frequencies intended ribbon and planar-diaphragm devices, but the properties of these differ considerably from the former.

3.1 Magnetic Force Effects

Figure 3.1 shows magnet assemblies commonly used today in speaker drive units. The magnet itself (stamped by grey) is usually of ferrite material and ring-shaped. To this is glued fast steel pole pieces, that direct the magnetic flux to flow through the air gap as effectively as possible. In quality drivers, a T-shaped center pole is sometimes employed (Fig. b), providing more symmetric flux distribution than in the Fig. 'a' case.

By using more effective and more expensive materials than ferrite, the magnet itself can be made smaller, enabling it to be positioned in the

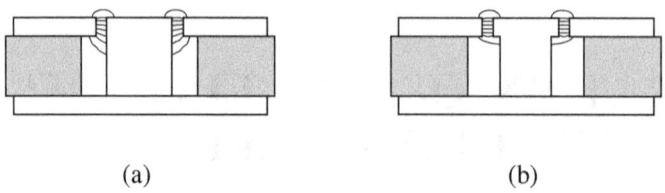

(a)   (b)

Figure 3.1.   Cross-sections of common magnet assemblies. a) Typical shape of pole pieces in inexpensive ferrite magnet drive units. Magnetic flux in the air gap is not symmetric. b) Better solution, that employs a T-shaped pole piece.

middle of the assembly. The stray field is thereby lesser, and the flux doesn't have to go through a hole in the magnet; a thing which it naturally won't do. The mechanisms pertaining to the operation itself are nevertheless the same, irrespective of the structural details.

The drive force F acting upon the voice coil in the air gap is obtained from the well-known basic formula:

$$F = Bli \qquad (3.1)$$

B is the magnetic flux density (in Teslas) acting perpendicular to the wire, l is the length of wire contained within the magnetic field, and i is the current in the wire. B is the flux density that exists when the current is zero. The current always causes its own magnetic field, that may react with adjacent iron, but the effect does not relate to equation (3.1). (So-called flux modulation is thus not evil per se but a necessary part of the force generation.)

The constants B and l generally always appear together, and their product is referred to as the force factor, whose unit becomes N/A or Tm.

When examining equation (3.1), the attention should focus on a very important fact regarding control of the voice coil: namely, the voltage between the ends of the wire does not appear in this equation at all. To this, then, pertains our first cause of wonderment:

The drive force actuating the voice coil is determined by the current flowing in the coil. Instead, the voltage across the coil does not even affect straightly the development of the force. Even for this reason, it is ill justified and downright amazing that everywhere it is taken as self-evident that an amplifier has to feed loudspeaker terminals by a voltage signal without

any respect to current.

If the driver impedance were a mere pure, constant-staying resistance, then, and only then, would it be all the same whether the driver were fed by voltage or current since these two would at every moment be directly proportional to each other. If the impedance were even linear and interference-free, then still voltage drive would defend its status since the differences between these two driving modes would be mostly related to frequency response shaping. In reality, however, the driver impedance is, as will be demonstrated later, everything but interference-free and linear, and hence, only by acting directly on current, can it be guaranteed that the force actuating the voice coil corresponds to the driving signal as accurately as possible.

Even without special knowledge on the behavior of impedance, the discrepancy between equation (3.1) and contemporary audio amplifier technology should ring some alarm bell within the inward parts of every intellectually honest individual interested in electrical engineering. Or, at least, this shortcoming should make pertinent people seriously call into question those inducements by which the present, physical rationales ignoring practice has gained its justification.

The analogue signal obtained from program sources, like e.g. CD players, is always a voltage signal, and there is nothing inappropriate in it. Signal processing in the front stages of amplifiers is also practical to be performed in voltage mode since using current signals in this context wouldn't provide any advantage. Hence:

In order to put the voice coil into motion, *somewhere* in the signal path must occur a conversion from a voltage signal to a current signal. The contemporary practice is exclusively that this pivotal conversion is left solely as the responsibility of the speaker driver. In the driver, however, the conversion always occurs *uncontrolled*, various electromotive forces being heavily involved in the development of impedance. Instead, when performed in the amplifier, this conversion can be carried out *controlled*, providing the speaker with direct drive, instead of indirect one.

Irrespective of where the deepest reasons for the continuation of the current unnatural tradition might be, to one who has somewhere listened an ordinary hifi set of say 10 000 currency units and compared this with a do-it-yourself set utilizing current-drive and costing a small fraction of the former, the difference in favor of the latter is so unambiguous that the universal consensus with which current-drive technology has been kept out of the reach of customers, can't be without denoting a continuous source of astonishment.

In addition to the force described above, the voice coil is also influenced by an unwanted magnetic force effect, stemming from the fact that the field induced by the voice coil current attracts iron contained in the magnetic circuit. The phenomenon can be called solenoid force.

Figure 3.2 illustrates the flow of magnetic flux in the "motor structure" of a loudspeaker. Fig. 'a' depicts the flux caused by the permanent magnet alone. Fig. b, in turn, depicts the flux generated by the current-conducting voice coil, without the influence of the magnet. A part of this flux circulates through the entire magnetic circuit, giving rise to a current-dependent component in the circuit's total flux.

Flux lines always have a tendency to travel the straightest route and as effectively as possible, so, in this case, the flux touring above the coil attempts to press the coil into the magnet assembly.

Thus, regardless of the current's direction, the solenoid force always tries to pull the voice coil backwards. The voice coil flux, however, has

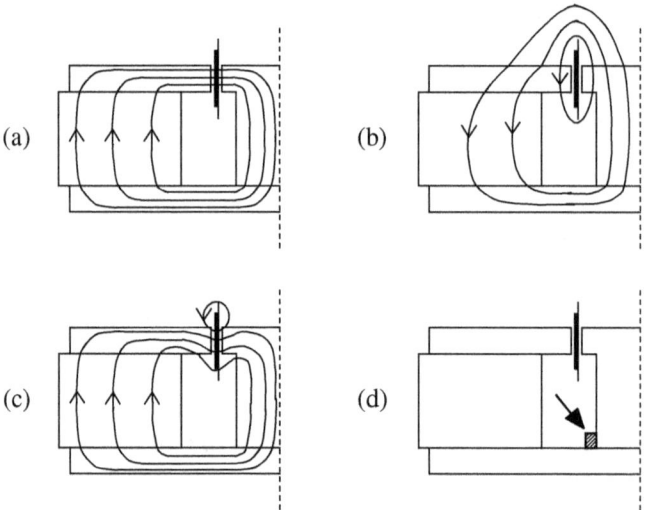

Figure 3.2.   a) Passage of flux lines in a magnetic circuit without current in the voice coil. b) Magnetic flux caused by voice coil current without the field of the permanent magnet. In this case, flux due to the voice coil tends to decrease the flux that loops in the circuit. c) Magnetic flux produced by the interacting voice coil and permanent magnet. d) Magnetic circuit equipped with a short-circuit ring. Transformer loading caused by the ring reduces the effects of inductance.

to proceed a big part of its way in the air, so, in practice, the solenoid force fortunately stays much smaller than the actual drive force. Nevertheless, the solenoid force is a significant cause of 2nd-harmonic distortion.

Figure 3.2c depicts the flux brought about by the magnet and voice coil together. In the vicinity of the voice coil, the flux lines are strained downward, so the drive force acts in this case upward. The flux touring through the bottom is now slightly weaker than in case 'a'.

The application of current-drive does not affect the magnitude of the solenoid force itself, but in the elimination of the adverse effects of flux modulation and the inductance relating to it, the driving method has a very essential import.

The solenoid force can, however, be significantly reduced by using, in the magnetic circuit, materials that are good electrical conductors. Some manufacturers have used a shorting ring around the center pole, as shown in Fig. 3.2d. The ring corresponds, in operation, to a short-circuited secondary winding of a transformer.

When the secondary of a transformer is loaded by some impedance, this impedance is seen on the primary side, in principle, multiplied by the square of the turns ratio. Hence, in this case, a resistive load is established in parallel with the inductance of the primary winding (i.e. voice coil), shorting part of the inductance's current. Thereby, the voice-coil-induced flux and the solenoid force associated with it reduce, although the inductance in itself does not decrease. The same advantage is also gained by using well-conducting magnet material, like neodymium, whose eddy currents effect a similar loading.

Irrespective of other structural details, it is usually worth striving for strong magnetic flux in the air gap since thereby one not only improves sensitivity and efficiency but also decreases the relative significance of the solenoid force.

3.2 Motional Equations under Current-Drive

It is easy to think that in the ideal case the displacement of a loudspeaker diaphragm should follow accurately the applied signal. In reality, however, it is not so and ought not to be since the instantaneous value of the acoustic pressure developed is not due to the diaphragm position. In the following, we will thus consider how the motion of the diaphragm is actually determined, assuming it is rigid and the operation of the system is linear.

Figure 3.3a shows a cross-section of a typical cone driver in a closed cabinet. The moving portion of the driver consists of the cone, the voice coil with its former that actuates the cone, and the dust protection cap. The suspensions of the cone may also be partially included in the moving mass, into which in practice also integrates a bit of the surrounding air.

The duty of the suspensions is to allow motion in the direction of the axis and, at the same time, prevent traverse motion; and they constitute a spring that tends to return the diaphragm to its rest position. In closed enclosure, the driver diaphragm is also loaded at low frequencies by the air spring of the enclosure. Manufacturers generally specify for their drivers the equivalent volume, with which the spring force caused by the enclosure equals the spring force of the cone suspensions.

Besides the mass and the springs, the mobility of the diaphragm is also affected by a third factor, mechanical resistance, that tends to slow down movement. This effect develops because the deformations taking place in the suspensions require energy, and this manifests as a counterforce proportional to the velocity of the diaphragm. Some retardation is

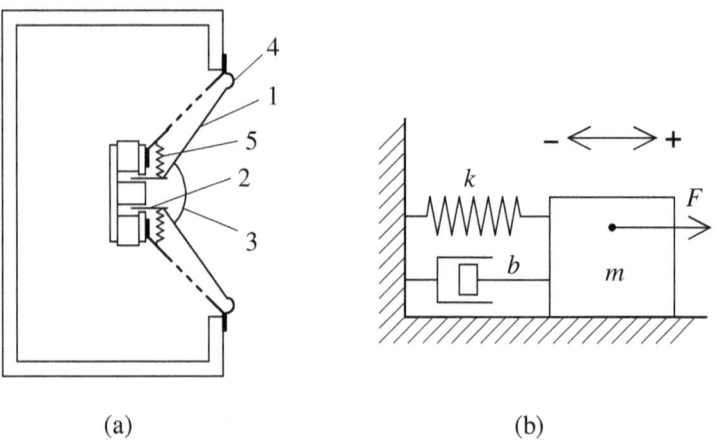

(a)   (b)

Figure 3.3.   a) Cone driver in closed enclosure. The diaphragm (1), voice coil former (2) attached to it, and dust cap (3) are able to move, hanged on the outer suspension (4) and inner suspension (5). The outer suspension is, in hifi speakers, usually of rubber, and the inner one is usually of folded and stiffened cloth. The air compressed between the dust cap and the magnetic pole is always led out via some route. b) Mechanical model corresponding to the speaker of Fig. 'a'. The mass piece is assumed to move on its track without friction. In addition to the mass inertia, the system's operation is governed by the spring constant k and damping constant b.

also caused by air flows in the interior structures of the driver and in the damping material used in the enclosure.

On these grounds, we may construct for the driver the mechanical model shown in Fig. 3.3b. The force F, produced by the voice coil, acts upon a sliding object that has mass m and that is attached to a spring and a damper. The spring constant k includes the effects of all springs. The mechanical resistance is represented by the damping constant b.

The force F becomes therefore divided into three components, which are:

- force that accelerates the mass, ma, where a is the acceleration

- force that stretches the spring, kx, where x is the displacement from the rest position

- force that moves the damper, bv, where v is the velocity

The positive direction of all motion-describing quantities has been defined as the same (in Fig. 3.3b, to the right). Hence, we may write:

$$F = ma + bv + kx \qquad (3.2)$$

Since velocity is the time derivative of distance and acceleration is, in turn, the time derivative of velocity, we further obtain, recalling formula (3.1):

$$m\frac{d^2x}{dt} + b\frac{dx}{dt} + kx = Bli \qquad (3.3)$$

which describes in a differential equation the dependency between displacement x and applied current i.

The transfer function corresponding to equation (3.3) can now be directly written, according to the principles explained in appendix sections B4 and B5:

$$\frac{X}{I} = \frac{Bl}{ms^2 + bs + k}$$

$$= \frac{Bl/m}{s^2 + \dfrac{b}{m}s + \dfrac{k}{m}} \qquad (3.4)$$

where X and I are interpreted as phasors. The result is a 2nd-order low-pass function, the behavior of which was discussed in section 2.1. Thus, a driver in free space or in a closed cabinet (in the following, F/C driver) constitutes a 2nd-order system, that is characterized by a certain natural frequency (resonant frequency) and a certain Q value.

By comparing expressions (3.4) and (2.5), it is seen that

$$\omega_0 = \sqrt{\frac{k}{m}} \tag{3.5}$$

The resonant frequency is thus determined by the spring constant and the moving mass.

The normal operation band of a driver is located above the resonant frequency. So, when the operation is desired to extend low, ω_0 should be aimed low. This is only accomplished by making the mass large and the suspensions flexible. Increasing the mass has, however, always the drawback that the overall sensitivity degrades, as can be noticed from the transfer function (3.4).

In the parameter lists published by driver manufacturers, the moving mass is usually labeled M_{ms}. Instead of the spring constant, they specify its reciprocal C_{ms}.

Further, by comparing expressions (3.4) and (2.5), it is seen that $b/m = \omega_0/Q$ which yields:

$$Q = \frac{\sqrt{km}}{b} \tag{3.6}$$

The Q factor of the system is therefore, under current-drive, inversely proportional to the damping constant (i.e. mechanical resistance) and depends only on mechanical parameters. Equation (3.6) gives so-called mechanical Q, that is generally labeled in data sheets by Q_{ms}. So:

The total Q factor of a driver operating on current-drive in a closed cabinet is solely constituted of the mechanical Q, that, in the case of free air, is represented by the parameter Q_{ms}. Instead, parameters relating to a driver under voltage drive, Q_{es} (so-called electrical Q) and Q_{ts} (total Q), are useless under current-drive.

The curve families representing the second-order low-pass function, like (3.4), have been shown before in Fig. 2.1. At frequencies much lower than ω_0, the displacement of the diaphragm is thus almost independent of frequency and approximately in phase with the current. In this region, the dominant counter-force is due to the spring.

At frequencies distinctly higher than ω_0, i.e. in the normal operation band of the speaker, the displacement decreases inversely proportional to the square of frequency, the dominant counter-force being due to the mass. It is perhaps a little surprising that, in this region, the displacement of the diaphragm is about in opposite phase with respect to the current. In other words, at the moment when the current applied to the plus-terminal of the driver reaches a positive peak, the diaphragm is in its rearmost position.

At the resonant frequency, the counter-forces caused by the spring and mass cancel out each other, and we are left only with the damper, that has a conclusive effect on the Q value.

Above, we have mostly discussed about cone drivers, but the equations presented also hold as such for dome-type high-frequency units. These differ structurally from that in Fig. 3.3a mainly in that the cone and outer suspension have been removed, and a dome of the same width as the voice coil acts as the diaphragm. Also, high-frequency drivers (tweeters) are always sealed from the back, so that pressure variations generated by the bass driver (woofer) wouldn't mess up the operation.

The mechanism shown in Fig. 3.3b is common also elsewhere than in loudspeakers, for a corresponding model applies, for instance, to the suspension of a car wheel. The wheel with its drum establishes a mass that is able to move attached to a spring and a shock absorber. The duty of the shock absorber, or damper, is here to lower the Q value to a reasonable level, so that the system doesn't exhibit resonance on a bumpy road.

3.3 Effect of Motional EMF

Loudspeaker drive units are usually assigned a nominal impedance (usually 4, 6, or 8 Ω), which provides, in practice, only a kind of average value of the actual impedance in the frequency region of interest. When measuring the DC resistance, the result obtained is usually about 75% of this nominal value. On the other hand, at the high and low extremes of the frequency range, the impedance can be many times higher than the nominal value. The reason for this are the two kinds of electromotive forces (EMF) induced in the voice coil: the motional EMF caused by the motion of the coil and the inductive EMF caused by the inductance of the coil.

Electromotive "forces" are voltages, by nature, although they are called forces. They can be represented, in the circuit, by built-in voltage sources, whose effects can be examined externally by measurement. Motional EMF is utilized, for example, in electro-dynamic microphones.

Electromotive forces always appear in series with the wire's resistance, so we may use, for the driver, an equivalent circuit shown in Fig. 3.4. The coil's DC resistance has been designated by R_c and the inductance by L_c. The inductive EMF is represented by voltage e_i. R_p represents the loading the eddy current and hysteresis losses introduce upon the inductance. R_p is not constant but increases with frequency.

Voltage source e_m represents the motional EMF, which can always be calculated from the formula:

$$e_m = B\,l\,v \qquad (3.7)$$

The motional EMF induced in the circuit is thus directly proportional to the velocity of the voice coil, v. Also here, B is the flux density the wire sees when no current is flowing. All EMF due to the current-induced flux is contained in e_i.

Electromotive forces by nature strive to oppose the factor that originally caused them. Consequently, both e_m and e_i have such polarity that they tend to reduce voice coil current, increasing accordingly the total impedance.

According to a very common conviction, the motional EMF (also called back EMF) of the driver should somehow be suppressed or eliminated by keeping the output impedance (or resistance) of the amplifier low, so that the amplifier acts as an ideal voltage source. The truth is, however, that motional EMF can by no means be suppressed into nonexistence or oblivion since law (3.7) can never cease to be valid and the relative proportion of EMF in the voltage across the driver can never be

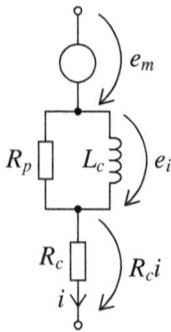

Figure 3.4. Electrical equivalent circuit of a speaker driver. The terminal voltage can always be divided into three components, which are: resistive voltage drop $R_c i$, inductive EMF e_i, and motional EMF e_m. Due to the electromotive forces, the driver's impedance magnitude is always greater than the mere DC resistance.

reduced, no matter what kind of output impedance is used. To the same pseudo-scientific illusory thinking, upon which the ascendancy of voltage drive in fact largely rests, also belongs a fiction that an output impedance as low as possible, relative to the driver impedance, "controls" the speaker and hence as though prevents extraneous oscillations.

According to equation (3.7), e_m is always present when the voice coil is in motion, irrespective of the cause of this motion. In a loudspeaker, the motion is due to the drive force, that in turn, is directly proportional to the current, as described by equation (3.1). In phasor form, it may hence be written: $E_m = Z_m I$ where Z_m is the electrical impedance (motional impedance) caused by the motional EMF. The appearance of the back EMF, therefore, does not in any way depend on e.g. whether a signal is starting or stopping or whether the diaphragm is returning toward the rest position or in which direction power is instantaneously flowing between the amplifier and speaker.

The model of Fig. 3.4 thus constitutes a linear impedance consisting of three internal components, whose relative magnitudes, at a given frequency, depend only on the driver's own parameters and the enclosure but not on external circuitry.

The amplifier sees the speaker driver always as a mere impedance load, that also incorporates the motional impedance. The motional EMF is thus always present as one voltage component in the closed circuit formed by the amplifier's output and the driver. Hence, this so-called back EMF by no means loses its significance when the amplifier's output impedance is low, but inevitably acts as an essential factor in the driver's voltage, as shown in Fig. 3.4, as long as the voice coil moves at all.

Moreover, when the system is linear, *all* effects of the motional EMF are incorporated in the motional impedance Z_m, and besides it, the motional EMF does not have any own, separate impact on the transient properties, that are often sought to be "controlled" by minimizing the output impedance. The transient reproduction properties of a linear system are completely determined by the frequency response properties through the Fourier transform, as presented in appendix section C4. Consequently, if the changing of some parameter (like e.g. the output impedance) doesn't have any apparent effect on the frequency response, the transient performance cannot change either.

The output impedance of the amplifier affects, of course, how voltage divides between this impedance and the speaker, but the signal filtering that occurs thereby, does not affect the proportions of the voltage components within the driver itself. In practice, the output impedance always appears in series with the driver, so, for the end result, it is the same whether the output's or voice coil's resistance is changed. In practice, even mere temperature changes in the voice coil have much greater effect than a usual output resistance of the order of 0.1 Ω.

To obtain a more specific view about the behavior of motional impedance, we make use of transfer function (3.4) and the differentiation rule of rotating phasors, (b7) in appendix B. Since velocity is always the derivative of distance, one may write, using phasors: $V = j\omega X = sX$. Recalling yet equation (3.7), we obtain as the motional impedance of an F/C driver

$$Z_m = \frac{E_m}{I} = \frac{BlV}{I} = Bls\frac{X}{I}$$

$$= \frac{(Bl)^2}{m} \cdot \frac{s}{s^2 + \frac{b}{m}s + \frac{k}{m}} \tag{3.8}$$

Z_m is thus, in form, a 2nd-order band-pass function, having the same ω_0 and Q as the mechanical system. Also, it is seen that the transfer function from current to velocity, V/I, is the same as expression (3.8) divided by the force factor Bl.

The shape of a second-order band-pass response has been portrayed earlier in Fig. 2.3, also showing the corresponding low-pass response. Consequently, if the displacement of a diaphragm behaves at constant current level like graph A, then velocity, motional EMF, and motional impedance behave like graph B.

Expression (3.8) has the same form as the impedance of the parallel resonance network, (2.18). The motional impedance of an F/C driver can thus be modelled in practice by a parallel connection of a coil, capacitor, and resistor; by adjusting L, C, and R to match the measured impedance peak.

The Nyquist plot of the motional impedance (3.8) is, consequently, a circle like in Fig. 2.6a. At low frequencies, Z_m approaches pure inductance; at high frequencies, pure capacitance; and, at the resonant frequency, Z_m is purely resistive. For the absolute value $|Z_m|$, we obtain from equation (3.8):

$$|Z_m| = \frac{(Bl)^2}{m} \cdot \frac{\omega}{\sqrt{\left(\dfrac{k}{m} - \omega^2\right)^2 + \left(\dfrac{b\omega}{m}\right)^2}} \tag{3.9}$$

We take as an example a typical, moderately small, unenclosed bass-midrange driver for which $Bl = 6$ Tm, $m = 0.008$ kg, $k = 1000$ N/m, and $b = 1.5$ Ns/m. These values yield a resonant frequency of 56 Hz and a mechanical Q value of 1.9. A curve of the absolute value of the motional impedance is shown in Fig. 3.5, also featuring a typical magnitude of voice coil resistance (6 Ω).

In the region of the resonant frequency, $|Z_m|$ is many times higher than the DC resistance R_c and becomes equal only at 140 Hz. At frequencies higher than this, $|Z_m|$ is about inversely proportional to frequency, so that still at 2 kHz, $|Z_m|$ is 6% of R_c. Hence:

Figure 3.5 proves erroneous such a view that motional EMF would only have impact in the resonance region and would be negligibly small at higher frequencies. Although the magnitude of the motional impedance is far from the resonance point less than the DC resistance, the motional impedance has, ne-

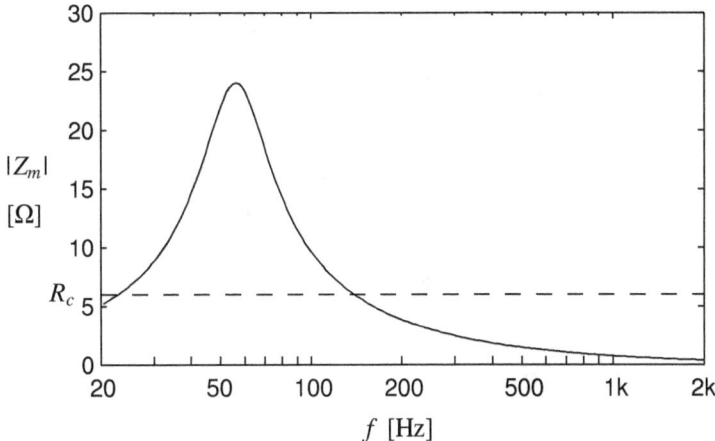

Figure 3.5. Magnitude plot of the motional impedance of an ordinary hifi driver with approx. 5-inch diameter. The dashed line, set at 6 Ω level, represents, for comparison, typical voice coil resistance. The motional impedance is not negligible even at the upper end of the operation band.

vertheless, yet very essential significance through the whole mid-frequency range.

When measuring the total impedance of a driver, this feature does not easily show up due to phase differences between the various components.

Moreover, the example driver described does not represent, in this regard, any worst case. According to equation (3.9), $|Z_m|$ is proportional to the square of the force factor Bl. When using higher force factors, as is customary when desiring high sensitivity, the motional impedance can be a lot higher than in Fig. 3.5. For example, an increase of 3 dB in the force factor already effects the doubling of $|Z_m|$.

When the driver is mounted in a closed cabinet, the spring constant k increases, effecting increase in ω_0 and Q, according to equations (3.5) and (3.6). Then, also the peak of the motional impedance moves to a higher frequency, increasing further the proportion of $|Z_m|$, especially in the lower mid-frequency range. In bass reflex enclosures, the motional impedance has, in turn, two peak points, of which the higher is usually near 100 Hz.

In high-frequency drivers, the effect of motional EMF is also strong although no attention is generally paid to it. The resonant frequency is usually of the order of 1 kHz, and the force factor is typically about 3 Tm while the moving mass is some 0.3 g.

Far above the resonant frequency, the motional impedance is nearly independent of k and b. Equation (3.8) then simplifies to the form:

$$Z_m \approx \frac{(Bl)^2}{j\omega m} , \qquad \omega \gg \omega_0 \qquad\qquad (3.10)$$

By using the values mentioned, the magnitude of the motional impedance of a tweeter is found to be e.g. at 4 kHz still as high as 1.2 Ω. The value is, in its order of magnitude, everything but such that could be flat ignored when sound quality is of any concern.

3.4 Inductive EMF

All EMF inducing in the voice coil can be considered being due to Faraday's law of induction, which states that the EMF generated in each coil loop equals the rate of change of the magnetic flux passing through the loop. So, the difference between motional EMF and inductive EMF is, in fact, only in the reason for which the flux changes.

In the case of motional EMF, this flux alternation stems from the fact that, as the loop moves, greater or lesser part of the permanent magnet's flux tours through the loop (see Fig. 3.2a). In the case of inductive EMF, in turn, the flux of the loop alternates as a consequence of alternation in the coil's current (Fig. 3.2b). As long as the operation is linear, these two mechanisms act independent of each other, according to the super-position principle.

Due to loss mechanisms arising in iron, the inductive EMF (e_i in Fig. 3.4) cannot be easily described by equations. Driver manufacturers generally specify some value for the voice coil inductance, but based on this, not much can be deduced about the behavior of the inductive EMF and the impedance due to it. The reason is in the eddy current losses, whose relative proportion in the total power losses strongly increases as frequency rises. The phenomenon of hysteresis, inherent in the magnetization of iron, also has its effect on the resistive loading that appears in parallel with the inductance (L_c in Fig. 3.4), since overcoming hysteresis requires work too. Both loss factors manifest through the transformer action described in section 3.1.

Inductance depends on coil structure in the following way:

$$L_c = \frac{N^2}{R_m} \tag{3.11}$$

where N is the number of turns and R_m is the reluctance, or magnetic resistance, seen by the flux. R_m opposes the flow of flux correspondingly as electrical resistance opposes the flow of current.

Relation (3.11) is valid for all kinds of coils, but its significance in this context is mostly advisory. Inductance depends on the square of the number of turns and how easily the flux of the voice coil is able to flow. Most part of the reluctance accrues inside the coil where the flux flows relatively narrowly. So inductance is, in principle, directly proportional to the cross-sectional area of the coil.

Figure 3.6 shows measurement results of the magnitude of the inductive impedance in a 6.5-inch woofer and in a one-inch tweeter unit. In order to prevent motional impedance from marring the measurement, the voice coils were fixed, with epoxy glue, firmly to the pole pieces. The results have been obtained by applying the conventional impedance measurement method, by measuring the voltage difference between the driver and a resistor equal in value with the voice coil resistance, the current being equal for both. Thereby, we are left with the mere induc-

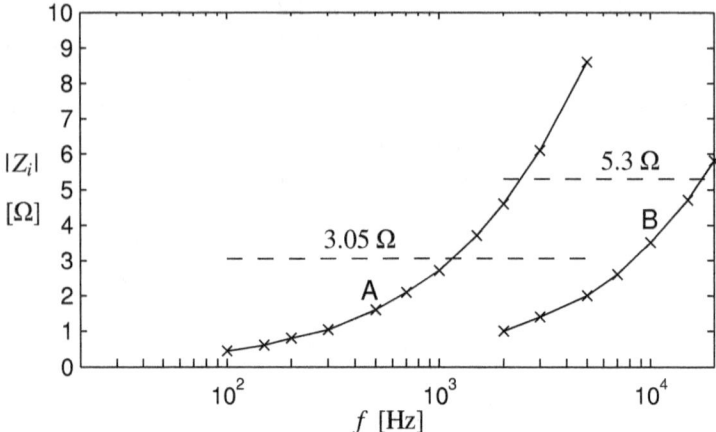

Figure 3.6. Magnitude of inductive impedance as a function of frequency in a single woofer (A) and tweeter (B). The drivers used in measurement were Vifa P17WJ-00-04 and Peerless 811815. Dashed lines indicate corresponding voice coil resistances. Both cases show the inductive impedance to be a significant factor even at the low end of the driver's frequency region.

tive voltage e_i, that introduces impedance Z_i.

Purely inductive impedance rises in direct proportion to frequency ($Z = j\omega L$), but the inductive behavior of a voice coil falls far behind this. The direction angle of Z_i does not reach 90° either but settles, at high frequencies, in the proximity of 50°.

It is seen, from Fig. 3.6, that in the woofer $|Z_i|$ exceeds the voice coil resistance 3.05 Ω already slightly above 1 kHz. Even when going to low frequencies, the effect of inductance does not become negligible as is often supposed, but in this case, e.g. at 100 Hz, $|Z_i|$ is still 15% of R_c. Also in the tweeter, $|Z_i|$ is remarkably high through the whole operation band, being yet at 2 kHz 19% of R_c. Thus, it must be noted:

The effect of inductance is not insignificant anywhere in the normal operation band of a driver, but the inductive impedance Z_i constitutes for all frequencies an essential component in the total impedance of the driver.

The diameter of the voice coil was in the measured woofer only 32 mm. In larger drivers, the voice coil and its inductance can be much greater. For example, in a typical 12-inch driver, the intersection point of $|Z_i|$ and R_c falls near 400 Hz, so that inductance begins to have effect even on the bass alignment. Whether the driver is 4- or 8-ohmic, is quite irrelevant in this regard because the ratio of inductance and resistance is, in practice, quite independent of the rated impedance.

3.5 Total Impedance

As we know, the total impedance of a voice coil is always the sum of the wire resistance, motional impedance and inductive impedance. In woofers, low frequencies are dominated by the motional impedance, center area of the operation band by the resistance, and the top end by the inductance. The frequency range covered by tweeters is generally narrower (about 1 decade), for which reason the motional and inductive impedances do not reach as dominant status as in a bass-midrange unit of a typical 2-way system. Yet even in tweeters, the motional impedance rises, when the resonant frequency is approached, to the same order of magnitude with the resistance; and at the upper edges of the hearing range, the inductive impedance reaches similar position.

The components of impedance differ in direction angle, so, in order to acquire an overall view, they must be analyzed in the complex plane. Figure 3.7 depicts, using Nyquist plot, the composition of the impedance of an F/C driver at different frequencies.

Based on earlier observations, the sum of the motional impedance Z_m and resistance R_c follows a circular locus, indicated by the dashed line, the diameter being equal to the motional impedance magnitude at the resonant frequency. This value is obtained from equation (3.8), as terms s^2 and k/m cancel out each other, leaving left $Z_m(\omega_0) = (Bl)^2/b$. The peak value of the total impedance, reached at the resonant frequency, is therefore (assuming the effect of inductance is negligible)

$$Z(\omega_0) = R_c + \frac{(Bl)^2}{b} \qquad (3.12)$$

As frequency increases, the inductive impedance Z_i begins to deflect the locus of the total impedance more and more in the direction of the positive imaginary axis and finally also in the direction of the positive real axis.

Figure 3.7a shows the situation below the resonant frequency where the total impedance is inductive while Z_i is too small to be drawn. Figure b shows the situation slightly above the resonant frequency where the total impedance is capacitive and the contribution of Z_i begins to

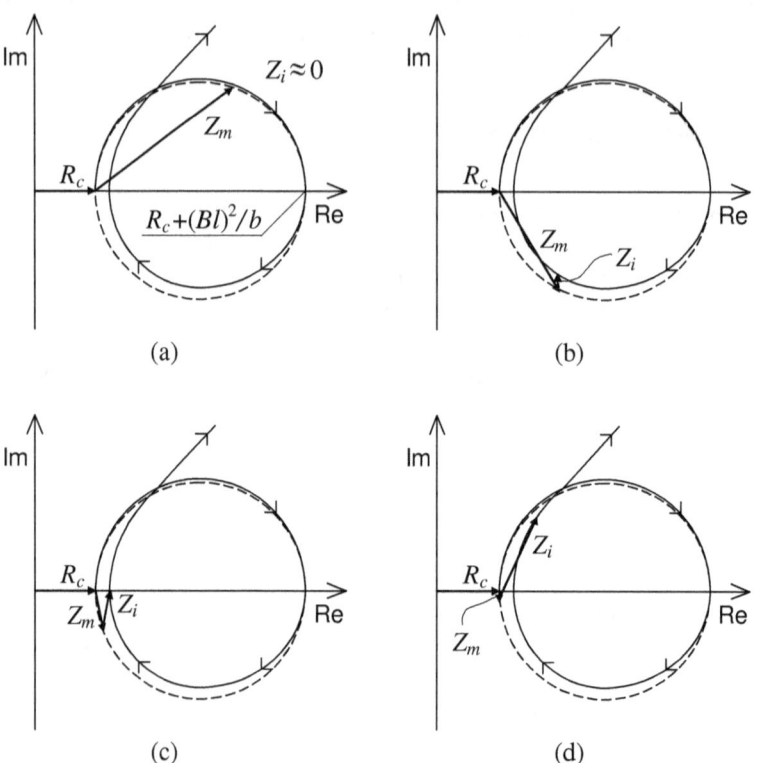

Figure 3.7. Impedance behavior of a driver in free air or closed cabinet. At low frequencies, the locus of total impedance follows a circular arc but deviates with increasing frequency, gradually, in the inductive direction. The diameter of the circle (dashed line) is $(Bl)^2/b$, where b is the damping constant. The composition of impedance is shown below the resonant frequency (a), somewhat above the resonant frequency (b), at the point of minimum impedance (c), and in the highest part of the operation band (d).

show up. In Fig. c, the total impedance is resistive and only a bit higher than R_c; Z_m and Z_i being almost opposite to each other. In the case of Fig. d, frequency is so high that Z_i makes the total impedance strongly inductive again, while Z_m is very small. Along with Z_i, the total impedance continues its growth and remains inductive up to very high frequencies.

The above described applies, by its principles, also to high-frequency drivers, but especially in structures equipped with a rear chamber, the behavior of the motional EMF can be more irregular in the resonance region, and besides the actual impedance tip, there may also appear other prominences. Ferrofluid, that is often used for voice coil cooling, in turn, strongly increases the damping coefficient b, making the impedance circle considerably smaller than shown in Fig. 3.7. Thus, ferrofluid lowers the impedance peak and reduces the Q value.

The essential behavior of the impedance magnitude of an F/C driver is shown in Fig.3.8a, the dashed line representing again voice coil resistance. The form of $|Z|$ corresponds to the model of Fig. 3.7. At f_c, that corresponds to the minimum point represented by Fig. 3.7c, $|Z|$ is generally yet 10-20% greater than R_c.

By looking at mere total impedance, one can easily get the impression that the effect of motional EMF would vanish at about f_c. Correspondingly, the effect of inductance would seem to be confined above frequency f_c. The truth shows up, however, from Fig. 3.7c, that shows how, at the mentioned frequency, Z_m and Z_i largely cover each other, having nevertheless considerable magnitude. It is even probable that in many high-sensitivity drivers, that are commonly used e.g. in orchestral and PA speakers*, the sum of $|Z_m|$ and $|Z_i|$ exceeds the voice coil resistance at *all* frequencies of the operation band.

Figure 3.8b depicts the behavior of the direction angle ($\angle Z$), corresponding to the magnitude graph of Fig. 3.8a. The angle has a positive (inductive) peak below the resonant frequency and a corresponding negative (capacitive) peak above it, as can be concluded from Fig. 3.7. The height of the peaks depends on the ratio of the circle's diameter and R_c.

It is of benefit, concerning amplifier operation, that the load impedance does not become excessively reactive at any frequency, that is, the absolute value of $\angle Z$ remains sufficiently low. This means, in practice, that the mechanical Q must be restricted if an F/C driver is to be easy matter for the amplifier. A moderate Q value also aids in the implementation of equalizer circuits for current-drive.

3.6 Motional Equations under Voltage Drive

In section 3.2, we derived, for an F/C driver, the transfer function from current to displacement, assuming frequency to be so low that mo-

---

* PA stands for "public address" and has originally denoted an announcement system. Nowadays the concept mainly involves amplification equipment used by performers.

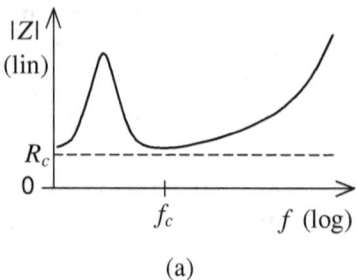

(a)

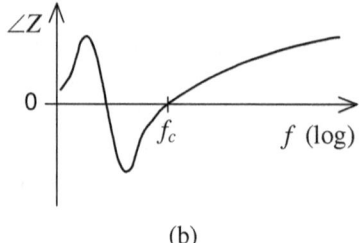

(b)

Figure 3.8. a) Typical impedance magnitude behavior of a driver. $|Z|$ is higher than voice coil resistance R_c at all frequencies. b) Direction angle curve corresponding to the magnitude curve of Fig. 'a'. $\angle Z$ changes its sign at the resonant frequency and at f_c, where $|Z|$ is approx. at minimum.

tion of the diaphragm can be regarded as being piston-like. Due to equation (3.8), we also know the behavior of the motional impedance and hence that of the total impedance, at the lowest frequencies (ignoring inductance, as is customary). By combining these pieces of knowledge, we find out the motion under voltage source feeding.

On these grounds, the total impedance is thus

$$Z = R_c + \frac{(Bl)^2}{m} \cdot \frac{s}{s^2 + \dfrac{b}{m}s + \dfrac{k}{m}}$$

from which we obtain, after a couple of modification phases:

$$Z = R_c \frac{s^2 + \left[\dfrac{b+(Bl)^2/R_c}{m}\right]s + \dfrac{k}{m}}{s^2 + \dfrac{b}{m}s + \dfrac{k}{m}} \tag{3.13}$$

This corresponds, in form, to the band-boost function of Table 1 (section 2.2) and approaches R_c at low and high frequencies, as is relevant.

When using voltage drive, the current conducted by the driver is fil-

tered according to the generalized Ohm's law, determined by the impedance Z. The transfer function from drive voltage to displacement, X/U,
becomes now, using equations (3.4) and (3.13),

$$\frac{X}{U} = \frac{X}{IZ} = \frac{Bl/m}{s^2 + \dfrac{b}{m}s + \dfrac{k}{m}} \cdot \frac{1}{R_c} \cdot \frac{s^2 + \dfrac{b}{m}s + \dfrac{k}{m}}{s^2 + \left[\dfrac{b+(Bl)^2/R_c}{m}\right]s + \dfrac{k}{m}}$$

$$= \frac{Bl}{mR_c} \cdot \frac{1}{s^2 + \left[\dfrac{b+(Bl)^2/R_c}{m}\right]s + \dfrac{k}{m}} \tag{3.14}$$

which is a 2nd-order low-pass function, as also expression (3.4).

By comparing result (3.14) to the standard form (2.5), it is observed
that the characteristic frequency is the same as also in the current-drive
case, i.e. expression (3.5). The Q factor, instead, changes totally. Using
formula (2.4), we obtain now:

$$Q = \frac{\sqrt{k/m}}{(b+(Bl)^2/R_c)/m}$$

$$= \frac{\sqrt{km}}{b+(Bl)^2/R_c} \tag{3.15}$$

So, in addition to the mechanical parameters, the Q of an F/C driver
under voltage drive (i.e. total Q factor) also depends on the force factor
and resistance.

Compared with the Q under current-drive, (3.6), expression (3.15)
always yields a lower value, due to the term $(Bl)^2/R_c$, that parallels the
mechanical damping constant b and represents an electrical damping
constant, or simply, electrical damping. $(Bl)^2/R_c$ is generally greater
than b, so the effect of electrical damping on the system's Q is conclusive.

Doubts cast upon the functionality of current-drive are most often
based just on the lack of such electrical damping and the usually over-
sized Q, that results as the consequence.

**Electrical damping is, however, a very paltry justification for
such a supreme and decisive choice of policy that the driving**

principle of loudspeakers presents, since the bass resonance only governs a small portion of the audio spectrum; and thus far, there has not been even proper effort to seek those solutions, based on active and passive processing as well as driver and enclosure techniques, by which the fundamental resonance can be reasonably controlled, without resorting to the filtering of current brought about by the motional impedance.

Formula (3.15) is also applicable when the amplifier does not act as an ideal voltage source but exhibits output resistance R_o. Then, one has to substitute R_c+R_o in place of R_c since all resistance appearing in series is equal in effect. When R_o becomes very high, one ends up with equation (3.6) since ideal current-drive corresponds to infinite output resistance.

By setting b to zero in expression (3.15), we are left with so-called electrical Q factor (designated Q_e), that thus corresponds to a condition where there appears no mechanical damping force. Q_e may be expressed in terms of the mechanical Q (designated Q_m) as follows:

$$Q_e = \frac{\sqrt{km}R_c}{(Bl)^2} = \frac{\sqrt{km}}{b}\cdot\frac{R_cb}{(Bl)^2} = Q_m\frac{R_c}{Z(\omega_0)-R_c} \tag{3.16}$$

where relation (3.12) has been used.

In practice, Q_m is found by determining the Q of the motional impedance (method described in section 13.1). After this, Q_e is obtained from equation (3.16) without additional measurements. When Q_m and Q_e are known, the total Q can be calculated, on the grounds of equation (3.15), from the formula:

$$\frac{1}{Q} = \frac{1}{Q_m} + \frac{1}{Q_e} \tag{3.17}$$

Alternatively:

$$Q = \frac{Q_m Q_e}{Q_m+Q_e} \tag{3.18}$$

3.7 The Origin of Sound

So far, we have considered the motion of the diaphragm and related factors, but the goal is, of course, to understand and know the behavior of the acoustic pressure, i.e. sound, radiated by the driver. In texts dealing with the generation and propagation of sound waves, it is customary to present the subject in terms of acoustic impedance, radiation impedance, and other rather abstract concepts and mostly for those who are already almost specialists. This kind of theoretical approach is, however, not very illustrative and not even necessary when seeking to characterize the pressure signal produced by a loudspeaker, for the subject can also be examined based on minimum-phase linear systems.

In the following, it is assumed that the loading on the driver diaphragm due to moving air mass can be ignored, which holds quite well unless the issue is about horn speakers.

It is known that the displacement of an ideal piston mounted in an infinite baffle* must become fourfold as frequency is halved, in order to keep sound pressure on the center axis constant. Thus, to accomplish flat frequency behavior, the displacement must be, in principle, inversely proportional to the square of frequency which just corresponds to the situation in a speaker driver when operating in the frequency region that is governed by the moving mass, that is, above the resonance region.

It is also known that, in the mentioned region, the pressure signal is, when leaving, in phase with the applied current, assuming that current regarded as positive tends to push the diaphragm forward. However, as the displacement is in opposite phase relative to current (as was noticed in section 3.2), the pressure must, in turn, be in opposite phase relative to the displacement.

On the above basis, it is evident that the transfer function from displacement to pressure corresponds, in the referred case of infinite baffle, to a double differentiator, so the pressure must, when generating, follow the second time derivative of the displacement, that is, acceleration. The baffle mentioned doesn't have to be planar; it is enough that the solid angle seen by the piston does not depend on distance at the wavelengths concerned. Thus, it can be stated:

The pressure radiated by a speaker diaphragm in its front is, as a main rule, directly proportional to the diaphragm's acceleration, not to the displacement or velocity, as is suggested in many interpretations. With sine wave, this implies, inter alia, that the leaving pressure reaches its maximum when the diaphragm is in its rearmost position, because the acceleration is

---

* Denotes a flat and rigid panel whose edges are in all directions so far as to not have any effect on the sound generated by a vibrator in the panel. The term is sometimes used incorrectly to denote a closed cabinet, which is, however, quite a different thing.

then at its positive peak value.

The oppositeness of phase between pressure and displacement may at first sound strange unless it is taken into account that air particles also have a certain mass that seeks to preserve its motional state. Simplified, this can be viewed as follows: When the diaphragm is moving backward, particles in the vicinity reach the same velocity and follow with. As the diaphragm displacement approaches its negative apex, the motion slows down and eventually turns in the opposite direction. The air particles, however, tend to continue their travel and thus accumulate against the diaphragm, establishing overpressure. Correspondingly, as the diaphragm is in its forward travel and then turns back, the air bulk doesn't follow with immediately but spreads, generating underpressure. (In the proximity of the sound source, pressure generally reaches its peak at a different time than velocity although in a wave propagating far away from the source these two are always in phase.)

A mere unmoving diaphragm, standing out from an infinite panel or the like, does not effect pressure in an open space. Even when moving with constant velocity, such a diaphragm does not effect pressure in an open space although air flow may occur. Only a diaphragm or surface in accelerating motion is able to compress or rarefy the air particles nearby and generate pressure.

The transfer function of a pure differentiator is mere s, as was also found in the context of equation (3.8). Consequently, the phasor of acceleration (A) is obtained by multiplying the phasor of displacement (X) by the factor s^2 (or by multiplying the phasor of velocity by s). Hence, based on equation (3.4), the dependency between diaphragm acceleration and applied current in an F/C driver becomes

$$\frac{A}{I} = \frac{Bl}{m} \cdot \frac{s^2}{s^2 + \dfrac{b}{m}s + \dfrac{k}{m}} \tag{3.19}$$

Acceleration thus follows a 2nd-order high-pass function that has the same resonant frequency and Q as the displacement and velocity have. The transfer function from voltage to acceleration is obtained in a similar way from equation (3.14).

The behavior of basic second-order responses is shown in Fig. 2.3. Earlier, it was already found that when curve A corresponds to displacement, curve B corresponds to velocity. Curve C corresponds then to acceleration, which, in the conditions mentioned, also represents sound pressure. The amplitude response obtained with a closed enclosure type thus falls below the resonant frequency with a 2nd-degree slope, or 12 dB per octave.

From Fig. 2.3b it appears that the phase of acceleration and pressure (curve C) precedes the applied signal more and more, as frequency is lowered. The value of Q also affects the picture, as shown in Fig. 2.1b. At the resonant frequency, this phase lead is 90° which can cause unfavorable surprises with tweeters and midrange drivers if the phenomenon has not been duly taken into account in crossover filter design. Particularly in woofers, this inevitable phase shift causes low frequencies to be reproduced in advance of higher ones, in terms of phase. For an individual driver, amplitude and phase are always tied together according to the principle of minimum phase.

Figure 3.9 illustrates the phase relationships of an F/C driver using phasor diagrams. Acceleration, velocity, and displacement always stay at the same angle relative to each other but turn with respect to current. Figure 'a' depicts the situation far below the resonant frequency where displacement and current are almost in phase, acceleration being a half cycle ahead of them. At the resonance (Fig. b), velocity reaches its maximum value and becomes in-phase with current. Far above the resonant frequency (Fig. c), in turn, acceleration is virtually coincident with current, displacement becoming very small. The motional EMF always accompanies the velocity phasor.

Besides frequency analysis, it is also informative to take a look at the

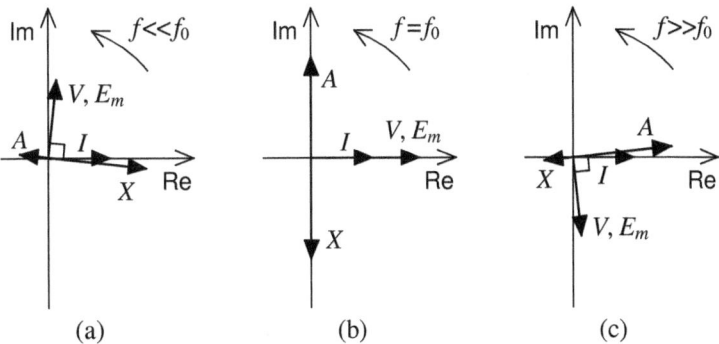

Figure 3.9. Phasor diagram of a speaker driver at three different frequencies (f_0 =resonant frequency), current I serving as the reference. As the phasors rotate counterclockwise, their real parts represent the instantaneous values. Acceleration (A) and velocity (V), and respectively, velocity and displacement (X) are, by definition, always perpendicular to each other. Phasor lengths have no significance in this context, except in comparison between the Figures.

time behavior of a driver, for the actual movement of the diaphragm, in transient reproduction, doesn't really coincide with common imagery.

We take as an example the test signal shown in Fig. 3.10a, consisting of two symmetric rectangular pulses, to represent current applied to the voice coil. (We ignore that, in practice, no signal source is able to generate discontinuities.) The driver is assumed to operate ideally for all frequencies, so that the acceleration of the diaphragm perfectly follows this signal.

The velocity of the diaphragm, that is obtained as the integral function of the acceleration, then varies according to the triangular shape of Fig. b, staying non-negative all the time (assuming the diaphragm was initially at rest). The displacement, that in turn is obtained as the integral of the velocity, thus behaves as shown in Fig. c. During the positive input pulse, the displacement increases along an upward-opening parabola curve, and during the negative pulse, the increase still continues, but now along a downward-opening parabola.

Thus, in this ideal case, the diaphragm stays, after the signal, permanently in a deviated position (solid line) because the frequency region reproduced by the driver has no lower limit. In practical drivers, however, there always has to be a spring that effects the return of the diaphragm toward the rest position (dashed line), thus limiting the lower cutoff frequency. The behavior of the displacement is even a bit surprising

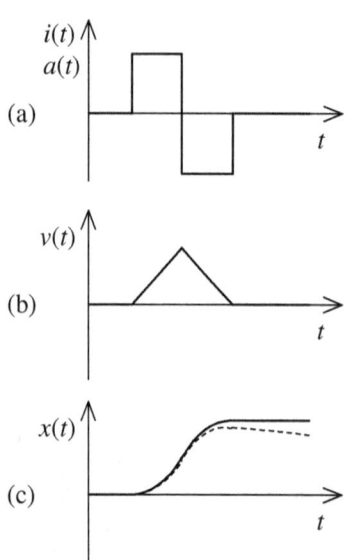

Figure 3.10. Time behavior example of an ideal driver. a) Test signal congruent with diaphragm acceleration. b) Velocity corresponding to the acceleration of Fig. 'a'. c) Displacement corresponding to the previous graphs (solid line) and one achievable with a practical driver (dashed line).

since, by looking at the curve of Fig. c, it does not readily come to mind that as the outcome is a sound signal shown in Fig. 'a'.

When discussing the transient properties, it is often worried how the diaphragm behaves when a signal suddenly stops, and how the consequent post-oscillations are curbed. To produce a sound, in general (e.g. in musical instruments), always requires, however, a vibrating surface or equivalent that has, at every moment, certain displacement, velocity, and acceleration. A sudden stopping of a sound would require, in practice, that these and also the derivative of the acceleration, the next derivative, and so on, should somehow be brought to zero at the same moment. This is an impossible occurrence which also shows up from Fig. 3.10 in that returning the diaphragm home requires new velocity and new acceleration. Hence, no signal originating from any practical sound source ever stops abruptly but always through exponential decay since only on the exponential function all its derivatives decay at the same rate.

According to a common thought, especially propagated by advertisers, the mass of the diaphragm should be as low as possible, in order that it could "follow" rapid signal changes and thus reproduce strokes accurately. However, based on Fig. 3.10c, one has to ask: For what is low mass needed here? In what way does the smallness of mass help, even when desiring to reproduce the unrealistic rectangular signal of Fig. 3.10a? The answer is simple:

The magnitude of the mass of the moving parts of a driver has directly nothing to do with how accurately the driver is able to reproduce various transient shapes. The mass only affects directly the overall sensitivity and resonance properties, and excessive lightness is even disadvantageous due to heightened resonant frequency.

The lightness of the diaphragm material might give some benefit, in terms of sound quality, in the sense that at high accelerations the tensive forces stressing the diaphragm would stay smaller, thus reducing deformations. On the other hand, by allowing greater mass, the rigidity of the diaphragm can be substantially improved, so the above aspect for small mass doesn't carry very far either.

3.8 Directivity and Horn Effect

As frequency increases and wavelength decreases, the sound field of an enclosed driver gradually narrows from omnidirectional to involve only the front side, finally concentrating more and more in the direction of the axis. The phenomenon stems from two factors: the confining effect of the cabinet's front panel and other possible restraining surfaces and, at high frequencies, also the fact that the diaphragm's width becomes significant with respect to wavelength. The former has essential impact also on the front axis sound pressure, whereas the latter only manifests as a weakening of radiation in the side directions.

The driving principle of a speaker is irrelevant with regard to directivity properties, but when pursuing current-drive, one has to pay more attention to the frequency behavior of the driver-enclosure-combination since the same equalization methods that are customary in conventional passive speakers, cannot be used.

The solid angle seen by the vibrating surface directly affects how great is the sound pressure produced by the driver in the space sector in question. The smaller the solid angle seen by the diaphragm at each wavelength, the wider the movement the air particles must exhibit in the range of influence of the diaphragm, in order to respond to the volume alternation caused by the diaphragm. The sound pressure obtained is thus, in principle, inversely proportional to that solid angle from which the diaphragm is able to collect air particles within a certain period of time; and as wavelength increases, farther and farther surfaces affect the determination of this sector.

Figure 3.11 illustrates the change of an enclosed driver from omnidirectional to half-space-radiating, assuming there is no other sound baffling surfaces nearby. The phenomenon is termed *baffle step* since there appears, in the frequency response of a speaker, a step-like rise of 6 dB, as the solid angle halves from 4π to 2π steradians.

It is natural to define the center frequency of the step as the point where the response has changed 3 dB. As a rule of thumb, the wavelength corresponding to this frequency (λ_{3dB}) is about 6-fold compared to the distance from the front panel edge to the center point of the diaphragm (Fig. 'a'). For example, a distance of 15 cm yields a wavelength of about 90 cm, corresponding to a center frequency of 380 Hz.

The shape of the baffle step itself is, in principle, well defined (solid line in Fig. 3.11b), and, for its electrical compensation, a 1st-order filter is enough. The change occurs almost completely in a frequency range of one decade. The actual behavior with a box-like enclosure resembles,

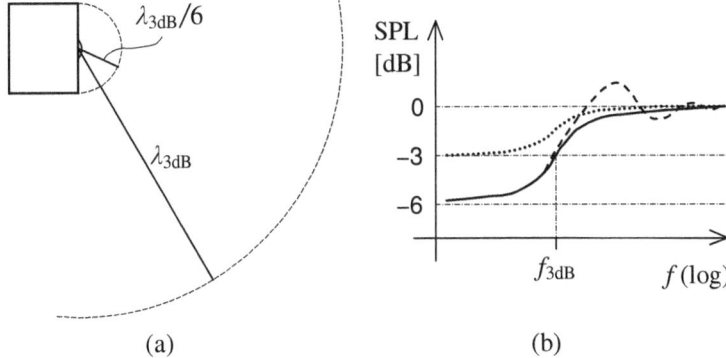

(a)   (b)

Figure 3.11.   a) The wavelength corresponding to the center frequency of the cabinet-induced baffle step proportioned to the width of the front panel. b) General behavior of the baffle step. Solid line: the effect of bare baffle step on the sound pressure measured from the front side (SPL = sound pressure level). Dashed line: a more realistic response, where is seen alternation caused by edge reflections. Dotted line: total response into all directions (power response).

however, more that indicated by the dashed line since the diffractive reflections oriented from the cabinet edges to the listening point cause boosts and cuts when merging with the directly coming radiation.

The dotted line represents the total response, that involves all directions and is also proportional to the total acoustic power and is therefore also referred to as the *power response.*

The baffle step appears only 3 dB high in the power response since the radiation oriented to the back half partially makes up for the lack introduced in the front half radiation at low frequencies. So, as the front pressure drops to half, an other sound field of corresponding intensity develops in the back side which doubles the total radiated power and thus increases the power response by 3 dB, compared to a case where the back radiation is not taken into account. (An increase of 3 dB always denotes a doubling of power and a multiplication of level by $\sqrt{2}$.) The power response doesn't exhibit undulations caused by edge diffractions because the reflections do not increase or decrease the total power of radiation at any frequency.

The increase in power response due to decreasing solid angle also implies an increase in power efficiency since the input power remains unchanged. The phenomenon is utilized in horn-loaded speakers, whose efficiencies are in the class of tens of percents, while a typical figure in a hifi driver radiating into half-space remains below 1 percent. The rise in radiation effectiveness also increases the stress exerted on the diaphragm which can introduce even distortion unless the driver is suited for horn loading.

At high frequencies, the radiation field obtained also narrows due to the driver's own directivity since radiation originating from different parts of the diaphragm arrives to a point aside from the center axis in different phase, so the sound attenuates the more the higher the frequency is and the farther one goes from the center axis direction.

Figure 3.12a depicts the outline of a typical frequency response of a cone driver up from the mid-frequencies in three different directions, in conditions corresponding to an infinite baffle, and using current-drive. The steep fall of sound pressure in the side directions also limits, for its part, the usable frequency band of the driver. Curves taken at 60° off-axis can also be used for rough estimation of the power response since this direction represents, in practice, best the whole half-space.

At high frequencies, the diaphragm doesn't act uniformly but breaks up into differently vibrating regions since the velocity of sound in the diaphragm material itself is generally not sufficient to keep the whole cone at the same phase. At the same time, there also occurs some disconnection of the diaphragm so that only the central part of the cone vibrates with the voice coil, the other part remaining as a passive flare. This shrinking of the effective area is largely compensated by a reduction in the moving mass, but, as frequency rises further, sound pressure

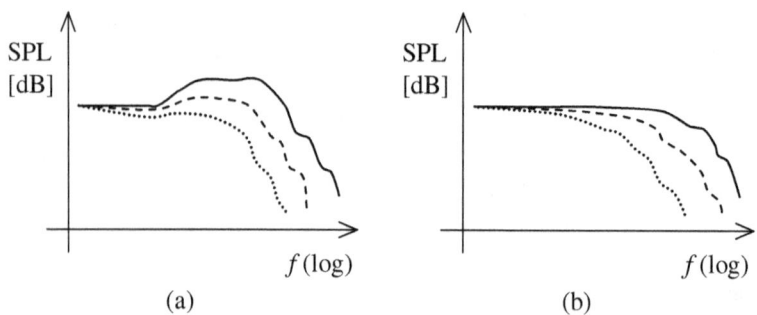

Figure 3.12. Effect of directivity on the performance of a cone driver at high frequencies. a) General behavior of the frequency response under constant current; directly in the front (solid line), at 30° angle (dashed line), and at 60° angle (dotted line). Because of the horn effect caused by the cone, the upper end of the reproduction region tends to be accentuated. b) Curve set corresponding to Fig. 'a' without the horn effect.

finally collapses also on the front axis.

At high frequencies, the horn action of the cone easily causes an extensive accentuation in the frequency response, as sketched in Fig. 3.12a. On voltage drive, the phenomenon is not as prominent due to the filtering effect of voice coil inductance. This accentuation of high frequencies limits the suitability of many present-day drivers, designed solely for voltage, to be current-operated. Without the horn effect, the response on current-drive would stay, in principle, flat up to the final fall-off (Fig. 3.12b).

The material and size of the diaphragm have a remarkable effect on how pronounced this high-tone boost appears. Polypropylene, that has good internal damping, seems to perform in this regard better than many other substances. In metal cones (aluminum, magnesium), the velocity of sound is so high that any horn effect virtually cannot arise, but due to strong resonation in the treble region, these are not very usable at least for passive current-drive applications.

With small cones, the rise in sensitivity is usually of the order of a few decibels and is compensated, at least partially, by the increasing directivity. In large drivers, however, the rise is often as high as 10 dB which calls for appropriate compensation or low enough crossover frequency.

3.9 Power Consumption

Of the quantities describing loudspeaker properties, for many it is power that first comes to mind. Consequently, in the marketing of stereo sets, increasingly bizarre power markings are employed to make an impression; but as a matter of fact, the power limits usually specified for both drivers and complete speakers are some of the most meaningless parameters and fully dependent on the way of definition.

Measurements of power handling capacity are generally performed using a noise signal of specific frequency distribution and crest factor, assuming yet that the load is a resistance equal to the rated impedance. The power ratings determined may perhaps give some reference relating to the amplifier power that can be recommended with a certain type of program material; but instead, they tell nothing about how loud sound the speaker is able to produce without being distorted, and often not even the maximum power tolerated by the voice coil.

It is also quite misleading that the power specified for a driver is not based on the power consumed by the driver itself but on the consumption of the whole speaker system, as a part of which the driver is assumed to operate, through a given type of crossover filter. Hence, there can be presented also for tweeters nominal powers of the order of 100 watts, and no manufacturer dares to differ from the lot and specify, even in addition, a more realistic power capacity, for the figure wouldn't be deemed very attractive.

All electric power taken turns into heat in various parts of the speaker, excluding the portion of acoustically radiated power, which also finally turns into heat when absorbing into room surfaces. In a speaker driver, the power dissipated divides into three components, which, in the case of current-drive, can be conveniently expressed as follows:

- Power dissipated in the voice coil resistance, $R_c i^2$, where i is the RMS value of current. This component only warms the voice coil, from which the heat transfers, with time, also to the magnetic circuit.

- Power consumed by eddy currents and hysteresis, $\text{Re}(Z_i)i^2$, which is thus proportional to the real part of the inductive impedance. This power manifests as warming in the magnetic circuit and has most effect at high frequencies.

- Power consumed by mechanical losses, $\text{Re}(Z_m)i^2$, which is greatest at the resonant frequency, where the real part of the motional impedance reaches its maximum. The warming effect mostly concentrates in the suspensions and in places exhibiting flow resistance.

Consequently, the total sine wave power is $\text{Re}(Z)i^2$, where Z is the total impedance.

All power consumed by the driver thus does not heat the voice coil although the first-mentioned component is dominant in most part of the reproduction band. Using Fig. 3.7, it can be concluded that the power component spent in mechanical work dominates the total power in the resonance region. The power loss of the magnetic circuit, in turn, does not necessarily rise to dominance in the driver's normal operating region but, while warming the pole pieces, can nevertheless increase the risk of voice coil overheating.

In practice, especially at bass frequencies, the limit of sound distortion is encountered much earlier than the limit of voice coil endurance since the powers corresponding to maximum displacement are, in the resonance region and below it, surprisingly low. Figure 3.13 shows a realistic example of this. The curves have been constructed by making

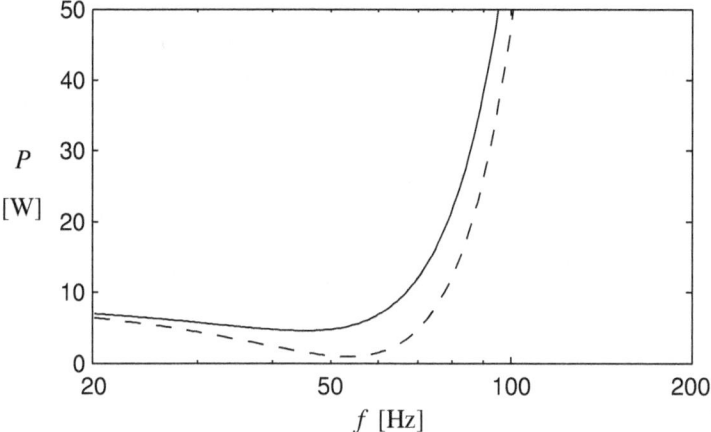

Figure 3.13. Power consumption vs. frequency in a typical 8-inch woofer in closed 40-litre enclosure while cone excursion is kept at maximum (5 mm in one direction). Solid line indicates total power and dashed line power dissipated by the voice coil. Parameters used: Bl = 9 N/A, m = 0.025 kg, k = 3000 N/m (equivalent volume 80 l, Cms = 0.001 m/N), b = 3 Ns/m, and R_c = 6 Ω.

use of equations (3.4) and (3.8) for a typical 8-inch woofer that has been mounted in a 40-litre enclosure and has a linear excursion of ±5 mm.

Rated power for this kind of driver is most often about 100 W, but the excursion-limited power consumption remains here, below and near the resonant frequency (55 Hz), below 10 watts, embarking on steep rise only at higher frequencies (solid line). It is also interesting that the power heating the voice coil (dashed line) is only 1 W at lowest.

Moreover, these powers are valid for sine wave, whose crest factor, i.e. the ratio of peak value to RMS value, is $\sqrt{2}$. In real-world sound signals, the crest factor is considerably higher which further reduces the average power demand in the bass region. Hence, the power handling capacity is quite a useless property at least for speakers intended for the mere low-bass range (subwoofers), that are often marketed by power figures of the kilowatt class.

An interesting phenomenon relating to power consumption is the improvement of the speaker's efficiency at low frequencies when the number of drivers increases.

When a second driver is added near the first and fed by the same signal, the total input power for sure doubles (ignoring negligibly minor changes in the motional impedance). The acoustic power radiated also doubles if the wavelength is so small that the phase difference between the waves sent by the drivers is, all directions regarded, random; that is, the waves are not mutually coincident, except in some directions, like on the front axis.

If, by contrast, the wavelength is so large that the waves sent by the drivers are in phase regardless of the direction, the pressure is in all directions double compared to the radiation of a single driver, denoting fourfold acoustic power. Thus, the efficiency doubles, at low frequencies, due to the presence of the second driver.

This being the case, one may well ask: if the efficiency of a loudspeaker seems to be, at certain frequencies, directly proportional to the number of drivers, can we not, by adding drivers, finally achieve an efficiency greater than one? The frequency can namely always be chosen so low that the condition for the wavelength is satisfied.

The explanation abides in that, as the number increases, the air loading exerted on the diaphragm of each driver also increases, adding to the effective moving mass and thus decreasing the diaphragm's motion, until finally, when the number is still doubled, the acoustic power only doubles instead of becoming fourfold, thus leaving the efficiency unchanged.

Using several drivers is nevertheless, from the standpoint of power consumption, generally a more advantageous solution than increasing the size of a single driver.

3.10 Microphone EMF

The operation of transducers working on the electro-dynamic principle is always reversible, that is to say, the same structure that acts as a loudspeaker also acts as a microphone, and vice versa. This microphone feature of loudspeakers (and correspondingly, the loudspeaker feature of microphones) is thus in action at every moment, whether we wanted it or not. The diaphragm motion arisen by external pressure alternation namely always causes its own EMF component, which summates with the motional EMF deriving from the loudspeaker action.

The mechanical model of Fig. 3.3b can also be used when the driver operates as a microphone. The only difference is that the force F is not generated by current but by the acoustic pressure acting on the diaphragm.

We will consider an enclosed driver that is not loaded electrically and that is small in size with respect to wavelength. Instead of force

(3.1), the diaphragm is now being exerted by force $F = -Sp$, where S is the effective diaphragm area and p is the pressure acting in front of the diaphragm. (Positive pressure gives rise to backward force.) Constant Bl in equation (3.3) is now replaced by $-S$ and current i by pressure p. The transfer function from acoustic pressure to displacement thus becomes, corresponding to equation (3.4),

$$\frac{X}{P} = -\frac{S/m}{s^2 + \frac{b}{m}s + \frac{k}{m}} \tag{3.20}$$

By the same procedure that was used to obtain equation (3.8), we can now derive the relationship between the EMF of the microphone (labeled E_p) and the pressure:

$$\frac{E_p}{P} = -\frac{SBl}{m} \cdot \frac{s}{s^2 + \frac{b}{m}s + \frac{k}{m}} \tag{3.21}$$

Thus, the frequency response of a sealed microphone element follows, in principle, a corresponding 2nd-order band-pass function as the motional impedance of the element.

How, then, is it possible to achieve flat frequency response with dynamic pressure microphones altogether? Very good result is indeed not easily achieved, but, by making the Q value very low (< 0.2) and by choosing the resonant frequency appropriately, quite a useful frequency range can still be obtained. The response is also usually modified with extraneous cavities.

Another major class of dynamic microphones is formed by so-called pressure gradient microphones, that in fact respond to particle velocity, instead of pressure, and correspond, by their principle, to a free or openly enclosed speaker driver. In this kind of structure, there is established, between the front and back surfaces of the diaphragm, acoustic cancellation, that in loudspeaker use causes the well-known 1st-degree roll-off in the front-measured frequency response, down from a certain cut-off frequency. In microphone use, however, the corresponding phenomenon is of advantage because the differentiator action it introduces turns the aforementioned band-pass response into high-pass, which often serves the purpose better.

Transfer function (3.21) can also be substantially simplified if frequency is confined to be distinctly above the resonace point. Conforming with formula (3.10), one can namely write:

$$\frac{E_p}{P} \approx -\frac{SBl}{j\omega m} \quad , \qquad \omega \gg \omega_0 \qquad (3.22)$$

The microphone EMF produced by an enclosed speaker driver is thus, in the mass-dominated region, in principle, inversely proportional to frequency. However, as wavelength decreases close to the dimensions of the device, equation (3.22) no more represents the whole truth.

4
THE CONSEQUENCES OF VOLTAGE DRIVE

When listening to speech or music through a conventional, even of high-quality, loudspeaker system, we quite clearly detect that the sound emanates from loudspeakers. A kind of electrical stamp or characteristic in the sound always reveals that one is experiencing an electronically reproduced image and not a genuine, live performance. This general impression of roughness, that could be called something like synthetic dressing, does not disappear even with the most expensive equipment and harms especially acoustic music since instruments do not sound as they do in actuality, and choral voices, for example, clot and become easily distorted. Amongst hifi hobbyists, an often used term is listening fatigue, which is related to lacks in sound quality but can seldom be linked to any measurable explanation.

What is the reason that, even in this age of communication technology, a more lifelike sound reproduction has not been achieved? Solutions have been sought mostly by increasing the number of sound channels and by digital processors striving to convey spatial information, but the factors essentially affecting the operation accuracy of the speaker driver itself, the electromotive forces, have been left without proper attention although some discussion relating to current-drive experiments has been seen in the past decades.

As it appeared in the previous chapter, the electromotive forces are a significant part of the driver's total voltage for all frequencies. In the following, we will examine what kind of destruction these parasite voltages and some related factors produce when they are freely allowed to merge with the applied signal. Let's therefore let the cat out of the bag.

4.1 Circulation of the Electromotive Forces

As was already found, an electro-dynamic transducer itself doesn't know whether it is intended to convert electrical signal to mechanical motion or do the reverse, so it serves both offices all the time. When dealing with a low-impedance source or load, however, these two functions will not remain separated from each other but intermingle in a way that is not acceptable from the standpoint of either goal. Taking into account how poor usually is the quality of the microphone signal generated by loudspeakers, it is truly unfortunate how indifferently its effects have been taken and how little is generally known about it.

The motional EMF induced in the voice coil, discussed in sections 3.3 and 3.10, follows at every moment the velocity of the voice coil, in principle, according to equation (3.7).

Figure 4.1 shows how this EMF (e) acts in the circuit formed by the amplifier output and speaker driver. Figure 'a' represents the situation on voltage drive where the EMF generated by the driver appears in series with the amplifier output voltage, thus affecting essentially the development of current. In addition to the current component u_o/R_c, representing the applied signal, there also flows an extraneous current component, e/R_c, that can in many respects be regarded as a disturbance factor. Because e is, in practice, almost of the same order of magnitude as u_o, current e/R_c has also a very pivotal role in the development of the drive force actuating the voice coil.

Figure 4.1b represents, respectively, a driver operated by a current

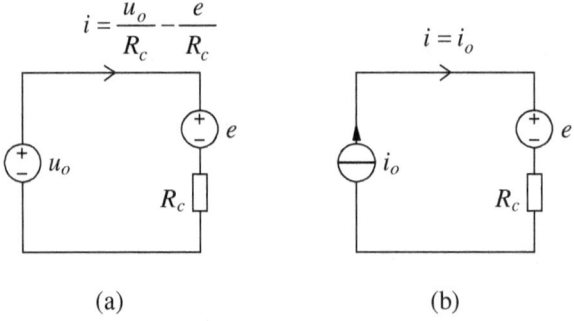

(a)   (b)

Figure 4.1. The composition of current in a speaker driver, with voltage source feed (a) and current source feed (b). On voltage drive, the current is affected, besides the applied voltage u_o, by the driver's electromotive force e. Instead, on current-drive, the resulting current is exactly what is fed (i_o).

source. Now, the EMF introduced in the voice coil is not able by any means to affect the flow of current, which is determined solely by the feeding source. As the current is forced, the EMF only manifests as an additional voltage component at the driver's terminals but is not involved in voice coil actuation.

Consequently, under voltage drive, the voltage applied as though first effects a certain current, that sets the voice coil and diaphragm in motion. This motion, in turn, induces an EMF, which causes its own current component, that is limited only by the voice coil resistance (R_c). This current, in turn, again actuates the voice coil, in which is now generated a new EMF, from which ensues new current, and so on. Thus, in better words:

A loudspeaker circuit operating on voltage drive exhibits a feedback effect where the EMF deriving from voice coil motion directly summates with the voltage applied to the driver, so that the resulting current is a mixture of the original signal and a spurious signal corrupted by the speaker's own mechanical, electrical, and acoustic properties and circulated in the feedback process.

A corresponding feedback mechanism is in action also with the inductive EMF.

Imagine whatever non-ideality or interference factor that strikes the operation of the driver, introducing an EMF. In the case of Fig. 4.1a, this EMF always effects a corresponding disturbance current because voltage source u_o appears as a short-circuit for other sources.

As the same disturbance EMF appears in the Fig. b circuit, the outcome is entirely different since the current source i_o is seen by the EMF as an open circuit, thus eliminating the generation of unwanted currents and keeping the voice coil operation immune not only to own spurious voltages but also to other adverse factors discussed later.

4.2 Microphone Feedback

A loudspeaker diaphragm produces sound by the same principles both forward and backward. However, inside an enclosure, the sound pressure is, at bass and lower midrange frequencies, many times higher than outside because the interior pressure is not able to spread to the ambience and because the interior surfaces close to the driver affect like horn loading, increasing the intensity of the back radiation. In terms of quality, however, the back radiation can never match the front radiation due to the blocking effect of the frame and the uncontrolled behavior of the cone's inner suspension. Inside the enclosure, the sound is usually also muddled by strong reverberation together with possible standing waves and other enclosure-based colorations.

In bass reflex enclosures, that have gained ascendancy today, it is not possible to use so much damping material that it would conclusively decrease the level of the lower mid-frequencies inside. In closed cabinets, it has not been customary to use great amounts of damping material either because it has been suspected to cut the volume. In supermarket grade sets, it is even common that there is no damping material at all or only a bit. Therefore, it is inevitable that, in all loudspeaker implementations of the present, a portion of the enclosure sounds penetrates through the cone, thus merging with the speaker's front radiation (Fig. 4.2a).

This passing through is limited only by the mass associated with the cone since electrical damping (section 3.6) has no significance when being far from the resonant frequency. The present-day pursuit of low moving mass has thus worsened the leakage of enclosure sounds though in lightweight low-cost speakers the resonation of the walls itself may be an even greater detriment.

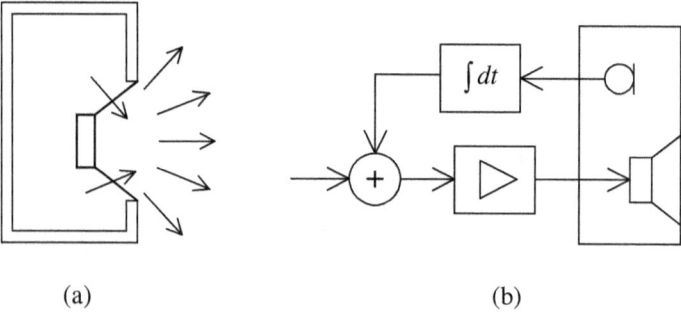

(a)   (b)

Figure 4.2.   a) Reflected back-radiation passing through the diaphragm and merging with the front-radiation. b) A feedback system that functionally corresponds to the behavior of a voltage-driven loudspeaker, as pressure alternations inside the cabinet affect cone motion. An amplifier that drives a blamelessly operating speaker driver is fed by a signal to which has been mixed a signal derived from a poor-quality microphone disposed in the cabinet. The integrator after the microphone represents only the frequency dependence of the EMF generated by the speaker's microphone action (equation (3.22)).

When hitting the cone and its suspensions, the enclosure sounds develop forces which the driver, in the manner of its own microphone action, converts into voltage. Especially in large drivers, these forces do not even distribute evenly on the whole diaphragm but can, depending on the wavelength, press different portions differently, thus introducing swinging movements also. Considering yet the above-described general quality properties of the enclosure sound, it would seem reasonable to take care, except of good damping, also that the remaining leak sounds are not able to modify the very signal that is being reproduced.

The using of current-drive itself does not eliminate the penetration of the enclosure sounds, but it prevents the indefinite EMF, generated by them, from riding on top of the value signal.

When using voltage-drive, this EMF acts, in the midrange, correspondingly as the feedback arrangement shown in Fig. 4.2b, where the output obtained from a poor-quality pressure microphone, mounted inside the cabinet, is summed to the program signal itself via a certain filter that follows the transfer function of the speaker's microphone operation and is thus presented as an integrator.

How many an earnest hifi listener, professional in the field, or even lay consumer would accept the connection shown in Fig. 4.2b to be actually built in his equipment? Probably not many would be delighted for such an effect if it were offered as a separate accessory. But when acting built-in in the driver's own circuit, a corresponding phenomenon is, however, operative all over the world as part of a standard practice assumed as a given.

The EMF generated by the enclosure noise can be measured by using a cabinet in which has been mounted two identical drivers, one being driven by sine wave and the other acting as a microphone. If it can be assumed that both drivers see the pressure, bouncing in the cabinet, in equal magnitude, both also incur equal enclosure-derived EMF. This is fulfilled fairly well especially if the drivers are disposed in their panel symmetrically because then they are in equivalent position also with respect to standing waves.

The accuracy of the measurement is though deteriorated a little by the fact that the enclosure sounds are able to leak from two passages while the purpose is after all to examine the enclosure noise EMF introduced by a single driver for itself. In any case, the experiment provides a good conservative estimate of the effect's magnitude and at least indicates accurately how two drivers disposed in the same chamber disturb each other when operating on voltage drive.

Figure 4.3 displays results from one such measurement. The experiment was carried out using an 11-litre home-theater center-speaker cabinet, in which was tightly mounted two 5-inch woofers intended for hifi usage. The measurement was confined to the midband ($\geq$200 Hz) because at bass frequencies the cabinet pressure is virtually independent of position and is incorporated in the speaker's designed operation.

The curves indicate the proportion of the enclosure-derived EMF (E_p) in the active driver's terminal voltage (U_o) in variously damped cabinets. The percentage indicates, at the same time, how effectively the enclosure sounds modulate the voice coil current under voltage drive, that is, how large is the feedback signal that enters the summing block in Fig. 4.2b.

The EMF measured without damping material (curve A) varies in the 200-350 Hz region from 3 to 5% but rises at the standing-wave frequency even above 11%. As frequency increases, the EMF decreases due to weakening of the enclosure noise itself and the integrator property of

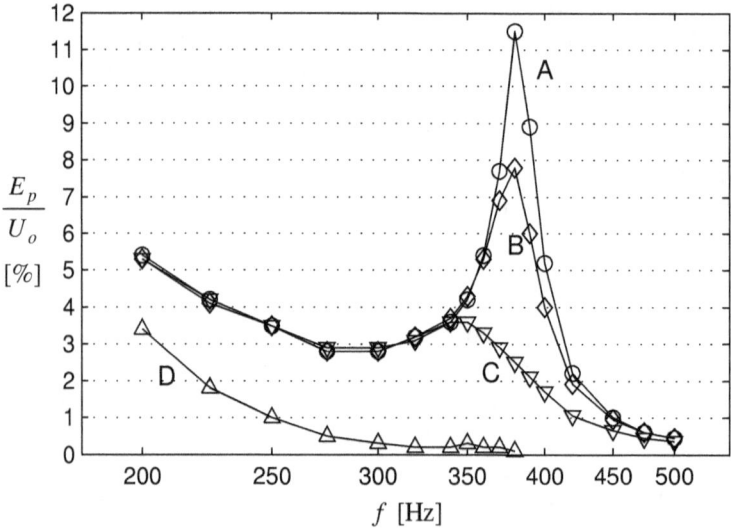

Figure 4.3. Magnitude of enclosure-derived EMF proportioned to the driver's total voltage in a 5-inch woofer (Vifa M13SG-09-08) using a cabinet of 11 litres with four different stuffings. A: without damping material; B: cabinet's rear wall covered by a thin layer of polyester fibre; C: cabinet half-filled with polyester fibre; D: cabinet full-packed with cotton cloth. The peak at 380 Hz (350 Hz for curve C) is due to the first standing wave, as the half of the wavelength becomes equal to the chamber's largest dimension.

the driver's microphone action (equation (3.22)), being nevertheless still at 500 Hz about half a percent. A thin absorbent material covering one interior wall is virtually of no help (curve B) although the strength of the standing wave somewhat relents.

Curves A and B represent typical enclosure damping in common-grade stereo and home theater sets as well as in PA speakers. In addition to the general reverberation, also the standing wave determined by the longest cabinet dimension is strongly involved in the EMF feedback and production of cone-penetrating interference.

When the cabinet is half-filled with polyester fibre (curve C), the standing wave already attenuates quite well, but ordinary reflections do not diminish at the low end at all, and even at the high end the reduction is minor. The filling degree represented by curve C is usual in quality grade bass reflex speakers but inadequate to essentially relieve the inside developing pressure.

It is possible to damp a cabinet also so that pressure at the mid-frequencies decreases conclusively. Curve D has been measured with the cabinet full-packed with ordinary sheet cloth so that only behind the drivers there is some empty space. The pressure-induced EMF has distinctly attenuated already at 200 Hz, diminishing yet in the region of 300 Hz to almost negligible quantities. (The measurement is slightly also affected by the outwardly arriving pressure.) A damping of this effectiveness is not known to be employed anywhere, but the experiment shows, however, that even the penetration problem is surmountable.

When interpreting the results of Fig. 4.3, one must yet regard that the drivers were quite far from each other, relative to the depth and height of the cabinet used. Thus, especially at the higher frequencies, the measured EMF may underestimate a little how the active driver itself experiences the case.

With the obtained results and driver data, one can also estimate the relative level of the penetrating back radiation.

As an example, we consider the situation at 300 Hz where the driver's impedance magnitude is about 7 Ω. The magnitude of the motional impedance is, in turn, using formula (3.10) and the rated parameters (Bl = 5.2 Tm and m = 0.0065 kg), 2.2 Ω. Further, the ratio of the motional EMF (excluding microphone contribution) to total voltage is the same as the ratio of the motional impedance to total impedance since both have the same current. Consequently, the motional EMF magnitude of the active driver is in this case $(2.2/7)U_o = 0.31U_o$. On the other hand, the enclosure-derived EMF, read e.g. from curve C, is $0.029U_o$, so the cone motion caused by cabinet pressure is 0.09 times as high as the drive-induced motion.

Therefore, the enclosure sound passing the cone is at 300 Hz only 21 dB weaker than the direct sound that is to be listened. Such a collateral sound should already worry any quality-interested listener. The help is, however, easily accessible since with the cabinet damping corresponding to curve D the levels ratio will be 0.003/0.31, or 40 dB, which is already a much more tolerable level difference.

Moreover, the sound penetration is also essentially affected by the effective area of the cone. As the operative pressure remains unchanged, a large diaphragm namely moves as much as a small one, assuming the ratio of mass and area is constant. Consequently, the strength of the penetrating sound is, in principle, directly proportional to cone size.

The effective area of the driver used in the experiment was only 80 cm^2. If the area were, in the same conditions, say 130 cm^2, which is a typical value in 6.5-inch drivers, the aforementioned 21 dB level difference would diminish to a paltry 17 dB and at the standing wave frequency even lower. Increasing the enclosure volume does, in practice, weaken the interior pressure, though, but this doesn't, however, solve the problem but rather shifts it to lower frequencies.

It is thus evident that even the current-drive principle cannot fully achieve its potential unless enclosure damping is also begun to be considered from new premises.

The feedback percentages presented in Fig. 4.3 are hair-raising even as such. However, the midrange sensitivity of the drivers used is, according to the manufacturer's graphs, only 87-88 dB (1W, 1m), corresponding to a typical level in small hifi drivers.

As sensitivity rises, the microphone feedback also essentially strengthens because, when the drive signal is kept unchanged, both the sound pressure and microphone sensitivity increase. An increment of 1 dB in sensitivity thus increases the microphone feedback by 2 dB (assuming that $E_p/U_o \ll 1$). Hence, if the drivers in the said test cabinet had a sensitivity of say 91 dB, which is still a usual value in many applications in this size class, the percentages shown in Fig. 4.3 would double which should already give some reason for pondering even for the most seasoned designer.

Also, this is not yet all since, in cone drivers intended for PA use, sensitivities even of the order of 100 dB are common. No wonder then that the ability of PA systems for faithful sound reproduction is even generally considered lacking, and the voice of a singer is indeed often only a ghost image of the actual one. The reasons are, of course, manifold, but the cabinet noise feedback approaching 20 percents is hardly the least of them.

The above-described phenomena are also present in the operation of tweeters. Only the frequencies considered are about 20-fold compared to the previous analysis.

In dome type tweeters, the back space of the diaphragm has usually been extended by a cavity drilled in the pole piece and has generally been damped better than bass enclosures. The diverse reflections, that hit the diaphragm from the back side, produce, however, EMF interference similarly as in cone drivers. The magnitude of the feedback is difficult to get measured, though, but there shouldn't be any reason why it would be essentially lesser than with woofers. Instead, the moving mass per area is in tweeters generally smaller, so the susceptibility to sound penetration and accompanied post-oscillations is even greater than in bass enclosures.

In small and cheap tweeters, there is often only a flat metal surface behind the diaphragm. As the space is very tiny, actual reflections are able to develop only at the highest frequencies, but at the same time, other pressure-induced problems and the resonant frequency increase.

4.3 Outward Microphone Coupling

Microphone effects are not limited merely to the enclosure-pressure-induced parasite signal although this constitutes the main part of acoustically coupled interference. Adjacent drivers namely introduce electromotive forces for each other also by means of outer pressure. The coupling is strongest between adjacent cone drivers, but the effect is also observable when the other participant is a little dome.

The subject was examined with two 8-inch drivers, each in its own closed cabinet, the sensitivity of the drivers being about 90 dB (1W, 1m). The cabinets were side by side, almost touching, so that the distance between the drivers' center axes was 26 cm, while the distance to the floor was 60 cm.

The EMF induced in the receiving driver was measured up to 200 Hz by means of current, the terminals being shorted which corresponds to the case of a voltage-driven speaker where electrical damping curbs diaphragm motion in the resonance region. At frequencies over 200 Hz, where loading no more affects the result, normal voltage measurement was used.

The outcome is shown in Fig. 4.4. The EMF magnitude introduced in

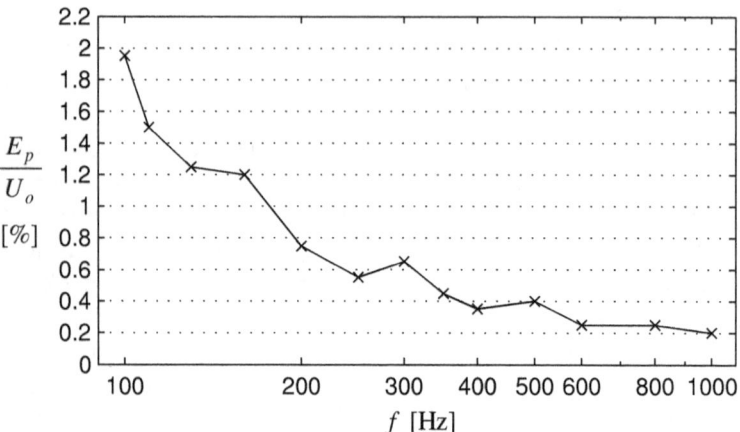

Figure 4.4. Example of the outwardly coupled microphone effect between two adjacent drivers. The curve represents the ratio of induced microphone EMF (E_p) to driver's terminal voltage (U_o) under voltage drive. The measurement has been made with ordinary 20-cm diameter drivers whose centers were 26 cm from each other.

the receiving driver, relative to the terminal voltage of the sending driver, extends at 100 Hz to almost 2 percents and decreases unevenly with increasing frequency, ending at 1 kHz yet to a level of 0.2%. Moreover, the drivers used were not as close to each other as possible in practice. If they were mounted almost in contact, the EMF percentages shown would yet have to be multiplied by 1.2. (Pressure is generally inversely proportional to the distance of the source.)

The measurement directly indicates the relative magnitude of this particular EMF component in parallel-connected drivers under voltage drive. The voice coil current is, of course, modified in the same proportion, thus establishing acoustic feedback from one driver to the other. Because the interference is conveyed outwardly, damping materials or interior structures have no effect on it.

Series connection does not bring improvement either since, although the impedance seen by each EMF roughly doubles, the EMFs acting in series keep the interference current approximately the same. The main difference is that in the series connection both EMFs affect the operation of both drivers.

Instead, on current-drive, these interference currents either are not generated or they cancel out. In a current-driven series connection, EMF effects only show up as voltage fluctuation like in a single driver. In a parallel connection, the drivers form a loop where EMF-derived currents are able to circulate, though; but since these currents flow in the drivers in opposite directions, audible interference effects remain slight. Also, it should be noted that in a current-driven parallel connection the EMFs of the drivers act against each other, so in many cases even the mentioned loop current stays low.

The sensitivity of the driver is a crucial factor also in the strength of the outward microphone coupling, as in the cabinet noise feedback. By rising the sensitivity from the above 90 dB to say 96 dB, the EMF ratio shown in Fig. 4.4 would in principle become fourfold though in practice the ratio does not exceed the acousto-mechanical coupling ratio. Consequently, in PA speakers this parasite signal reaches at low frequencies at least to the order of 5%.

A pressure wave applied on the cone from a side direction does not necessarily effect uniform force on the whole surface, but, depending on the wavelength and also diaphragm shape, different areas may vibrate in different phase and magnitude. The motion transferred to the voice coil is thus not only axial, but other modes of vibration may also occur, like with the back radiation. Extra hue is also introduced by the delay due to the spacing of the drivers, being of the order of a millisecond, and by the edge effects of the front panel. Although the EMF introduced on sine wave is yet sort of sinusoidal, one cannot expect very applaudable transient properties from this kind of a microphone.

Passive motion in the speaker diaphragm due to external pressure fluctuations is not yet detrimental in itself. In some room surfaces there may also be felt vibration when playing low tones, and this is also relevant. Detrimental is, instead, that the vibration of some parts is coupled backward in the signal path, thus creating very complicated near-echo effects.

As with adjacent drivers, coaxially disposed drivers are also susceptible to pressure fluctuations from one another. In coaxial drivers, where the treble unit has been built inside the coil former of the woofer, there easily takes place remarkable acoustic coupling between the units.

This has significance at least in the region of the crossover frequency where the signal of the sending unit has not yet attenuated much and the receiving unit is still capable of effective operation. Because, on voltage drive, the impedance seen by a driver is in many crossover filter circuits rather low for all frequencies, requisites for the generation of EMF-derived currents are good also here.

Figure 4.5 shows the strength of the microphone coupling occurring from the treble unit to the bass unit in a coaxial fibreglass-cone driver.

Coupling also occurs in the other direction but mostly only in the region of the treble unit's resonant frequency.

The EMF introduced in the bass unit is more than 1% of the treble unit's voltage in quite a wide frequency range. (The measurement has been confined to 7 kHz, where the response of the bass unit approximately extends.) Below 1.7 kHz, the EMF ratio starts to fall off which is due to normal frequency response behavior of the treble unit.

The ratio does not, however, decrease with increasing frequency, as in the previous example, but takes on even a rise. At least part of the reason is in the cone disconnection phenomenon, that in loudspeaker action causes the vibration to concentrate at high frequencies to the root of the cone. Hence, when wavelength is of the order of the cone radius, forces acting on different hoops are not able to cancel each other, as would happen in a totally rigid cone. The measurement may also include some coupling derived from the mutual inductance of the voice coils.

In the test, it was further observed that the established microphone voltage strongly depends on the cone's displacement. By pulling the cone forward a few millimetres, the EMF percentages shown in Fig. 4.5 even doubled, and correspondingly, by pushing backward, the values approximately halved. Therefore, a voltage-driven coaxial device not only develops unnecessary shadow signals from one side to the other, but these parasite products are further yet modulated heavily and non-linearly in congruence with the cone motion caused by bass tones.

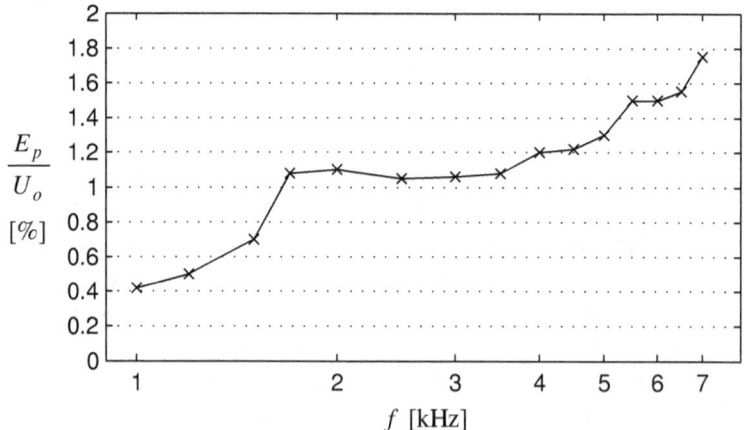

Figure 4.5.   Microphone EMF induced in the bass unit, proportioned to the treble unit's voltage, in a domestic coaxial driver (Seas G17RE COAX/TVF). Measurement covers the frequency range that both units are able to reproduce.

As a whole, one can thus note:

In loudspeaker systems employing several adjacent cone drivers and in coaxial drivers, rather common in hifi use; there is generated due to outward microphone coupling considerable electromotive forces, which under voltage drive produce in the voice coils interference currents of audible level. With lower driver sensitivities, these currents are, at their effective frequencies, roughly estimating, only 40 dB weaker; and in PA applications only 30 dB weaker; than the actual signal. Instead, under current-drive, these detriments are avoidable because the total current of the drive units can then be kept accurate, irrespective of any microphone effects.

4.4 Feedback of Mechanical Disturbances

In section 3.3, we derived a beautiful expression for the motional impedance of an F/C driver and, likewise, for the relationship between velocity and current (equation (3.8)), taking as the basis the ideal model of Fig. 3.3b. If the behavior of the motional EMF were indeed so definite and linear for all frequencies, a voltage-operated driver would be able to perform the required voltage/current-conversion in this regard blamelessly, and the differences with respect to current-drive would be considerably lesser. However, besides the microphone effects, there exists a whole lot of other mechanisms that deviate the motional EMF from this regularity.

At least the following non-idealities are able to generate in the voice coil extraneous mechanical vibrations, regardless of signal level:

- Reflections from the cone rim. Despite the matching of material properties, a portion of the transverse wave, propagating in the cone, returns back toward the center. At a certain frequency, the surround may even vibrate in antiphase to the cone, resulting in a dip in the frequency response.

- In a dome diaphragm, the returning of the mechanical wave back to the joint of the coil former

- Loose mass and reflection effects of the cone's inner suspension

- Modification of the effective mass due to waving and disconnection of the diaphragm.

- Bell modes developing in the cone at certain frequencies, causing the diaphragm to divide into sectors that vibrate in different phases. These deformations make the cone intermittently higher, thus introducing fluctuation in the voice coil position.

- Air currents through a perforated coil former and through the air gap of the magnetic circuit

- Stirring of ferrofluid around the voice coil

- Flexing of the voice coil adhesives and coil former. At the high end of the frequency range, cone excursions are so tiny that even a slight compliance in the glue layers can introduce response alterations and even hysteresis.

- Air loading required for acoustic radiation. Especially on the part of backward radiation, this loading may include indefinite attributes.

Reflections and resonances induced by cone suspensions (Fig. 4.6a) are, of course, harmful in themselves, irrespective of the driving practice, since they blur the driver's phase behavior and bring about roughness also in the amplitude response. Here is one reason why it is worthwhile to keep the crossover frequency low. In tweeters, a corresponding phenomenon is that the wave leaving the dome root propagates via the apex back to the root (Fig. 4.6b), from which possibly occurs yet reflections.

However, in voltage-operated drivers, the adverse effects are not

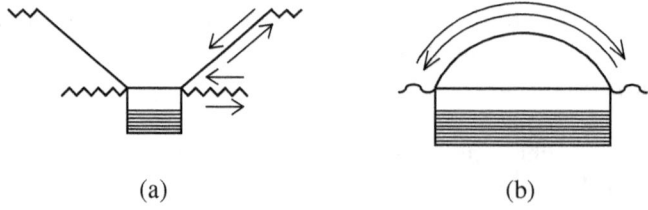

(a)   (b)

Figure 4.6.   a) Mechanical reflections caused by the suspensions introduce delayed vibrations to the voice coil. b) Mechanical wave propagating in a dome diaphragm ends up back to the voice coil with related consequences.

limited to this, but the damage is yet put in circulation through the feedback caused by the voice coil's motional EMF. Thereby, these myriad mechanical reverberative interferences are mingled as part of the voice coil current and hence as part of the upcoming sound signal. Considering all the above listed factors, the current of a voltage-operated driver always incorporates a diverse batch of inappropriate mechanically derived components, that corrupt sound quality also at those frequencies where microphone effects do not reach to.

Due to internal losses of diaphragm materials, it is usual that, as wavelength becomes small enough, the wave propagating in the diaphragm substantially attenuates before arriving at the rim. Respective reflections then, of course, lose their significance. Due to the reduced mass loading the voice coil, however, the motional impedance then does not fall off in inverse proportion to frequency as would otherwise happen, according to formula (3.10), but remains at a considerably higher level. Consequently, also the ability of various interference forces to effect motion in the voice coil is magnified in the same proportion.

Everyone who has been dealing with pressure air knows that there is no flow without a sound. The pressure developing under the dust cap is often conducted through holes, in the coil former, into the space under the inner suspension, and from there, the flow continues out through the porous suspension textile. It is also very common that this pressure does not have any arranged exhaust, the only route going then between the voice coil and pole pieces. The blow sounds produced by these currents do not necessarily carry to the listener directly, because of the intervening speaker diaphragm. However, in a disconnected voice coil they easily bring about vibrations, that, again, are passed on electrically.

In major part of tweeters currently being produced, the air gap has been filled with magnetic liquid, whose purpose is to cool the voice coil and to perform as a damper in the resonant frequency region. It is, however, difficult to imagine how such a fluid could keep completely steady or move linearly with the coil in all situations. Slopping is characteristic for liquids, and because a filling that covers the voice coil cannot be negligible in mass, some kind of noise inevitably results.

Concluding from the extent of usage, the benefits of ferrofluid have, however, been deemed greater than the drawbacks which is naturally not a wonder, on voltage drive, for the generated interference sounds are easily left, even when reinforced by the EMF feedback, in the shadow of other recycling products. However, when moving to current-drive, the sense of ferrofluids ought to be assessed anew, for a direct way of driving can bring out much that has been impossible to discern before. Unfortunately though, driver manufacturers do not always even state directly which types are equipped with the fluid and which are not.

The loading exerted on the diaphragm by the surrounding air shows up, above all, as extra mass, that changes at the high end of the reproduction band into mechanical resistance. In the back side of the cone, however, air has to move relatively narrowly, encountering sharp-edged obstacles and the partially transparent inner suspension. At low frequencies, also the damping material properties affect the acoustic operation of the cone.

In these conditions, there easily occurs strong diffractions and non-linear pressure behavior; and on voltage drive, also the vestiges of such effects integrate as part of the driver's current.

Thus, the voice coil is always subject to several remarkable mechanically or pneumatically coupled interference forces, due to which the behavior of the motional EMF is not at all so simple and well-defined as to make it judicious to keep this voltage component as a filter in a conversion between signal quantities, as is done today.

A speaker driver, with all its non-idealities, is always a weak link in the sound reproduction chain, and current-drive does not change this fact either. However, because of the existing practice of operation, the chain incorporates *two* weak links since the problems limiting the performance of the driver's moving parts are also reflected in the voltage/current-conversion, that has been left as the responsibility of the speaker.

The goal has been only that the amplifier is able to supply *enough* current in various conditions, never minding what that current in fact consists of.

It is possible to measure the behavior of the motional EMF by using two identical drivers, of which one's voice coil has been glued immovable. Both drivers are fed by equal sinusoidal current via own series resistors. The motional EMF, that is thus present in one driver but missing from the other, is then found out simply by measuring the difference of the voltages developed. The differences occurring in the inductances and resistances of the drivers limit the absolute accuracy of the result, though, but the variations as a function of frequency are, nevertheless, shown up clearly.

Figure 4.7 shows the magnitude of the motional impedance in the mid-frequency region, determined this way, in the same 4-Ω woofer

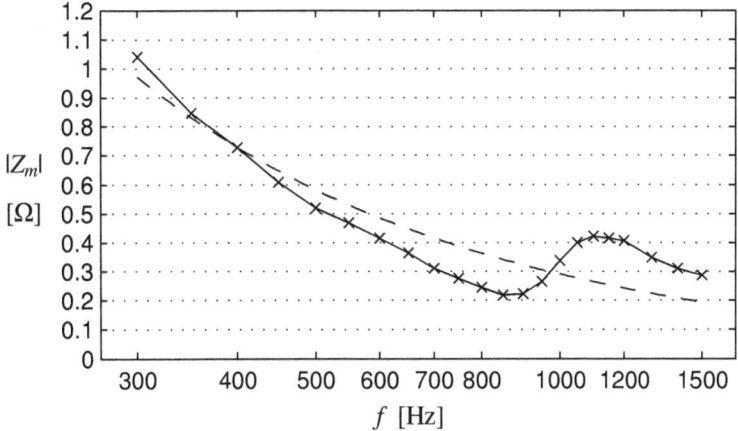

Figure 4.7. Solid line: motional impedance magnitude as a function of frequency in an unmounted 4-ohm bass-midrange driver (Vifa P17WJ-00-04). Dashed line: dependency that is inversely proportional to frequency.

that was earlier used in the measurement of inductive impedance (Fig. 3.6). The dashed line represents, for comparison, the ideal dependence given by equation (3.10), scaled to give the best fit.

The measured $|Z_m|$ remarkably deviates from the inversely proportional frequency dependence, and at 1.1 kHz, there occurs a hump, that has no counterpart in the driver's frequency response. It is difficult to say which of the above-described effects are responsible for this resonation, but in any case, the result indicates how uncertainly-based the build-up of the motional EMF generally is. As frequency increases further, $|Z_m|$ stays higher than its theoretical value, as expected.

At the peak, $|Z_m|$ extends to 0.42 Ω, which is over 10% of the rated impedance and about 14% of the DC resistance. At 300 Hz, the respective readings are 26% and 34%. It must yet be recalled that the relative proportion of the motional impedance increases as the factor $(Bl)^2/m$ increases (equation (3.10)).

Such a parasite voltage seems, nevertheless, to have gained acceptance although when e.g. amplifier distortions are concerned, even figures of a few tenths of a percent are often regarded as unforgivable.

In the impedance curve of many cone drivers, there can be seen, in the midrange or lower treble range, one or more narrow-sized humps. These peaks stem just from the fact that the voice coil may, to some extent, live its own life although the mass of the moving parts, taken as a whole, would behave more regularly.

The driver used in the above test had not been selected in any way, and virtually no bumps can yet be seen in the impedance curve. The EMF variation shown in Fig. 4.7 is therefore yet mild, compared with many other, even high-quality, drivers.

4.5 Phase Modulation

Above, we have considered small-signal phenomena, whose occurrence is not dependent on loudness. At high signal levels, there are also involved deformations due to hard acceleration, and force factor variation, that causes a side-effect we will discuss next.

As cone excursion becomes large enough, the amount of voice coil turns in the efficient magnetic field starts to decrease, and the amount of turns outside the air gap correspondingly increases. In drivers where the voice coil is distinctly taller or shallower than the air gap, this declining of the force factor essentially begins only after a specified excursion limit (parameter X_{max}), but in practice, due to the stray field (Fig. 3.1), the fall starts even before. Instead, in cases where, in order to maximize sensitivity, the voice coil length has been made to match the air gap, the range of constant force factor becomes rather diminutive.

This Bl variation as a function of displacement gives rise to well-known harmonic distortion and accompanying intermodulation distortion. Less known is, however, that the resulting drive force (equation (3.1)) involves on voltage drive, in addition to the direct Bl error, also distortion arising from the impaired motional EMF.

We will consider a voltage-operated woofer that is fed by two different frequencies, chosen so that one effects large excursion and the other is many times higher. Ignoring amplitude-related distortions, we look what happens to the phase of the latter signal.

Let us assume that the higher frequency is close to the instance of the driver's impedance minimum where the composition of impedance may originally be like in Fig. 4.8a (see also Fig. 3.7). The motional and inductive impedances have almost equal magnitudes, making the total impedance resistive. Hence, at this frequency, the current of the driver is initially in phase with the applied voltage.

However, as the force factor decreases in consequence of the low-frequency signal, the impedance changes since Z_m is proportional to the square of Bl, and as the displacement reaches its peak, the situation will be that shown in Fig. 4.8b. Due to the shortening of the Z_m vector, the direction angle of the total impedance has increased which is directly

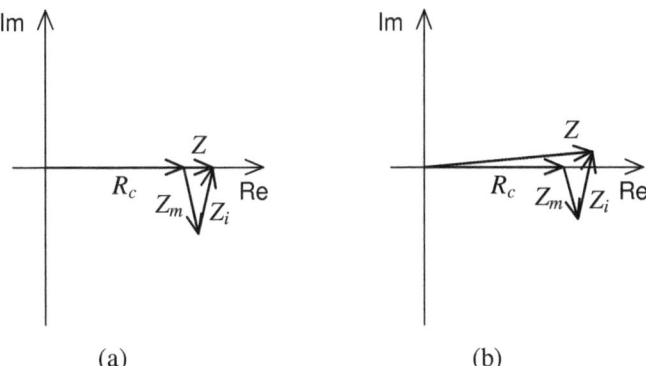

Figure 4.8. a) Impedance diagram of a woofer at the example frequency while the diaphragm is in its middle position. Total impedance Z has the direction of the real axis. b) Corresponding diagram while the cone is far from the middle position. Due to reduced motional impedance, Z has turned slightly inductive. (Z_i assumed constant).

reflected to the phase of the driver's current in that frequency region.

The Bl variation due to large displacements always causes in the motional EMF and consequently in the total impedance aberration that shows up, in the mid-frequency region, mostly as modulation of the impedance's angle in congruence with the displacement. On voltage drive, this modulation is directly transferred to the reproduced signal, giving rise to phase noise, or jitter, and thus adding significantly to nonlinearity during loud bass tones. Instead, on current-drive, this impedance interference is only conveyed to the driver's terminal voltage where it is of no harm; and hence, nonlinearities are limited to the distortion that directly stems from the Bl variation itself.

Figure 4.9 illustrates the effect of the phase modulation in a voltage-operated driver with sine waves having a frequency ratio of 1:10 (e.g. 50 Hz and 500 Hz). At the displacement peaks, the current in question lags the ideal sine shape and reaches it again as the voice coil passes its rest position. During cycles 1, 3, and 5, the wave frequency is approximately normal. Instead, in cycle 2, as the impedance is turning, the frequency becomes a little lower; and in cycle 4, the frequency is, corres-

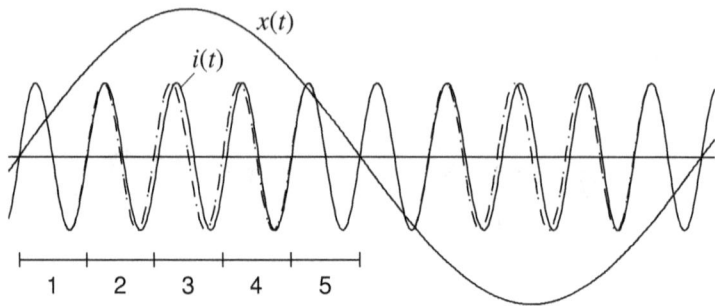

Figure 4.9. Phase modulation of a sinusoidal current ($i(t)$) under voltage drive due to large sinusoidal displacement ($x(t)$) having one tenth of the current's frequency. Dashed line represents a pure sinusoid. For clarity, the phase shift is shown somewhat exaggerated.

pondingly, higher than normal. The outcome is, thus, rapid frequency wavering, which denotes the scattering of pure spectral components into a band of some span.

At 500 Hz, $|Z_m|$ is typically about 20% of $|Z|$. If the decline in the force factor is say 10%, $|Z_m|$ will be 0.81 times the original, bringing about, with the mentioned impedance ratio, an increase of over 2° in the angle of $|Z|$. Assuming that this change takes place during about one cycle, as in Fig. 4.9, the relative frequency deviation of this cycle will be about 2°/360°, that is, at least 0.5%.

Frequency wavering, that resembles the above described, is also introduced by the Doppler effect, as the rather high diaphragm velocity, caused by bass signals, modulates the frequency of higher tones. For example, with an excursion of ±3 mm and a frequency of 50 Hz, the maximum velocity is about 1 m/s, corresponding to 0.3% of the velocity of sound. This detriment has sure been acknowledged, and it is even sought to be minimized, but why has this not happened to the impedance-derived frequency aberration?

4.6 Inductance Modulation

Common for the feedback effects described in the previous sections is that they occur by means of an EMF generated by extraneous or uncontrolled voice coil movement. However, in the following, we shall find that distortions and noises produced by the inductive EMF are not at all secondary in significance. The destructive effects of the inductive EMF and motion-based EMF in fact complement each other since the proportion of the former always increases with frequency, whereas in the latter the trend is reverse.

As the voice coil moves relative to the pole pieces, its effective magnetic resistance changes because, foremostly, the length that the current-induced flux lines travel in the air is dependent on the voice coil's position (see Fig. 3.2b). The deeper the coil goes into the magnet assembly, the lower is the reluctance and the higher is the inductance, according to equation (3.11). However, there is large variation in the sensitivity of the inductance, depending on the properties and shaping of the magnetic circuit.

The unsteadiness of the inductive impedance shows up, of course, directly in the total impedance, resulting again in consequences unhappy from the standpoint of voltage drive:

The displacement caused by low frequency components introduces for higher frequencies, by means of voice coil inductance, remarkable impedance magnitude fluctuation, that on voltage drive is directly reflected to the driver's current. As a result, the amplitude of the middle and treble tones becomes modulated in congruence with the total displacement. This is about as significant a source of intermodulation distortion than the *Bl* variation and yet fully preventable by moving to current-drive.

In addition to the amplitude modulation, the inductance variation also introduces phase modulation, correspondingly as the motional impedance variation in Fig. 4.8. This effect is most substantial in the lower midrange where the inductive impedance is practically perpendicular to the total impedance, thus contributing to the jitter of the voltage-operated driver.

Figure 4.10 shows measurement results of impedance behavior as a function of displacement in some, rather small bass-midrange drivers, at a frequency of 2 kHz. The displacement was effected mechanically and set by means of a measurement scale. (As the frequency is sufficient, the disturbance in motional impedance due to touching can be ignored.) The devices measured are not anything exceptional in this regard but fairly typical.

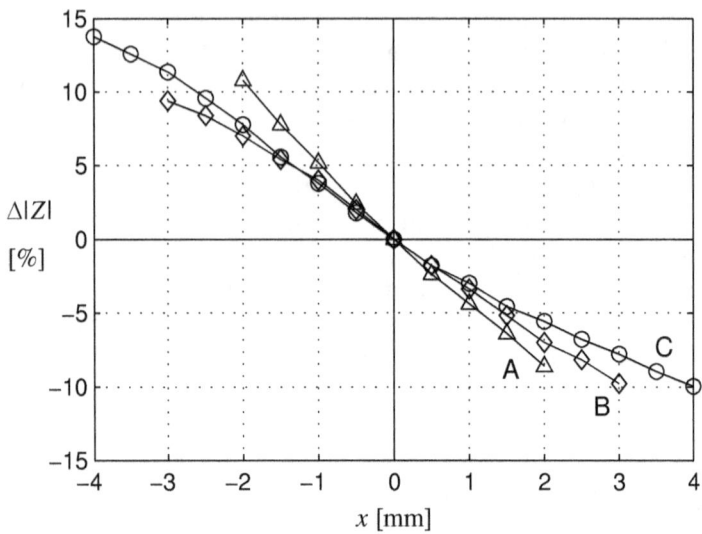

Figure 4.10. Relative change in impedance magnitude as a function of displacement, at 2 kHz frequency, in three different hifi grade bass-midrangers. A: Vifa M13SG-09-08; B: Seas P14RC4Y/DC (1 coil in use); C: Peerless 833429. The measurement has been extended, in each driver, to the rated Xmax limit.

The impedance variation occurring within the linear excursion range extends in all samples to the order of ±10%, curve C rising even up to 14%. Also, it is noteworthy that the rate of change can be higher at negative displacements, that is, when the cone is positioned backward. Increasing frequency from the 2 kHz does not, however, increase the slope in the way one could expect, based on the ratio of inductance and resistance.

When a signal's amplitude is being modulated, there is always established new, extraneous frequency components, i.e. distortion, whose magnitude directly depends on the strength of the modulation. (A fact about which the users and makers of various compression and level modification devices seem to be ignorant or indifferent.)

We will consider a little what happens to a sinusoidal signal current (i_s) having amplitude A and angular frequency ω_s. Let the interference-free current be thus

$$i_s(t) = A\sin(\omega_s t) \qquad (4.1)$$

Suppose now the amplitude alternates around its original value sinusoidally at frequency ω_m ($\omega_m \ll \omega_s$). Then, one can write for the amplitude-modulated current:

$$i_{sm}(t) = A[1 + m\sin(\omega_m t)]\sin(\omega_s t) \qquad (4.2)$$

where m is the modulation index, less than unity. By making use of the trigonometric relation (c5), in appendix C, and the rule $\cos(-\alpha) = \cos(\alpha)$, equation (4.2) can be expressed in terms of single frequency components:

$$i_{sm}(t) = A\sin(\omega_s t) + \frac{mA}{2}[\cos((\omega_s - \omega_m)t) - \cos((\omega_s + \omega_m)t)] \quad (4.3)$$

The modulated signal current thus contains, in addition to the original sine wave, two distortion components, the frequencies of which are $\omega_s - \omega_m$ and $\omega_s + \omega_m$. Because these side frequencies are not multiples of the applied frequency, the distortion is nonharmonic by nature.

Figure 4.11 depicts the amplitude spectrum relating to the inductance-derived intermodulation distortion. On the sides of the signal frequency ω_s, there appears two equal-sized modulation products, whose amplitude is $mA/2$. When the modulating signal occupies a band, instead of just a single frequency (ω_m), the whole band is correspondingly symmetrically reflected around ω_s.

With program material containing bass frequencies, the distortion due to inductance modulation easily rises to great heights. If the variation in impedance is e.g. ±10%, which would seem to be typical, as the excursion reaches its rated value; m is correspondingly 0.1, so the magnitude of the distortion components will be $A/20$, or 5%, of the original wave. When two 5-percent distortion components are added together, in

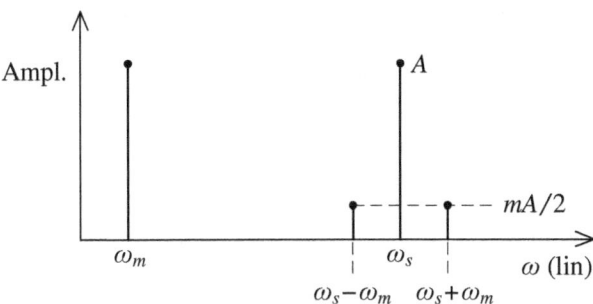

Figure 4.11. Line spectrum illustrating amplitude modulation on the basis of equation (4.3). Frequencies $\omega_s - \omega_m$ and $\omega_s + \omega_m$ represent intermodulation (IM) distortion.

the normal way, quadratically, one obtains as the total distortion a full 7%.

Unlike *Bl*-based modulation, inductance modulation becomes already significant at relative small excursions since the impedance variation is quite linear and occurs in both directions. Hence, when the motion of the diaphragm of a voltage-operated driver becomes visible for the eye, the distortion is typically already of a noticeable level. Inductance modulation is also a generally known shortcoming, but the contemporary audio equipment industry hasn't found an affordable solution for the matter.

4.7 Inductance Nonlinearity

The magnetic circuits of speakers are generally made of steel materials that have high enough magnetic saturation level. Although the used steel varieties are relatively soft, the turning of the elementary magnets, required for a change in the flux, always causes some internal friction, due to which the flux does not change directly in accordance with the magnetizing current but drags a little behind it, bringing about a nonlinearity phenomenon called hysteresis. Even if the pole pieces were magnetized, with the permanent magnet, into saturation (which is, however, generally not the case), this does not prevent the fluctuations of the flux since the flux component introduced by the voice coil is, especially in the proximity of the air gap, about perpendicular to the permanent flux.

The inductive EMF of a winding always follows the actual rate of flux change, that is affected, in addition to the magnetizing current and eddy currents, by the lag introduced by hysteresis. So, this perpetually altering delay is always included in the relationship between the voice coil's inductive EMF and current, and this relationship in turn governs, in a wide frequency range, the dependence between the driver's total voltage and current.

How acceptable would it generally be deemed that, in series with the (voltage-operated) driver or elsewhere in the signal path, there would be used an inductor wound on a mere bar of iron? The practice could evoke justified criticism, and the properties of ferromagnetic materials would readily be retrieved to the mind. However, when a comparable coil is acting in the driver itself, the problem is not seen, and all the

magnetic nonlinearities are able to step in, like from a Trojan horse.

In certain quality drivers, there has been attached, on the boundaries of the air gap or around the center pole, shorting rings of copper to level the impedance and to reduce distortion. While decreasing the fluctuation of the flux, such rings also keep down, to some extent, the effects of hysteresis; but this is, however, an unnecessarily imperfect and costly solution to a problem that disappears with the correct driving method. These secondary current conductors are, anyway, hardly of any harm, even on current-drive.

The cunning of hysteresis abides in that the effects do not necessarily show up in the measurement of harmonic distortion. The reason is that, as frequency increases, the ends of the hysteresis loop (flux density vs. magnetizing current) round, and the shape of the loop approaches an ellipse; meaning that a sine wave is, luckily, mapped almost into a sine wave despite the indefinite delay mechanism that acts in between. Nevertheless, the behavior is not linear, and the impact on more complicated signals can be rather harmful, on voltage drive.

The non-ideality of the voice coil's magnetic circuit becomes tangibly apparent by measuring the impedance at several different signal levels. The inductance is then found to distinctly increase with the measurement current. The effect is not explainable by the usual decrease in permeability because then the direction of change should be opposite. Instead, the reason may relate to the mutual interactions between the voice coil flux and the permanent magnet flux, which is much stronger in comparison.

Figure 4.12 shows measurement results of the impedance behavior as a function of current at 2 kHz frequency, in the devices familiar from Fig. 4.10. For perceptibility, the measurement data has been proportioned to a reference value obtained at 3 mA.

In one sample out of three, the impedance increases even over 10%, and in two samples a good 5% as current becomes hundredfold from the 3 mA. The change is not linear nor logarithmic but pretty much irregular. Most evidently, the rise yet continues at higher currents; but there, measurement is hampered by the temperature dependence of the voice coil resistance.

The measurement frequency chosen (2 kHz) is most often included in the band served by the analyzed drivers. As frequency increases from the mentioned, the impedance variation also increases a little yet. The current-dependence of inductance is also observable in tweeters, even though to a considerably lesser extent.

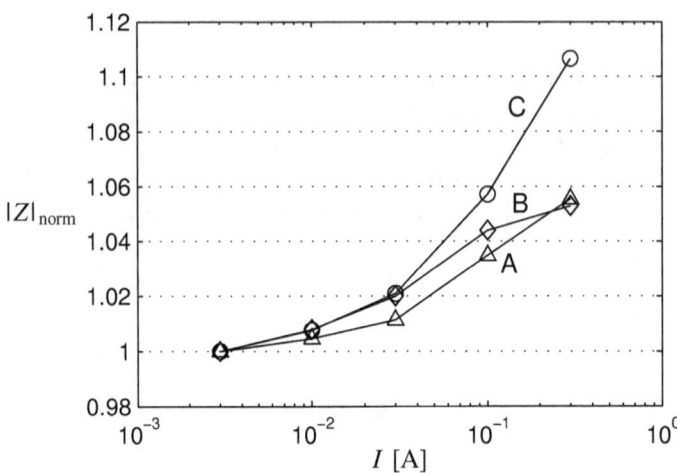

Figure 4.12. Measured current-dependences of impedance magnitude at 2 kHz. The drivers (A, B, C) are the same as in Fig. 4.10. The impedances have been normalized with respect to the lowest value. The measurement has been performed by comparing the voltage across the driver to the voltage across a resistor connected in series. Warming of the voice coil has not been affecting the results.

What kind of an imprint does a nonlinearity effect of the order of 10 percents make in the sound quality of a voltage-driven loudspeaker? In the frequency region governed by inductance and at rather high signal levels, the effect may well be characterized as catastrophic, but our culture has become so accustomed to a certain electrical timbre that even the contribution of this distortion mechanism has not prompted the acknowledgement of verities and reassessment of conventions.

The results of Fig. 4.12 must not be interpreted only by thinking that they would effect a mere reduction of the dynamics by an amount of decibels that corresponds to the impedance change. The gain factor of any system cannot namely alter without introducing, at the same time, a quantity of nonharmonic distortion products proportional to the rate of the change. In this case, the impedance variations may occur quite rapidly, compared to the reproduced frequencies, and accompanied by hysteresis effects, so possibilities for the nonlinear aberration of transient waves are abundant.

The magnetic roughness of iron also manifests, to some extent, as harmonic distortion in the relationship between the driver's voltage and current although the determination of this distortion does not yet give a view of the true severity of the detriments. At low frequencies, the obtained result is also affected by the nonlinearity of the motional impedance.

Figure 4.13 shows a quite typical distortion behavior between voltage and current, measured from a 5-inch bass-midrange driver at a power level of 1 W. Here, the input was current and the output was voltage, but the distortion should be the same also in the other direction.

In the upper midrange, the harmonic distortion approaches one percent and keeps also in the lower midrange close to a half percent. This much of harmonic distortion is thus removed from the speaker, when moving to current-drive (Distortion measured from the sound does not, however, necessarily reduce by the same numbers since distortion percentages arising from different sources are not directly additive.) The overall harmonic distortion of a voltage-operated driver is, at the same power and frequencies, generally of the order of a percent or two, so the contribution of the voltage/current-conversion in this number is appreciable.

Mills and Hawksford [1] have performed some interesting distortion comparison between voltage- and current-drive with the low-frequency driver of a Celestion SL600 loudspeaker. In measurements made with a current of 1A, in peak value, current-drive resulted, at the frequencies examined, in substantially lower distortion. At 100 Hz, the 2nd harmonic was reduced by 9 dB and the 3rd and 4th harmonics both by 3 dB. At 3 kHz, in turn, the total harmonic distortion was reduced even by 26 dB. Tweeters were also found to benefit from moving to current-drive

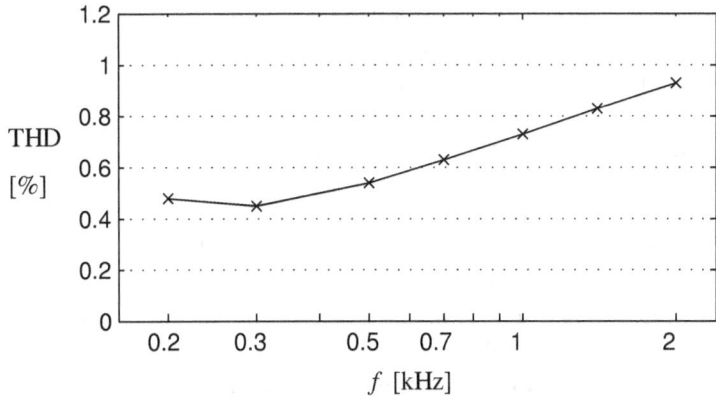

Figure 4.13. Total harmonic distortion (THD) of the voltage/current-conversion, measured from the example driver Seas P14RC4Y/DC (1 coil in use). The voltage was 2.83 V, corresponding to a power of 1 W into a nominal load of 8 Ω.

by an amount of 3-7 dB.

The changing of flux in iron occurs when the microscopic clusters or areas, formed by the elementary magnets, turn their magnetic orientation. These changes in orientation do not, however, occur smoothly with the magnetizing field (current) but stepwise in leaps, giving rise to rapid small bounces in the flux density. The interference signal thus arising is referred to as the Barkhausen noise, according to the discoverer, and it has significance e.g. in the operation of tape recorders.

Most part of the Barkhausen noise's power is located at ultrasonic frequencies, but the phenomenon may also have some significance on the definition of loudspeakers, due to the program-dependent interference EMF induced in the voice coil. In ordinary use, this noise can hardly be discerned since it is left, in its order of magnitude, well in the shadow of other adverse effects. In demanding hifi listening, however, the situation can be different, and only by current-drive it can be ascertained that the recording precision of the CD format (not to mention SACD) will actually be available to the last bit.

4.8 Resistance Variations

The wire metal used in voice coils is almost invariably copper or aluminum, the resistance of both increasing with temperature at a rate of 0.4% per Celsius degree. The allowed temperature rise of the voice coil may be again, depending on the endurance of the coil former and adhesives, even around 200 degrees. Hence, it is fairly realistic to estimate that, when operating near the high end of the safe power area, the resistance can increase even 50% which requires a 125-degree rise in temperature. How this resistance change affects the sound level and frequency balance, depends completely on the driving practice.

Excluding the extremes of the reproduction band, the magnitude of a driver's impedance is only a little higher than the voice coil resistance. A 50% increase in the resistance thus roughly denotes a 40% increase in the impedance, giving rise to a 3-dB decline in the voice coil current and sound level of a voltage-driven device.

As a consequence, the gain continuously alters in congruence with the signal level, with a delay determined by the associated thermal time constants. In large woofers, this time constant may be some tens of seconds, but in tweeters, only of the order of seconds. (A time constant represents, when a 1st-order model is applied, the period during which the step response has reached 63% of its final value.) If the signal levels of the left and right sides differ, at times, from each other, the setting of channel balance will also exhibit fluctuations.

Instead, when the current of the driver is forced to follow the input, the rise in resistance appears only as a corresponding rise in voltage, and sensitivity variations are prevented. This ought to mean something for those who wish to preserve the dynamic variations of music intact.

In multiway systems, the drivers serving the different bands do not necessarily warm by the same amount and especially not at the same speed, so a voltage-driven speaker easily develops variation also in the frequency response. Wide-range modifications in the frequency balance, of the order of decibels, are already audible and may further impair the plausibility of reproduction.

On voltage drive, the rise in resistance also affects the Q value of the woofer. According to equation (3.16), the electrical Q factor (Q_e) is directly proportional to the resistance R_c. Further, because in general $Q_m \gg Q_e$, the total Q factor increases, due to formula (3.18), almost in the same proportion as Q_e and R_c. If R_c increases say 50%, as was noted above, the increase in the total Q value is then typically over 40%, having already a considerable impact on the bass alignment. For example, in closed enclosures, the Q has often been set to 0.7, but after a rise in temperature, the value may be over 1, introducing boominess in the bass response and increasing the risk of driver failure.

Instead, on pure current-drive, the Q value depends only on mechanical parameters not subject to the voice coil temperature. Even when using passive bass equalization, the variation in response due to voice coil heating remains yet considerably lesser than on voltage drive.

Voltage drive introduces interchannel imbalance even without any warming taking place. Resistance, like inductance also, has a certain manufacturing tolerance, that gives rise to sensitivity deviation from unit to unit. If the accuracy of impedance is e.g. ±5%, the consequent difference in sensitivity between two drivers can be 0.8 dB at worst. Thus, current-drive also improves the matching of loudspeakers by eliminating certain uncertainty factors. In principle, of course, sensitivity differences can be canceled with the amplifier's balance knob, though, but with multiway speakers, the case is not that simple.

The only argument speaking for voltage drive here could be that the increasing resistance tends to decrease power accordingly, whereas on current-drive, the power correspondingly tends to rise somewhat. However, in respect of the basic duty of loudspeakers, it is more appropriate to seek to keep the level relationships steady and realistic than to limit power dissipation at the expense of sound quality. Also, current-principle offers the possibility to use elegant overload protection methods that cannot be applied in a voltage-driven system.

4.9 Circuit Element Nonlinearities

The weaknesses of voltage drive are not limited merely to the side-effects occurring in the driver itself, but often other circuit components are also able to deliver their possible nonlinearities to the load current. The effective mechanism is similar to that arising from the voice coil's electromotive forces. The distortion factors of the series-acting voltage drops as well as those of the parallel-flowing side currents become, in a conventional system, inevitably part of the driver's final current.

Figure 4.14 shows a two-way speaker with 1st-order filtering. Inductance L prevents high frequencies from passing to the woofer, and capacitance C, respectively, prevents low frequencies from passing to the tweeter. Z represents the interconnect impedance.

When the circuit is fed by a voltage signal and Z is small, the nonlinearities of the inductance L are reflected to the woofer's current the more, the higher is the voltage across the coil relative to the applied voltage, that is, the higher the frequency.

Instead, when a current signal is used, the nonlinearity of the inductance is able to manifest only as an extraneous voltage component, that does not introduce any current to the input wires. Therefore, the only possible flow route for the distortion current is the loop formed by the bass and treble arms. The total impedance of this loop is, however, in major part of the operating frequency range, so high that audible distortion remains very minor, compared with the voltage-driven case.

The principle of current-drive reduces in many cases conclusively the transfer of the nonlinearities of the series inductor

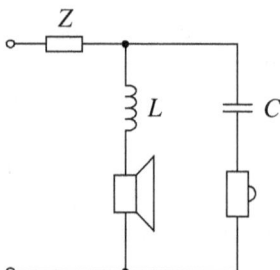

Figure 4.14. Simple two-way system, that can be used on both voltage- and current-drive. Z represents the impedance formed by cabling and connections.

to the sound signal. Hence, it is possible to use ferrite-core inductors also in many hifi applications where otherwise only air-core inductors would qualify. The physical size and DC resistance of coils can thus be considerably reduced.

In totally uncompromising implementations and other uses of inductance, it may nevertheless be justified to rely only on ironless coils.

Likewise, the nonlinearity effects possibly relating to the capacitance C are able, in the voltage-driven case, to transfer to the current of the series-connected driver. The reflection gets stronger as the capacitor's voltage increases, that is, as frequency decreases. Just as with the coil, the use of current-drive also eliminates this distortion effect almost entirely.

When the amplifier operates on one-sided supply voltage (e.g. battery-powered equipment), a large series capacitor is generally employed at the output to block DC voltage from the speaker. On voltage drive, this electrolytic capacitor also plays a role in the composition of current. A series capacitor is also needed in a current-output amplifier, but the circuit can be realized so that possible impurities are not transferred to the load.

The harmonic distortion of capacitors is generally negligibly small, but from the standpoint of sound quality, a more essential and cunning factor is yet the dielectric hysteresis relating to their insulator materials; as a phenomenon, similar to the hysteresis encountered in ferromagnetic materials. The higher the dissipation factor (tangent of the loss angle) of the capacitor, the more there is also room for hysteresis-derived delay between the voltage and charge. Although the nonlinear nature of this delay does not usually show up in sine wave measurements (except in ceramic capacitors), it may, nevertheless, produce some detriment on real signals.

Particularly in non-polarized (bipolar) electrolytic capacitors, occasional distortion is also introduced due to their internal, structural leakage diodes, that become conductive at times. Every time the capacitor's voltage reaches a new maximum, capacitance increases remarkably and returns again to the normal value when the voltage has started to fall. The phenomenon has been described in more detail in section 11.3.

The resistance and inductance accruing from the speaker leads and connections along the way can be incorporated into a single series impedance, as in Fig. 4.14. As long as this impedance stays constant, irrespective of time and signal level, its effects are limited, even on voltage drive, typically to a minor power loss and slight frequency response modifications. However, the contact resistances of connectors and possible switches can exhibit, due to aging and contamination, unreliable and even rectifying behavior.

Not much is required to bring about disturbances of significant level. If the speaker impedance is e.g. 4 Ω, the combined signal-dependent uncertainty of all contact resistances needs to be, in principle, only 4 mΩ to effect a distortion of 0.1% in a voltage-driven speaker. With connections having oxidation, it is easy to achieve even much larger divergences, so the need for keeping the current controlled is, also in this case, apparent. Often the output circuit is equipped with selection, soft-start, or protection relays and sometimes even fuses, which themselves are nonlinear. With current-drive, all such factors can be forgotten, as long as the voltage drops arising from them remain sort of moderate.

4.10 Drive at Low Frequencies

Thus far, it has become plain how overwhelming and epochal the benefits current-drive offers are, especially from middle to treble range. The principle of pure current-feed can well be applied for bass frequencies also, but in this region, the qualitative superiority of the method, with respect to voltage feed, is not as substantial.

In the region of the driver's resonant frequency, the inductive EMF is typically only a few percents of the total voltage, so the improvement gained from the elimination of the inductive interferences is at these frequencies not as significant as elsewhere. Further, at the greatest wavelengths, the pressure inside the enclosure is virtually static and establishes for the cone a relatively controlled and even counterforce. If the enclosure is rigid, internal microphone effects of detrimental nature are then not so easily generated, although the driver's sensitivity to pressure variations increases when the resonant frequency is approached. (A reflex type speaker exhibits two such resonances.) As the dimensions of the diaphragm are negligible relative to the wavelength, also the moving parts behave, in many respects, more ideally and more piston-like than in the other regions. Consequently, also the motional EMF generated in the voice coil reflects better the actual motion of the cone, being thus less subject to the mechanical interference factors. Regarding yet that the ear is quite insensitive to distortions in the bass region, serious harm will hardly be introduced if, in order to level the frequency response, the principle of pure current-drive has to be traded a little in the lowest octaves.

Some insight into the differences between the two driving methods can be gained by considering the dependence of the Q value on various factors. Both expressions of Q ((3.6) and (3.15)) have the same numerator, that contains the spring constant and mass; but on current-drive, the denominator consists of a mere damping constant b, whereas on voltage drive, the denominator is dominated by the term $(Bl)^2/R_c$. If the stability of the Q is taken as the criterion, the mutual superiority of the driving methods is then mostly dependent on which one behaves more steadily: b or $(Bl)^2$. A universal answer can hardly be given since the variation of both factors is largely dependent on driver realization. The nonlinearity of b can, however, be somewhat reduced by copious use of damping material.

As a merit of voltage drive can be counted, however, that, at those frequencies where the motional EMF is distinctly the dominant factor in the driver's voltage (concerning closed enclosures), one can write, as a rough approximation: $u_o \approx Blv$, that is to say, the velocity of the diaphragm tends to directly follow the applied voltage. This, as if an automatic motional feedback, has effect, however, only in the region of the impedance peak and does not compensate the force factor variation.

According to a common notion, electrical damping, that only voltage drive affords, would be even indispensable to prevent the diaphragm from somehow vibrating on its own and to keep it in "tight control". The view is also closely related to the one that the moving mass should be low with respect to the force factor, in order that the speaker would act "quickly" and stop suddenly when needed.

Common to these myths is that it has not been understood that sound is generated due to acceleration and that frequency response (including phase) and transient response directly derive from each other. If the ability of the driver to perform correctly in the time domain were indeed dependent on the aforementioned issues, hardly ever would we get hearable any recognizable or tolerable sound.

The only control effect that electrical damping can produce is in that if some external force (like a reflection from somewhere) tends to move the diaphragm at its resonant frequency, a small Q value helps to brake this movement. Instead, in other frequency ranges, electrical damping has no significance at all. Hence, if the bass frequency response of a current-driven speaker is somehow made equal to that obtained on voltage drive, then also the bass transient responses are, in principle, identical in both cases.

The mechanical Q value of drivers is generally so high that current-drive cannot be directly applied without some kind of response shaping. However, the heightened current-sensitivity of the resonance region can also be put to use if the Q value is not too high for passive compensation. With a suitably optimized RCL network, the response peak can namely be equalized quite neatly, without attenuating the lowest frequencies at all. Consequently:

By current-feed and passive equalization, the bass response of a driver in a closed enclosure can be made to extend considerably lower than conventionally is possible. And that is not all since, by filling the enclosure with an efficient enough damping material, the lower cut-off frequency of a current-driven speaker can be made, in practice, at least as low as is possible with a bass reflex enclosure of the same size.

The bass reflex function, at least as such, is not suitable to be used with current-drive. Nothing is lost, however, since in the above manner one can achieve a full-bodied and precise bass reproduction, without reflex ports and all the problems related to them.

---

[1]   P. G. L. Mills and M. O. J. Hawksford, "Distortion Reduction in Moving-Coil Loudspeaker Systems Using Current-Drive Technology", *Journal of the Audio Engineering Society*, vol. 37, March 1989, p. 129-148.

5

THE PRINCIPLES OF CURRENT-DRIVE

Moving to current-drive not only denotes a change in the operation principle of the power amplifier but a completely new way of thinking in all that has to do with the use of crossover filters and different passive equalizers as part of the signal chain. Even if the amplifier performed like an ideal current source, this is not yet enough, for the loudspeaker system must also be designed to support the current-principle so that the actual operation mode of the drivers does not shift back toward voltage drive because of the circuits on the signal's way.

An ordinary amplifier as if attempts to force its output voltage to always carefully follow the program signal, without any respect to what the current is, as long as its magnitude only stays within some allowed limits. A current-output amplifier, by contrast, as if attempts to force its output current to carefully follow the program signal, without any respect to what the voltage will be, as long as it only stays within some allowed limits. By following this principle of dualism between current and voltage, it is easy to draw a conclusion that the circuit topologies for a current-drive speaker would simply be obtained by converting the corresponding voltage drive topologies so that inductances are replaced by capacitances, capacitances by inductances, parallel connections by serial ones, and serial ones by parallel ones. This is, however, not the proper way to proceed in general, as we will find out in the following.

5.1 Thévenin and Norton Equivalents

All might agree that a loudspeaker drive unit has to be operated by an electrical signal. However, immediately next to this, a choice must be made concerning the nature of that electrical signal, that is, whether we seek to control the voltage or current of the load, since the simultaneous controlling of both is impossible. A completely ideal current-drive, as well as a completely ideal voltage drive, cannot be realized, so, in practice, one always has to settle for a compromise that lies somewhere between these two extremes. In which mode the driver eventually operates, is determined by the ratio of the impedances of the feeding source and the driver.

In Fig. 5.1a, a driver is fed from a voltage source (E) through a series impedance (Z). If Z is very small with respect to the load impedance Z_L, the load current I_L is determined almost solely by Z_L, and the load voltage U_L follows the feeding source E; so the driver operates, in this case, under voltage drive. (When comparing impedances, the absolute value, i.e. the length of the vector, is considered.)

In turn, if Z is very high with respect to Z_L, Z_L forms only a very small part of the circuit's total impedance. The load current I_L is then almost independent of Z_L and its variations, so the driver operates under current-drive. A high series impedance also requires a high source voltage if the current is to be kept sufficient; so, in practice, it is not economical to implement current-drive on the principle of Fig. 'a'.

If Z is of the same order than Z_L, it can be said that the current is determined in half by both impedances, and the variations of Z_L are reflected to the current with a half strength, compared with the case where $Z = 0$. The operation mode may then be regarded as a half way between voltage- and current-drive.

In Fig. 5.1b, the voltage source has been replaced by a current source and the series impedance by a parallel one. Looking from the load, the

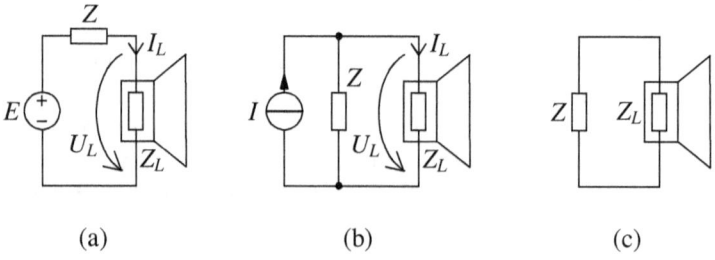

(a)   (b)   (c)

Figure 5.1.   a) Circuit feeding a speaker driver (impedance Z_L), presented using a voltage source (E) and series impedance (Z). b) Circuit feeding a speaker driver, presented using a current source (I) and parallel impedance (Z). c) Impedance seen by the driver is the same in both cases, i.e. Z.

circuits of Figures 'a' and b function totally identically if $I = E/Z$. Thus, either of the two models may be used, depending on the need, when analyzing the operation mode of the driver.

If Z in Fig. b is very low compared to the load impedance Z_L, voltage U_L is determined almost solely by Z since the load current is then very low. The operation of the driver is thus voltage-driven. In turn, if Z is very high compared to Z_L, the significance of Z becomes diminutive, and the driver operates under current-drive.

Therefore, in both models, a high value of Z denotes current-drive and a low value, respectively, voltage drive. Instead, the magnitude of the sources is irrelevant to the driving mode, so they can even be considered to be zeroes. A zero-valued voltage source corresponds to a short-circuit and a zero-valued current source, in turn, a break; so, irrespective of which model is used, the impedance seen by the driver, that is, the impedance that determines the driving mode, is always Z (Fig. 5.1c).

The models of Fig. 5.1 can be used to represent a network, made up of linear elements, also generally, not only in the case of a single source and a single source impedance. Any network, consisting of resistances, capacitances, inductances, and voltage and current sources, can namely always be substituted, between any two nodes, by a series connection of one voltage source and one impedance, referred to as the *Thévenin equivalent* of the network. In the substitution, one may also as well use the parallel connection of a current source and impedance (or admittance), referred to as the *Norton equivalent*.

Figure 5.2 provides an example of the application of the Thévenin and Norton methods. Figure 'a' shows a simple network that is examined between nodes A and B.

The source voltage (E_T) of the Thévenin equivalent (Fig. b) equals the voltage obtained from the network when not loaded and is in this case easily calculated from the voltage divider formed by Z_1 and Z_2.

The Thévenin impedance (Z_T), then, equals the total impedance of the network in question (between A and B) and can be determined by setting all sources to zero, that is, by replacing voltage sources by short-circuits and current sources by breaks. Consequently, Z_T is in this case the parallel connection impedance (product divided by sum) of Z_1 and Z_2, added by Z_3.

The source current (I_N) of the Norton equivalent (Fig. c) equals the current obtained from the network when short-circuited and can be determined either directly or using the relation $I_N = E_T/Z_T$. The Norton impedance Z_N always equals the corresponding Thévenin impedance since the impedance seen between A and B must be independent of the

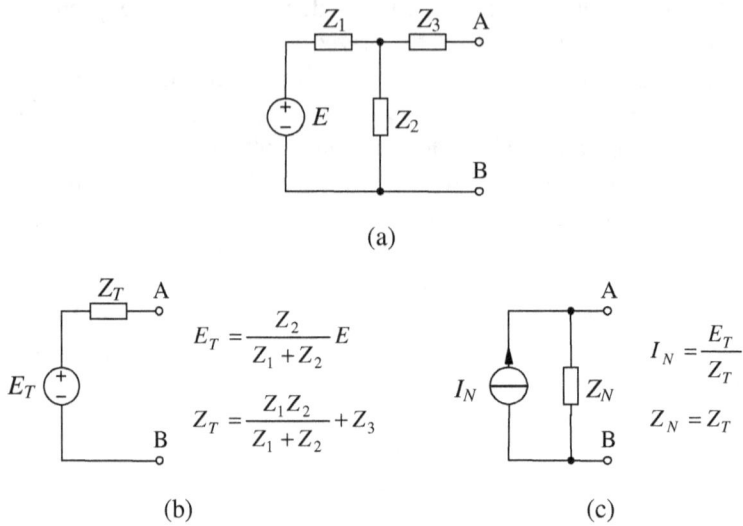

(a)

(b)   (c)

Figure 5.2. a) Simple example network, whose behavior, as seen from nodes A and B, is of interest. b) Thévenin equivalent which, at the given values of E_T and Z_T, substitutes for the network of Fig. 'a'. c) Norton equivalent which, at the given values of I_N and Z_N, substitutes for the network of Fig. 'a' (and the given Thévenin equivalent).

chosen presentation.

> **The requisite for current-drive is therefore that the Thévenin impedance (Norton impedance) seen by the driver, formed by the amplifier and passive circuits, is high enough relative to the driver's own impedance, for those frequencies the driver is able to reproduce.**

This requirement, in practice, considerably restricts what kind of filtering circuits can be used in a current-drive speaker. Especially circuit arms connected directly in parallel with the driver can easily lower the impedance too much.

Someone might ask whether current-drive is then, in this respect, inferior to voltage drive since a corresponding requirement doesn't seem to limit conventional speaker design. The answer is an absolute *no* since there is full dualism between the driving methods, and corresponding restrictions would sure also apply to voltage drive if someone, for some reason, desired to keep the source impedance low for all frequencies reproduced. Traditionally, however, not so much attention is paid to the matter; and so the actual operation mode of the drivers varies, in many cases even largely, nevertheless staying for the most part close to voltage drive.

5.2 Evaluation of the Driving Mode

In order to evaluate the fulfillment of current-drive more accurately than sketchily, we need some characteristic figure by which the suitability of different circuit solutions can be compared. Quite a natural choice for this is the magnitude ratio of the Thévenin impedance to the load impedance. The quantity is called here *current-drive index* and abbreviated CDI.

$$\text{CDI} = \frac{|Z_T|}{|Z_L|} \tag{5.1}$$

The higher the CDI value, the closer we are to ideal current-drive. A value of 10 has to be already regarded as good, and increasing the index from this doesn't bring substantial improvement any more. A value of 5 may yet be regarded as very satisfactory, but even this is not generally achieved for all frequencies. Still at a value of 1, which corresponds to a half current-drive, the sound quality is conclusively different than at a value of 0, that denotes pure voltage drive.

The total impedance constraining the EMF-derived currents is always $Z_T + Z_L$ (see Fig. 5.1c). Thus, for example, with an index of 1, the magnitude of the EMF currents is, in principle, half of that appearing on voltage drive; and with a value of 5, the EMF currents are about $1/6$ of the voltage-driven case, where $Z_T = 0$. Consequently, the attenuation ratio of the interference currents will be $1/(\text{CDI}+1)$.

However, the value of the current-drive index alone does not yet tell all about the attenuation of the EMF-derived interferences, for the sensitivity properties of the driver and the difference in the impedances' direction angles also affect the picture. For example, a mere increasing of the voice coil resistance decreases the CDI although the detrimental EMF currents actually weaken. The proportion of electromotive forces in a driver's voltage varies in accordance with sensitivity and inductance properties; so, in principle, drivers having low sensitivity and inductance will do with lower current-drive index than those having high sensitivity and inductance.

The attenuation ratio $1/(\text{CDI}+1)$ only holds exactly when the direction angles of Z_T and Z_L are equal. Generally however, they are not, and this may lead, especially at low index values, to a too optimistic estimation. In current-drive applications, however, one must have $|Z_T| \gg |Z_L|$, in which case the error introduced by ignoring the angles is no more significant.

The Thévenin impedance and thus also the current-drive index can be quite easily examined with circuit simulator programs. The load is then replaced by a 1-A current source, and the voltage developed across this source is monitored in AC analysis. The voltage directly indicates the Thévenin impedance seen by the load. For the index, it may often suffice that the load impedance is assumed to be constant.

Based on the above, there of course easily arises a question whether drivers can be connected in parallel without disturbing the current-drive principle. The issue was already touched in section 4.3, and the parallel connection does work, especially if the devices are identical and similarly positioned relative to the listener.

Figure 5.3 shows a parallel connection of two drivers and a Norton equivalent feeding it. Sources E_1 and E_2 model the electromotive forces of the respective drivers.

Let us first assume that $|Z_N| = \infty$. Due to the superposition principle, the currents introduced by the sources E_1 and E_2 can be treated independently of each other. E_1 generates, thus, loop current component I_1 and E_2, respectively, component I_2. (The source current I_N is not affected by these.) However, current I_1 flows in opposite directions in the two voice coils, and so does I_2. Hence, when the drivers are identical, the sonic effects of these current components are canceled out, at least in the front direction. In the side directions, instead, cancellation does not necessarily occur if the spacing between the drivers is significant with respect to the wavelength. Further, especially if the drivers have been mounted in

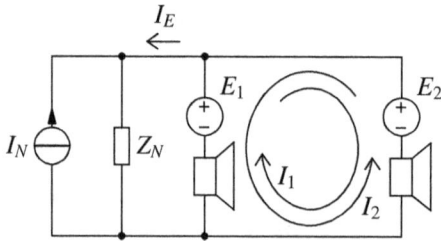

Figure 5.3. Interference current components produced by parallel-connected drivers. Loop currents I_1 and I_2, generated by the electromotive forces E_1 and E_2, are almost harmless due to canceling. Interference current I_E, that flows through the Norton impedance Z_N, is, instead, harmful, as in a single driver.

their chamber symmetrically, sources E_1 and E_2 as well as currents I_1 and I_2 resemble each other very strongly, making the net current in the loop actually rather minor.

If the Norton impedance cannot be considered infinite, it passes an EMF-derived current I_E, which is determined, according to the Thévenin theory, by the mean value of E_1 and E_2. Although the detriments of the mentioned loop currents are canceled quite effectively, I_E is as harmful as with a single driver. The current-drive index and attenuation ratio can be defined for the parallel connection by using, for the load impedance, the impedance of the parallel connection, which equals the impedance of one driver divided by the number of the drivers.

Because the effects of the loop currents are, nevertheless, not fully eliminated, it is recommendable to use, in current-drive speakers, series connection instead of parallel connection.

The equivalent circuit of a driver consists of three series-acting elements, as was presented in Fig. 3.4. Consequently, one may say, in a certain sense, that absolute voltage drive cannot really even exist. What, in fact, is the goal in the present operating philosophy, if anything at all? Is the intention of voltage drive to control the resistive voltage drop or perhaps the motional EMF or maybe even the inductive EMF? Whichever the target might be out of the mentioned alternatives, it cannot be reached since two of the components always establish a series impedance for the third. If, therefore, one desired to voltage-drive say the resistive part, the inductive part alone already causes that it is, in effect, not sensible to even talk about voltage drive.

5.3 Effect of Sensitivity Parameters

As was already noticed, the CDI is not an absolute measure of quality but just a comparison number that tells how much the EMF currents are attenuated with respect to pure voltage operation. The proportion of the EMF currents namely varies, depending on driver parameters, much even on voltage drive.

Table 2 indicates how the changing of the driver's different sensitivity parameters affects the development of EMF-derived sounds, given that the current-drive index is preserved constant in all cases. The Table shows the coefficients by which the pivotal pertinent quantities change as driver sensitivity is increased by doubling the effective wire length l, doubling the flux density B, halving the moving mass m, or doubling the radiation area S. When increasing the wire length, it has been assumed

Table 2. Effect of sensitivity doubling on the quantities related to the generation of EMF-derived sounds, assuming CDI stays constant. SPL =sound pressure, Z =driver impedance, e =electromotive force, i =EMF current, F =force. Subscript m refers to motional EMF and subscript i to inductance EMF. Columns represent the doubling of wire length ($l = 2x$), doubling of flux density ($B = 2x$), halving of mass ($m = \frac{1}{2}x$), and doubling of area ($S = 2x$).

	$l = 2x$	$B = 2x$	$m = \frac{1}{2}x$	$S = 2x$
SPL	2	2	2	2
Z	≈ 2	≈ 1	1	1
e_m (=Blv)	4	4	2	1
i_m	2	4	2	1
F_m	4	8	2	1
SPL$_m$	4	8	4	2
e_i ($\approx Ldi/dt$)	4	≈ 1	1	1
i_i	2	1	1	1
F_i	4	2	1	1
SPL$_i$	4	2	2	2

that resistance increases accordingly.

For example, the numbers in the column $l = 2x$ have been obtained as follows: As the wire length doubles, the force factor doubles; so, as the current remains constant, the sound pressure (SPL) doubles. Because both the diaphragm motion and force factor double, the motional EMF e_m becomes 4-fold. The impedance Z approximately doubles, and, as the CDI is held constant, the impedance seen by the driver is also doubled; so current i_m, caused by the motional EMF, doubles. This 2-fold current, together with the 2-fold force factor, causes the motional-EMF-derived drive force (F_m) and the respective sound pressure (SPL$_m$) to become 4-fold. Consequently, the ratio of SPL$_m$ and SPL becomes degraded by a factor of 2.

As the number of turns doubles, the inductance becomes, in principle, 4-fold, making the inductance EMF e_i also 4-fold. As the impedances are doubled, current i_i, caused by this EMF, hence doubles. Correspondingly as above, this causes the inductance-derived drive force (F_i) and consequent sound pressure (SPL$_i$) to become 4-fold, hence doubling the ratio of SPL$_i$ to SPL. The results imply that when the wire length is increased, the CDI should be increased even slightly more, in proportion, in order that the level of the EMF-derived sounds relative to the total sound pressure would remain unchanged.

From the next column, one can see that when the flux density doubles, ratio SPL_m/SPL becomes even 4-fold, while ratio SPL_i/SPL does not change. Thus, due to interferences deriving from the motional EMF, it is well to avoid using excessively high values of B. In practice, though, also the saturation of the magnetic circuit limits the possibilities to substantially increase the flux density.

When the moving mass is halved, ratio SPL_m/SPL doubles, while ratio SPL_i/SPL remains again unchanged. The result also implies, for its part, that striving for the lowest mass is not beneficial in terms of sound quality.

Increasing the area, by contrast, does not change the relative levels of the EMF sounds in any way. In practice, though, it is difficult to enlarge the area without increasing the mass as well.

Consequently, rising the sensitivity of a driver always increases especially the motional-EMF-derived detriments and correspondingly the demand of CDI. The goals of sensitivity and efficiency thus counteract the goal of sound quality if the EMF currents cannot be totally suppressed.

5.4 Multiway Systems

The operation band of a single drive unit is, also on current-drive, often inadequate to cover all required frequencies. In the implementation of multiway systems, however, one must take into account, in addition to the frequency response and matching between the drivers, also the requirements imposed by the current-drive principle.

In general, a crossover filter can be of parallel or series type and can be fed, in general, from a voltage or current source. Thus, there arises four different approaches, that are outlined in Fig. 5.4.

Figure 'a' represents the presently employed realization practice, in which a parallel-mode crossover filter is driven by a voltage source. The parallel mode has probably been adopted mostly because the arms of different drivers operate independently of each other which makes manual filter design easier. Adjusting the response of the woofer does not, therefore, influence the operation of the tweeter, and vice versa.

However, when voltage drive is desired, the series-mode filter (Fig. b) would be in fact, functionally, a more logical alternative. In the circuit of Fig. b, the sum of the low-pass and high-pass voltages ($U_l + U_h$) namely always equals the applied voltage (U_o), irrespective of the cir-

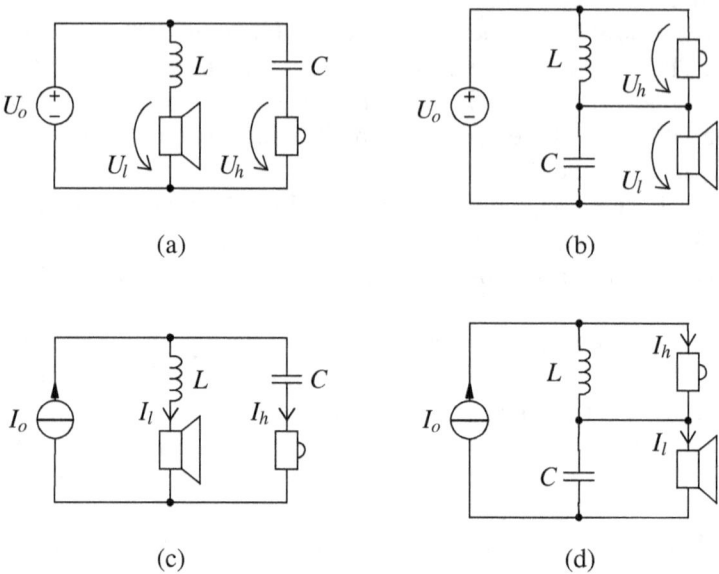

Figure 5.4. Possible realization principles of a simple two-way system. a) Parallel-mode filter with voltage drive. b) Series-mode filter with voltage drive. c) Parallel-mode filter with current-drive. d) Series-mode filter with current-drive.

cuit's tuning. The frequency response (in the front direction) would thus be, in principle, always flat over the crossover region if only the responses of the drivers were, under voltage drive, flat and equal. The power response, i.e. the overall response that takes account of all directions, can yet be adjusted by modifying the ratio of C and L. The circuit of Fig. 'a' does not have such advantages, and the voltages U_l and U_h have to be optimized separately.

Also from the viewpoint of impedance, the principle of Fig. b would do better than that of Fig. 'a' if the pursuit of low impedance were reasonable in the first place. In Fig. 'a', e.g. the impedance seen by the low-frequency driver consists solely of the series inductor and rises above the crossover frequency to fairly high values. In Fig. b, the corresponding impedance consists, instead, of the parallel connection of the capacitor, inductor, and high-frequency driver. The shape of this impedance resembles that shown in Fig. 2.5 (solid line), and its magnitude does not rise very high at any frequency.

Based on the dualism, we may now expect that, as the series-mode filter suited better for voltage drive, on current-drive things get reversed, and this indeed is the case. In the design sense, the realization principle of Fig. d is simpler since the low-frequency and high-frequency blocks operate independently. The other assets, then, are on the side of the Fig. c alternative.

In Fig. c, the sum of the low-pass and high-pass currents ($I_l + I_h$) always equals the applied current I_o, irrespective of the values of L and C. Correspondingly as in the case of Fig. b, the frequency response on the front axis will thus be automatically flat if only the driver responses are, under current-drive, flat and equal. In practice, though, the fundamental resonance of the high-frequency driver often causes that the driver responses are not very congruent, especially in terms of phase; but anyway, the benefit compared to Fig. d is definite.

Also here, the ratio of L and C can be used to adjust the power response around the crossover frequency and, to some extent, also the filtering steepness. The property is useful especially when the power response would otherwise exhibit a dip due to the directivity of the low-frequency driver.

Figure 5.5a illustrates the behavior of the currents I_l and I_h in the circuit of Fig. 5.4c at two different L/C ratios, with resistive load impedances. The dashed lines represent a conventional-like tuning, for which it is characteristic that the currents have attenuated at the crossover frequency by 3 dB (by factor $1/\sqrt{2}$) and that their phase difference is at this frequency, as well as elsewhere, 90°. This is achieved when all impedances are equal at the crossover frequency (labeled ω_c), that is, when $\omega_c L = 1/(\omega_c C) = R_L$ where R_L is the load resistance.

The impedance seen by the drivers consists of the series connection of the inductor, capacitor, and the other driver (resistance), i.e. a series resonance network, and is practically the same for both drivers. The current-drive index corresponding to the 90-degree tuning is shown by a dashed line in Fig. 5.5b. At the crossover frequency (resonant frequency), the value touches unity, and rises quite gently as frequency increases or decreases.

The current-drive index, and the filtering steepness likewise, can, however, be improved by increasing the circuit's reactances, that is, by raising the inductance and lowering the capacitance. The solid lines in Fig. 5.5 represent a setting in which $\omega_c L = 1/(\omega_c C) = \sqrt{3} R_L$, that is, the impedances (reactances) of the inductor and capacitor are at the crossover frequency $\sqrt{3}$ -fold relative to the load resistance. With this tuning, the currents at the crossover frequency are unattenuated and separated by 120°. Total radiation in the front direction remains unchanged, but the power response increases at the crossover point by 3 dB, compared

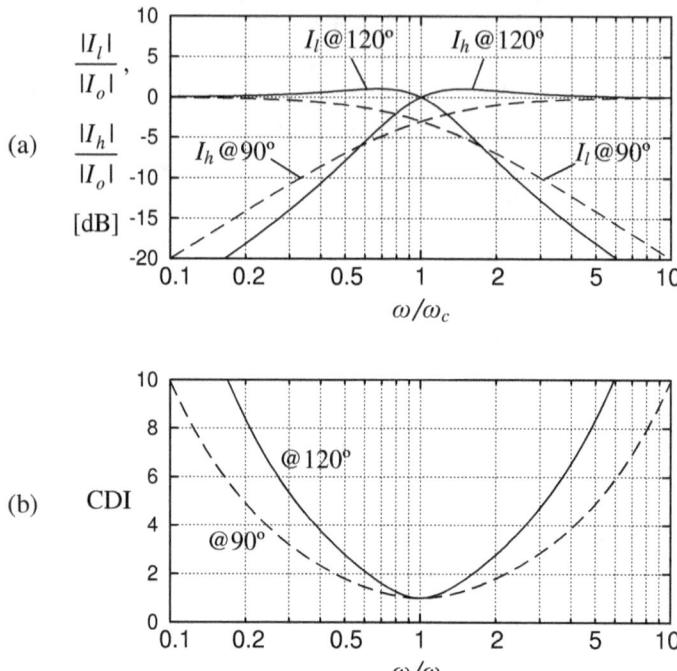

Figure 5.5. Behavior of the crossover filter of Fig. 5.4c on two different align-ments, assuming driver impedances are equal and resistive. The degree mark-ings refer to the phase difference between currents I_l and I_h at the crossover frequency. a) Currents I_l and I_h as a function of normalized frequency using conventional-like flat tuning (dashed lines) and a higher-impedance tuning (solid lines) that raises the power response at the crossover frequency by 3 dB. b) Current-drive indexes corresponding to the alignments of Fig. 'a'. (The index is the same for both drivers.)

with the previous case.

The current-drive index is now distinctly better than previously al-though the surround of the crossover point still seems problematic. A significant benefit is also that attenuation in the driver's stop-band can be made over 4 dB higher than in the 90-degree tuning. However, it is not useful to increase the phase difference noticeably from the 120°, for the power response and also the total impedance may ascend too much, and the design easily becomes too critical in terms of the drivers' mutual positioning and the listening direction.

From the standpoint of realizing current-drive, the outcome at the crossover frequency is also, fortunately, not so poor as could be concluded from the mere index value. Cancellation of EMF-derived sounds namely comes to aid also in the system at hand, in a like manner than in the parallel connection of Fig. 5.3. Also in this case, the EMF-derived current components circulate in opposite directions in the two arms, so the cancellation effect works quite well, provided, of course, that the current-driven frequency responses of the drivers are, in the crossover region, similar enough and that the acoustic source points of the drivers are equidistant from the listener. The latter condition requires, in practice, the tweeter unit to be positioned somewhat behind the woofer.

A working two-way system is thus fully possible but requires the phase relationships of the drivers to be regarded in a way that is not customary, in general. To reinforce the cancellation effect, the amplitude properties of the drivers should preferably match at least for an octave on both sides of the crossover frequency. To keep the sum frequency response right also requires that the partial responses must not become antiphase too close to the crossover frequency. For these reasons, the resonant frequency of the high-frequency device should be preferably at least two octaves lower than the crossover frequency. Existent tweeters are, however, not designed from this standpoint; and so their resonant frequencies are most often too high for the purpose.

It can be asked whether it is, in general, necessary to strive for high current-drive index outside the driver's passband? The answer is affirmative at least for two reasons. Steep filtering cannot be used without sacrificing the source impedance in a wide frequency range; so, of necessity, the operation of a driver also extends to a significant degree into the transition band, by which reason current-drive is also needed there. Secondly, the electromotive forces are not confined to those frequencies by which the driver is fed but typically include very manifold distortion components. Respective distortion currents will also be audible in the driver's stop region although the applied signal itself would not reach there.

Always a sure means of raising the level of current-drive is, of course, to use series resistance for each driver. This is expedient especially when the power consumption of the speaker is not of major concern. In any case, one may seek a suitable compromise between wasted power and the ideality of operation.

The principle of Fig. 5.4c is also suited for realizing three-way systems. The middle-range driver is attached to the circuit in the usual way through a series inductor and capacitor. The need to use three bands is, however, lesser than on voltage drive since voice coil inductance does not limit the frequency range of a current-operated driver. Also otherwise, the three-way alternative should be deliberated carefully because there are two crossover frequencies requiring care regarding the current-principle.

As a general rule for multiway applications, considering both the frequency response and current-principle, one can state:

All current that the amplifier feeds should flow through some speaker driver so that each driver is properly capable of reproducing those frequencies that to a significant degree are passed through it. In other words, side currents flowing past the drivers and driver currents not producing sound should be kept as low as possible. In addition, mutual differences in the drivers' phase behavior must be known, so that phasing in the transition region can be mastered and enclosing designed successfully.

Due to uneven frequency responses, the above current rule often has to be traded, but these exceptions should be tailored so that the current-drive principle does not suffer unduly.

Of course, one can get totally rid of the passive crossover by using a separate power amplifier stage for each driver and performing the necessary filtering functions before amplification. With this, active, principle, the current-drive index is determined solely by the quality of the amplifier, but as the drawback the amplifier and loudspeaker have to be designed for each other and, in practice, integrated together. Due to inconveniences relating to use and fewness of manufacturers mastering both fields, active speakers are unlikely to become soon prevalent; but for hobbyists, the method provides an opportunity for almost perfect current-drive irrespective of the number of divisions.

5.5 Suitability of Drive Units

The properties of existent speaker drivers, intended without exception for voltage drive, are quite often not very ideally suited for current-drive. For lack of better ones, they can nevertheless be used, very selectively, hoping that in the future the situation would be better in this respect.

Frequency response graphs specified for drivers are generally measured with a voltage of 2.83 V, corresponding to a power of 1 W into 8 Ω resistance. In terms of current-use, these measurements should be taken, correspondingly, with a current of 0.354 A. Voltage-based results can, however, be used by reading them together with the impedance graph. Fortunately, manufacturers customarily present the impedance curve in the same plot with the frequency response which makes it easier to find out the current-driven response.

Figure 5.6 illustrates the determination of the current frequency response from the voltage frequency response and impedance. The impedance axis must also be logarithmic (as it usually is) and preferably use the same pitch as the sound pressure axis.

The voltage response curve and the corresponding current response curve intersect at frequencies where $|Z| = 8\ \Omega$. As the impedance rises, the test signal of the current response increases relative to the test signal of the voltage response, so the difference between the curves changes by a corresponding amount of decibels. Consequently, the shape of the current frequency response is a direct combination of the voltage frequency response and impedance. For example, if the voltage response is constant over some interval, the current response in that interval follows the shape of the impedance curve.

As already noticed in section 3.8, cone drivers tend to accentuate the upper end of their operation range due to acoustic disconnection of the voice coil and the horn effect of the cone. The present striving for low mass also in the voice coil and its former yet reinforces this effect, that in a voltage-operated driver is largely compensated by the inductance-caused decline in current. In respect of the current frequency response,

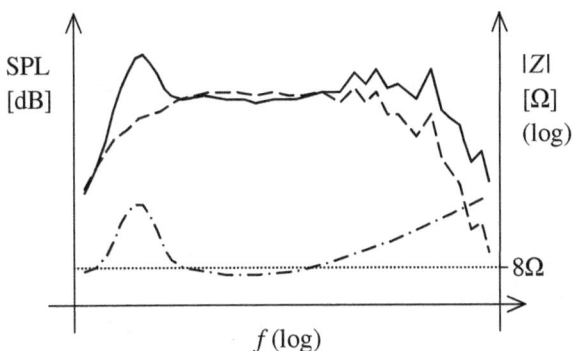

Figure 5.6. Diagram illustrating the dependence between current frequency response and voltage frequency response. The difference, in decibels, between the current response curve (solid line) and the voltage response curve (dashed line), determined at the same nominal power; equals the departure of the impedance (dash-dot line), in decibels, from its nominal value.

it would thus be well to give up this style that aims to minimize all moving masses, for by increasing the voice coil mass relative to the cone mass, the radiation sensitivity of the disconnected voice coil and hence the accentuation of the high frequencies could be somewhat constrained. Hardening of the structure might provide other advantages also, even though efficiency is slightly reduced.

The accentuation of the upper frequencies can be corrected to some extent also by directing the driver a little aside from the listener (see Fig. 3.12a). The best way for this is to point the front panel somewhat upward. An additional benefit is that the power response does not fall although response in the listening direction is equalized.

As the impedance increases with frequency, the difference between the above-described frequency responses also increases. Due to this, the reproduction band of a driver extends on current-drive always higher and ends more gradually than on voltage use.

As the low-pass filtering caused by voice coil inductance is left off, current-drive enables to use one-way principle in many such applications where one would usually have to employ a tweeter and a crossover filter in order to cover the highest frequencies.

A rise in the driver's cut-off frequency say from 8 kHz to 12 kHz (however defined) can be many times a decisive improvement by which the partitioning of frequencies and pertaining fitting problems can be avoided.

A large part of currently available woofers are not properly suited for current-drive use because of too high Q_m value. The resonant frequencies are also often unnecessarily high compared with what would be reasonably possible.

As we know from chapters 2 and 3, at the resonant frequency the current sensitivity of a driver becomes Q_m-fold relative to the normal level and, at maximum, even a little higher. In order that any correction would not be needed, the mechanical Q of an enclosed driver should be roughly 0.7 or at least less than 1. This is not generally achieved with currently available gear, but making such drivers would by no means be impossible if the subject were delved into.

If the Q value is less than 2, the response peak can be leveled by a passive equalization network. In this method, the current-drive index has to be traded, but mainly in the bass region only where the matter is not of such importance as elsewhere.

Active equalization, instead, can yet be employed with considerably higher Q values and without compromise concerning the driving mode. Here, the Q magnitude is primarily limited by how accurately the resonant frequency is known and how well it stays constant as the signal level and temperature vary. (Methods for resonance compensation are discussed more closely in chapter 8.)

Now, there might arise a question if the damping constant of a driver could be increased and the Q value thus decreased by using a second voice coil and the current introduced in it for braking? The coil would develop a velocity-dependent EMF in accordance with equation 3.7; and by using resistive load, also the current and the braking force caused by it would be directly proportional to the velocity.

Without doubt, this way one could sure bring about the needed extra damping and the desired Q value, but the EMF-derived current flowing in the brake coil would be, for sound quality, as detrimental as other similar currents, so the method is not useful. In fact, from the standpoint of the driving mode, it makes no difference whether the loading resistance is connected across the brake coil or the actual voice coil.

All EMF currents associated with the voice coil structure are able to act as sources of interference, so it is worthwhile to also pay some attention to the coil former and its possible eddy currents.

The most common former material today is aluminum, that conducts electricity fairly well. The former cylinder generally includes a narrow gap to avoid establishing a shorted secondary winding to the moving system. Despite this, the structure permits eddy currents to appear, as shown in Fig. 5.7.

In the area of the voice coil, there is induced an EMF corresponding to one turn, giving rise to current flow via the upper part of the former, where the magnetic fields do not reach much. A conductive coil former is thus, in terms of the current-drive principle, problematic at least to some extent, but in practice, the significance of these currents doesn't

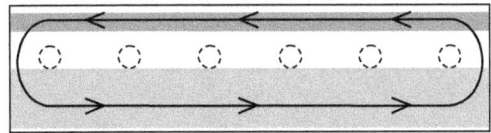

Figure 5.7. Typical voice coil former piece unrolled. The light grey strip denotes the area covered by the winding, and the dark grey strip the fixing point of the cone. Eddy currents are able to tour at least via the route marked by the arrow line.

seem very great.

Eddy currents always tend to slow down the motion that originally caused them, so their effect shows up as a slight increase in the mechanical resistance (damping constant) and hence a decrease in Q_m. However, electrical damping is not a right way to affect the Q value, as was already noticed when discussing the brake coil.

A conductive coil former reacts to both motion- and inductance-derived flux variation. The magnetic flux caused by the eddy currents always tends to counteract change in the original flux (not the flux in itself), so the first-mentioned flux is, in principle, in phase quadrature with the latter. For inductance, this means an increase in the resistive voltage component relative to the inductive component, that is, a decrease in the impedance's direction angle.

Figure 5.8 shows the effect of an aluminum coil former on the impedance angle of a 30-mm voice coil in the treble region, without the magnet assembly. At 2 kHz, the difference between the coils with and without the former is yet diminutive, but in the upper end of the hearing range, the angle difference increases to 7 degrees which already requires that the magnitude of the eddy-current-caused flux must be a good 10% of the original inductance flux.

Such eddy currents may exert significant extraneous forces on the coil former, and the presence of the magnetic circuit does not change this principle much either. For the highest frequencies, it would thus be justified to use, for confidence, insulating coil former material; but in

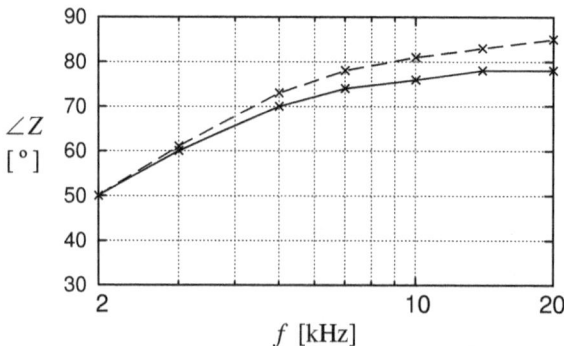

Figure 5.8. Direction angle of voice coil impedance of a 4-Ω cone driver, measured apart from the magnet assembly, with aluminum coil former (solid line) and without the former (dashed line). Coil diameter was 30 mm, coil height 14 mm, and former height 25 mm. At high frequencies, eddy current losses in the former decrease the angle, i.e. make the impedance more resistive.

practice, after other requirements, alternatives available in this respect are presently rather scarce.

In the previous section, it was already found that tweeters must be required to have low resonant frequency. The smallest values in currently available types fall around 500 Hz. However, the need would be to get the value somewhere in the range 300-400 Hz in order that rather low crossover frequencies would yet function appropriately. This can be achieved by using a sizable rear chamber, increasing the mass (along with force factor), and making the surround more flexible. No property should degrade due to these measures, and power handling could be even increased.

Further, it is desirable that the mechanical Q value of the upper-frequency driver would not exceed three or fall below unity. A high Q_m causes a pronounced response peak at the resonant frequency which complicates the managing of the total frequency response (The same problem also occurs on voltage drive.) A low Q_m again causes that the phase lead in the pressure of the upper-frequency driver with respect to its current easily becomes too large, thus making the acoustic source points of the drivers stray too far from each other (in the listening direction). Unfortunately, only few manufacturers have a practice to specify Q values for their tweeters; so, in the evaluation, one generally has to resort to the impedance curve or own measurement.

The main reason for the source point difference is, of course, the different distance of the voice coils from the plane of the mounting panel. The phase lead caused by the resonance of the upper-frequency driver acts in practice in the same direction; so in order to match the inherent phases at the crossover frequency, the front panel may have to be inclined quite a lot. Consequently, it would be helpful if the diaphragm of a dome-type tweeter were somewhat behind the mounting plane, meaning mild horn-shape. When using actual horn tweeters, inclination may not be necessary at all.

5.6 Microphones and Phono Cartridges

The front end of a sound reproduction chain always involves a microphone, that also often uses the electro-dynamic principle. A microphone capsule resembles, in basic structure, much a speaker driver and operates reversely with respect to it; so it is justified to ask if microphones are also subject to some unwanted feedback effects, that are dependent on the impedance seen by the voice coil? Presumably, it is not desirable that the quality of microphone signals constitutes a limiting factor as the reproduction flaws of loudspeakers diminish.

Figure 5.9a shows a way of realizing the commonly used, so-called pressure gradient microphone. A voice coil made of hair-thin wire is attached to a thin plastic diaphragm, that is exposed to sonic pressure from both the front and back sides. At low frequencies, the pressure developed behind the diaphragm is almost as great as in the front, and acoustic cancellation is strong. However, as frequency rises, the back pressure rolls off, and the net force acting upon the diaphragm correspondingly increases which, in principle, keeps the velocity of the coil even.

In Fig. 5.9b, a microphone is represented by an equivalent circuit, corresponding to Fig. 3.4, and the terminals are attached to an impedance (Z_a) representing amplifier input. As the voice coil moves due to the pressure, an electromotive force e_p is generated, giving rise to current i, through the circuit. R_c is most often of the order of 200 ohms, and e_i represents the inductance EMF like in speakers.

A typical value for Z_a is nowadays some 1 kΩ, but the standard impedance of 600 Ω is also used. The lower Z_a is, the higher is the voice coil current and the higher is e_i, that includes magnetic interference effects similarly as in speaker transducers. Especially the Barkhausen

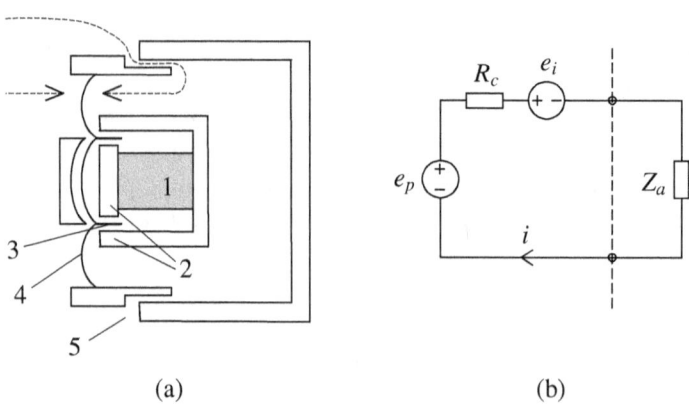

(a)   (b)

Figure 5.9.   a) One realization concept of the electro-dynamic microphone. 1) magnet, 2) pole pieces, 3) voice coil, 4) vibrating diaphragm, 5) side opening. Arrows depict the pressures acting on the front and back sides of the diaphragm. b) Equivalent circuit of a loaded electro-dynamic microphone. e_p =EMF produced by diaphragm motion, R_c =voice coil resistance, e_i =EMF produced by inductance, Z_a =amplifier input impedance.

noise may be even much more significant than in speakers because the signal levels are low.

At high frequencies, the magnitude of the inductive impedance can be, for example, one-fifth of R_c, that again can be one-fifth of Z_a. Voltage proportions are the same, so, in the present practice, the total effect of e_i on the output voltage is, at high frequencies, typically several percents, in fact for nothing.

In addition to flux modulation, current flowing in the voice coil naturally also introduces a force effect, as in loudspeakers.

We will estimate next the magnitude of this force with respect to the pressure-generated force in an ordinary pressure gradient microphone. Let us assume for the sensitivity 2 mV/Pa, the effective diaphragm area 2 cm^2, and the force factor 5 N/A, which are fully realistic values.

Suppose the microphone is exposed to an acoustic pressure of 1 Pa (corresponds to a sound level of 94 dB). If this pressure is passed only on the front surface of the diaphragm, as can be expected to be the case for the highest frequencies, the force acting on the diaphragm is 1 Pa · 0,0002 m^2 = 200 µN. However, as frequency decreases, the net pressure affecting the diaphragm also decreases; so at some relatively low frequency, the net force can well be e.g. 1/10 of the former, or 20 µN.

With the mentioned sensitivity, voltage developed at the terminals of Z_a is 2 mV, which, Z_a being 1 kΩ, yields a current of 2 µA. In the voice coil again, this current establishes a force, which is, according to equation (3.1), 5 N/A · 2 µA = 10 µN. Hence, according to this rough estimation, the magnitude of the force acting on the diaphragm due to the load current is, depending on frequency, even 5-50% of the pressure-induced force, and nothing prevents the ratio from being even greater. Thus, the loudspeaker action of microphones is, in practice, very significant, like the microphone action of loudspeakers.

The very light-structured diaphragm of microphones is, however, not intended for speaker operation, so these backward forces can possibly introduce extraneous deformations and thus degrade overall precision. The only remedy is to use high enough load impedance (10 kΩ at least). Possible loss of high frequencies due to cable capacitance should be dealt with, though.

The principle for dynamic microphones that corresponds to the current-driving of loudspeakers is that the current taken from the microphone must be kept as low as possible. Only this way can the interference factors deriving from the mechanical feedback and inductance be made negligible.

Some dynamic microphones incorporate a built-in transformer to increase the output voltage. Due to the transformer, the voice coil then sees the amplifier's input impedance divided by the square of the turns ratio; that is to say, if the transformer magnifies the voltage e.g. by ten, the impedance loading the voice coil is, in principle, $Z_a/100$. Consequently, when a transformer is used, the impedance loading the secondary has to be truly high if the voice coil current is to be kept moderate. With a too low loading impedance, even the voltage gain is lost because the voltage obtained for the primary collapses.

The operation of transformers in itself always involves non-idealities, as all iron-containing magnetic circuits, and it is also quite irrelevant whether the transformer is operated by voltage or current. Hence, these devices should be avoided everywhere in the signal path, especially when making quality recordings.

Also concerning microphones, it is often imagined that the smallness of the moving mass would denote good transient properties. However, the conception has no physical grounds any more than in the case of loudspeakers, neither are they generally even rendered. The mass affects directly only the microphone's sensitivity and resonant frequency, and the operation of the transducer itself is always minimum-phase. Instead, the many openings, cavities, and acoustic resistances, used for modifying the frequency response and directional pattern, may, to some extent, impair the time domain response because they involve delays of non-minimum-phase nature.

Capacitor microphones, whose operation is based on variations in the capacitance between the diaphragm and its back electrode, do not exhibit similar loading-induced side effects as the dynamic principle; and with the capacitor principle, it is possible to realize high-quality and high-sensitivity microphones, that, however, always require a built-in buffer amplifier and a supply voltage. Perhaps one reason for the highly-esteemed status of capacitor microphones can be found just in the above-described overloading condition of dynamic microphones.

Loading can cause unwanted effects also in other electro-dynamic transducers, like phono cartridges, guitar microphones, and playback heads of tape recorders. LP players are still produced and used despite the ascendancy of digital storage.

A phono cartridge, like a microphone, always generates an EMF that is proportional to the mutual velocity between the coil and the magnet. Consequently, the EMF introduced in the coil is the derivative function of the signal engraved on the disc, so the output voltage is highest at the highest frequencies.*

The recommended load impedance for moving-coil (MC) cartridges is typically only 100 Ω. With such a loading, the current flowing in the coil can be, especially at high frequencies, already of such magnitude that the resulting force, counteracting coil motion, can hamper the ability of the cartridge to follow fast signal alterations. Moreover, the effect of inductance is at least of the same order than in microphones.

When picking up concert music, it has become customary to use a multitude of microphones and record different instrument groups from close distance to enhance resolution and stereo imaging. The tracks are mixed afterwards and adjoined by manifold effects until the producer is satisfied with the result. Only a handful of audiophile labels aims to minimize the amount of microphones and to avoid multitrack techniques and artificial post-processing.

Could the prevailing state of affairs have somehow been influenced by the fact that the mixing engineers listen to their productions through their studio's voltage-driven monitor speakers, that, of course, are unable to render a truthful picture of the voicing, and then the shortcomings are tried to be fixed by introducing more microphones and more processings?

Current-drive technique reveals well the weaknesses or excessive manipulation of recordings, by which reason one does not necessarily bother to play all purchased renditions many times.

5.7 The Secret of Tube Amplifiers

Amongst audiophiles, there still lives a flourishing interest in amplifiers implemented with electronic tubes although in comparison of measurable performance quantities, solid state appliances generally outdo their nostalgic competitors hands down. In the sound of tube amplifiers, there is known to be, at best, pleasant characteristics, that are not easily described by words and that transistor equipment generally does not afford. How are this kind of observations explainable?

One reason may be, of course, the "warm" distortion produced by valves, that differs in quality from that caused by transistor stages. Also, the magnetic and other non-idealities of the output transformer, needed in tube amplifiers, may pass their own imprint on the sonic character.

---

* In the signal engraved on the disc, frequencies below 1 kHz have been boosted 12 dB relative to frequencies above 1 kHz. The differentiator action of the cartridge together with the RIAA correction of the preamplifier results in that the final response is flat.

The most significant differences are, however, found in the output impedance.

The output impedance of transistor amplifiers is typically less than 0.1 Ω, which denotes pure voltage feed for the speaker. In tube amplifiers, instead, the output impedance varies rather widely; from tenths of an ohm to even more than five ohms (with 8 Ω loading). A source impedance of even a couple of ohms is able to weaken the speaker's EMF currents so that the effects are observable; and as the value exceeds 5 Ω, the speaker may function at some frequencies even halfly current-driven.

It is indeed interesting, although not surprising, to notice how, in various sources, the listening evaluations published for amplifiers do correlate with the measured output impedance. When the impedance exceeds 3 Ω, positive comments e.g. about sound clarity and spatial impression increase even remarkably; and with appliances exceeding 5 Ω, "better than ever" style characterizations are already common, provided, of course, that the speakers used are suitable and that their frequency response is not too much confounded due to the frequency dependence of their impedance.

In actual current-drive use however, we can do without tubes unless one is then seeking certain distortion effects, as in guitar amplifiers. The designing of equipment according to the principles of current-drive enables one to travel the whole road along which there has been taken, in some top class amplifiers, a few steps.

Given hence that increasing the impedance enriches listening experiences, decreasing the same should respectively degrade them. This also seems to be the case, at least in light of those publicly accounted experimentations that deal with negative output impedance.

In subwoofers, it can be even of benefit that the voice coil resistance becomes almost cancelled by a negative output resistance since in this way one is able to quite directly control the velocity-proportional motional EMF; but otherwise, the using of negative resistance has understandably not gained popularity.

6
COMPENSATION OF THE RISING RESPONSE

Due to the phenomena presented in section 3.8, the current frequency response of an enclosed drive unit rises between the bass and treble regions generally so much that necessary correction is not accomplished merely by directing the speaker somewhat aside. This being the case, one has to resort to electrical compensation, that passively implemented always charges the current-drive index to some extent.

6.1 RCL Equalization

In order to effect attenuation in a speaker fed by current, a part of the current must be conducted past the driver. The first solution that comes to mind is, of course, to add an RC network in parallel with the driver, as e.g. in Fig. 6.1a. The accuracy of compensation would be, even with this circuit, sufficient for most cases, but the price to be paid is too hard: at high frequencies, the resistance appearing in parallel with the driver drops the current-drive index so low that the method has to be regarded as an outruled alternative unless the attenuation need is truly minor.

Figure 6.1b shows a usable attenuation concept, where the bypass arm is not connected to the driver directly but by means of a series impedance that contains inductance. At low frequencies, L_1 is conductive while C_1 corresponds to a break, so the driver receives full feed. At high frequencies, in turn, C_1 is conductive while L_1 equals a break, and thus the driver only gets a specified portion of the total current, as determined by the resistors. In the transition region, the steepness and shape of the attenuation curve can be adjusted by the choice of C_1 and L_1.

For high frequencies, the current-drive index is determined by the

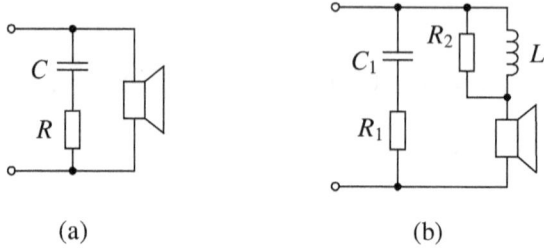

(a)   (b)

Figure 6.1.   a) Poor method for attenuating high frequencies. The operation mode easily drifts on the side of voltage drive. b) Better way to realize attenuation. At high frequencies, the driver sees impedance R_1+R_2.

sum of the resistances R_1 and R_2, that thus should be as high as possible. However, the impedance load of the amplifier cannot be increased unduly; so a suitable compromise has to be found between the index and, on the other hand, the voltage loading the amplifier and the power lost in the resistors. The tuning can be performed with a circuit simulator rather comfortably, but in order to get accurate results, the driver's own inductance must be taken into account by modelling it with an appropriate equivalent circuit.

Figure 6.2 displays the behavior of the attenuation circuit, using an example tuning that yields 6 dB attenuation and assuming an 8 Ω resistance in place of the driver. The compensation need arising from the baffle step of the enclosure is, in practice, generally less than 6 dB, so the rest of the obtained attenuation may be attributed to mitigate the horn effect. The correction function of Fig. 'a' can be fully sufficient for smallish drivers if the internal damping of the cone is good. It is better to use compensation rather too little than too much so that the current-drive index would not suffer unnecessarily.

Figure b shows the impedance seen from the circuit terminals, which is about a mirror image of the correction function. As frequency increases, the impedance rises here from 8 to 16 Ω. The result can be considered very temperate since, in comparison with a bare driver, the impedance and power consumption increase mostly only in the middle region. For high frequencies, in turn, the impedance stays, in practice, even lower than the driver's own, heavily inductive impedance.

The CDI (Fig. c) reaches its minimum at low mid-frequencies, where C_1 and L_1 tend to resonate. With the used 8 Ω nominal load, the index saturates at high frequencies to a value of 7. With a real driver, how-

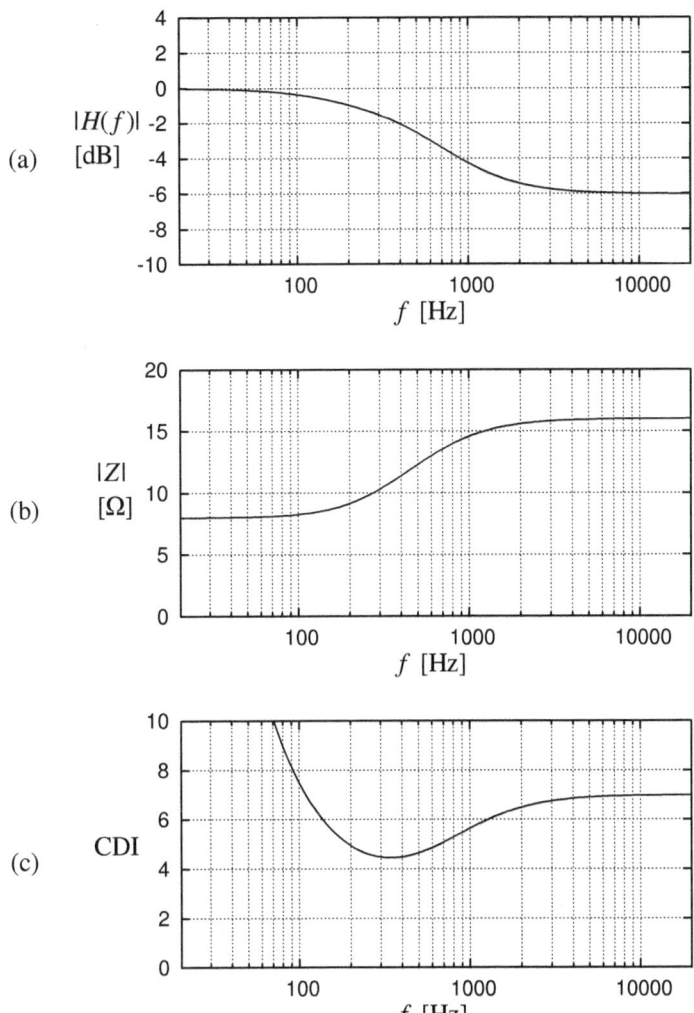

Figure 6.2.   a) Attenuation function produced by the circuit of Fig. 6.1b, using values $R_1 = 32\ \Omega$, $R_2 = 24\ \Omega$, $C_1 = 30\ \mu F$, and $L_1 = 4$ mH and assuming 8 Ω load. b) Corresponding impedance magnitude as seen from outside. c) Corresponding current-drive index.

ever, the rise in impedance due to inductance causes that the index is left somewhat lower than that. If the driver is to reproduce treble also, a small inductance may yet be added in series with C_1 in order to improve CDI for the highest frequencies.

A capacitor of a few tens of microfarads is yet quite easily implemented with plastic dielectrics, without having to resort to non-ideal electrolytics. A coil of the order of 4 mH, again, is not yet too large to be realized air-cored, but even the use of ferrites is not in this application so problematic as generally under voltage drive.

The presented attenuation concept can also be applied in multiway systems. Figure 6.3a shows a recommendable basic topology of a two-way speaker, in which the attenuation network of Fig. 6.1b has been conjoined with a parallel crossover filter. In addition, the treble arm has a series resistor, whose purpose is to augment the CDI of both drivers.

Figure b shows an example of the current dividing functions obtainable with the Fig. 'a' circuit, the crossover frequency being 2 kHz. Here, the attenuator has been tuned according to the previous example; so the ratio of the attenuator's total current, I_{l0}, and the current obtained by the driver, I_l, is as shown in Fig. 6.2a. It is best to define the crossover frequency to the intersection of I_{l0} and I_h since this point should conform to the actual crossover of pressures if the compensation function has been correctly chosen.

Both the driver's and attenuator's behavior is minimum-phase in the whole operation range, so, when correcting the driver's amplitude response, the phase response becomes rectified as well. Therefore, current I_{l0} corresponds, in principle, to the actual pressure behavior of the driver in terms of both amplitude and phase.

It is yet worth noting that the filtering steepness of current I_h is distinctly greater than generally in 1st-order filtering which stems from the impedance properties of the attenuator.

The tweeter impedance was assumed to be 6 Ω, which is a very common rating in domestic use. Hence, the 6 Ω used also for R_3 does not yet increase overall power consumption unduly; minding that, generally, only a minor part of sound signal power falls in the treble region. (Series resistors are also otherwise customarily employed in equalizing driver sensitivities.)

The index for the lower-frequency driver (Fig. 6.3c) is around three and a half for the whole midrange. Compared with Fig. 6.2c, the decline is substantial, but on the other hand, the result is not yet very bad since even at this level, the attenuation ratio of EMF currents is still 13 dB.

The tweeter index is also only moderate at the crossover frequency but quickly improves as frequency rises. The indexes can, of course, be bettered by increasing the resistance values, but in this, one can proceed only so far as the rise in voltage and power consumption allows.

The earlier-described cancellation mechanism of EMF-derived dis-

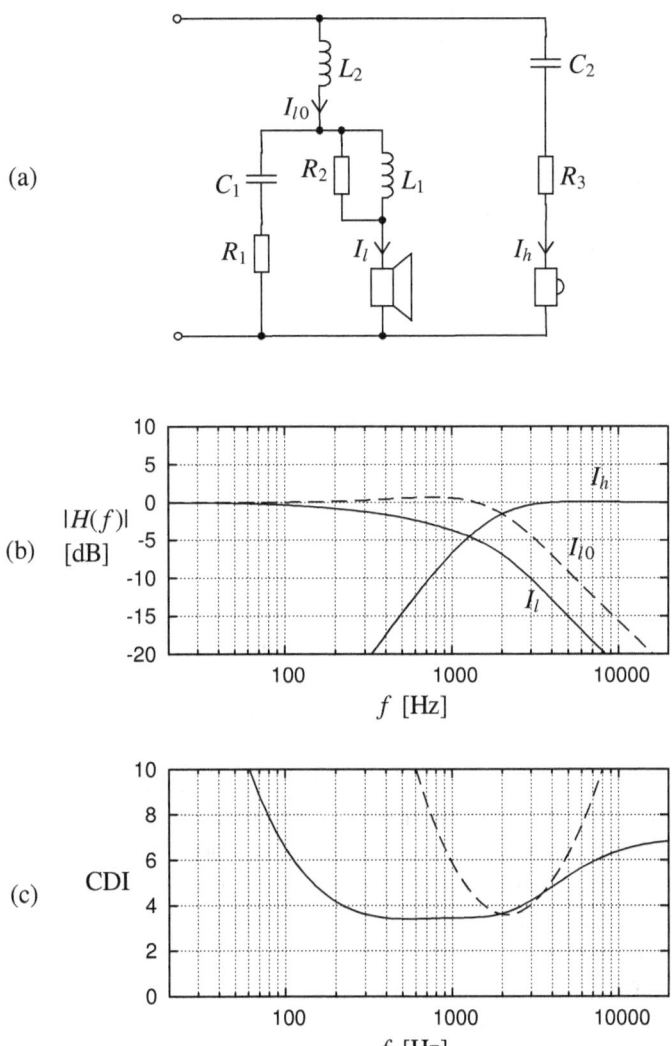

Figure 6.3. a) A two-way system employing the attenuation scheme of Fig. 6.1b. b) Behavior of currents in the Fig. 'a' circuit with an example tuning that yields 2 kHz for the crossover frequency. Settings are the same as in Fig. 6.2; additionally: $L_2 = 1.2$ mH, $C_2 = 4$ μF, and $R_3 = 6$ Ω. Woofer impedance was 8 Ω and tweeter impedance 6 Ω. c) Corresponding current-drive indexes for the woofer (solid line) and tweeter (dashed line).

turbances also acts in the system of Fig. 6.3a. Imagine that the tweeter develops, for whatever reason, an EMF pulse whose frequency content falls in the region of the crossover frequency, where the CDI is at minimum. This EMF introduces to the tweeter arm a certain current pulse, that also flows as such through inductor L_2 to the attenuation network. If the current frequency responses of the tweeter and the attenuated lower-frequency driver are congruent in the said region, as intended, the sound pulses produced by the drivers are similar but in opposite phase, canceling each other. At least concerning the tweeter, the attenuation of EMF interferences can thus be in actuality better than the CDI curve suggests.

In the other direction, however, the case is not as favorable. Imagine the same EMF pulse now develops in the bass-midrange driver. At the frequencies in question, C_1 corresponds, proximately, to a short and L_1, proximately, to a break, while the series connection of L_2 and C_2 in turn corresponds, proximately, to a short. The remaining, almost resistive network does direct the current pulse, that flows through the woofer, for the most part via the tweeter arm, but a portion of the current returns also via R_1. Because the woofer is also more sensitive than the tweeter, the cancellation is left in this case only partial.

Instead, interference attenuation in the lower midrange is not improved by the cancellation effect at all because, normally at these frequencies, the driver responses no more resemble each other. In fact, even the opposite may occur if the tweeter's phase response is, due to resonance, suitably twisted.

With small or narrow enclosures, the baffle step compensation may have to partially extend into the region served by the tweeter. Despite this, the tweeter does not necessarily need own correction network, for the rise in current response due to proximity of resonance may already suffice to cover the need.

In presently available consumer tweeters, both current and voltage sensitivities are, generally, a few decibels higher than in bass-midrangers of corresponding quality. The tweeter's output level is intended to be lowered in place with resistors. From the standpoint of current-drive, this is regrettable since the more the level has to be lowered, the more the driving mode is compromised. (The same also holds true when voltage drive is pursued.) Also otherwise, it is better not to have excessive sensitivity in the driver since the proportion of EMF sounds and hence also the need of CDI increase as the force factor is increased or the moving mass decreased.

If a directly suitable tweeter cannot be arranged, a shunt resistor may be added to the circuit of Fig. 6.3a; from the junction point of C_2 and R_3

to the lower terminal of the driver. An attenuation of some two decibels is yet possible to be realized without severe decrease in the CDI.

The element values used as an example are only directional and not intended as such for any practical application. Optimization always has to be performed with a circuit simulator, after the driver impedances have been modelled using the principles outlined in chapter 7.

6.2 Dual Coil Equalization

The above-described compensation scheme may degrade the mid-frequency EMF current attenuation too much, especially if the compensation need is large. Therefore, we need for the response equalization a method that interferes less with current-drive. This is made possible by the use of two voice coils.

Figure 6.4 shows the basic topologies for attenuating high frequencies with a two-part voice coil. The topology of Fig. 'a' yields a 6-dB step, assuming that the voice coils A and B have equal force factors. If the attenuation need is smaller, a resistor can yet be connected in parallel with the inductor L_1. If the attenuation need is greater than 6 dB, the variation of Fig. b can be used, which contains, over the former, one capacitor and resistor more.

The operation is based on that, at low frequencies, the signal current is passed as a whole through both voice coils, whereas at high frequencies, only coil B gets current, coil A being left passive. At low frequencies, L_1 is conducting, while C_1 and C_2 are virtually currentless. At high

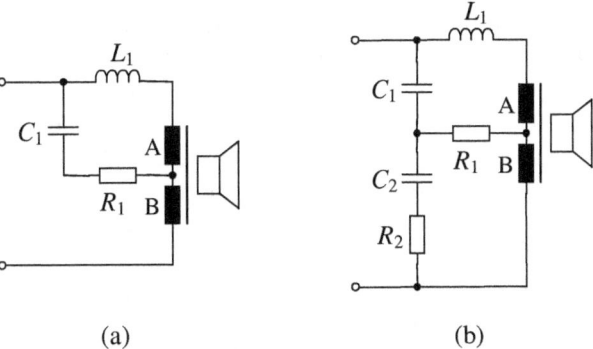

(a)   (b)

Figure 6.4. Compensation networks for the rising response, using a dual coil driver. a) Topology for attenuating high frequencies by 6 dB. b) Topology for attenuating high frequencies more than 6 dB.

frequencies, instead, L_1 and voice coil A are almost currentless, while the current flows via C_1. In Fig. 'a', all current is always passed through coil B; whereas in Fig. b, part of the current is also directed, at high frequencies, past coil B.

Resistor R_1 is chosen so high as the rise in total impedance allows. In case b, the desired high-frequency attenuation is set by R_2. The compensation function is shaped by trying diverse values for the other elements. In simulation, it must be regarded that the voice coil currents are in different phase, so their sum must be evaluated in vector form.

The benefit of the operation principle is that the impedance seen by coil B is in case 'a' infinite and also in case b, in practice, very high. The impedance seen by coil A is also relatively high, except in the resonance region of L_1 and C_1, where the impedance is at lowest equal to R_1 or, in the Fig. b case, little lower.

Figure 6.5 displays example curves from both circuit alternatives. The solid lines represent version 'a' and the dashed lines version b, that is set here to 10 dB attenuation. To facilitate comparison, version 'a' has been tuned, in terms of the correction function and total impedance, to conform with Fig. 6.2.

With two or three variables, the shape of the compensation curve cannot be determined all freely, of course, but the attenuation steps of Fig. 'a' are, in steepness, suitable for most purposes. Also here, the resistances have been chosen so that, with the nominal 4 Ω voice coil impedances, the total impedance (Fig. b) does not exceed 16 Ω.

The CDI minimum for voice coil A is in the 'a' case 3 and in the b case 3.7 (Fig. c). However, in the latter, small EMF currents are also passed through coil B which degrades a little the factual interference attenuation of alternative b. Taken as a whole, the driving mode is thus quite independent of the amount of high-frequency attenuation used.

How great, then, the index of coil A must be compared to an unsplit voice coil (like e.g. that in Fig. 6.1b), in order that the EMF detriments would remain at the same level? The answer is not so self-evident but can be found, for example, by the reasoning presented in the following.

First, it should be noted that the electromotive forces are in both coil halves always equal despite the fact that at high frequencies coil A is no more participating in program signal reproduction. Despite their different roles, coils A and B are thus in equal position in terms of their susceptibility to EMF currents.

Let us assume, then, that the same driver is used both in Fig. 6.4a and Fig. 6.1b circuits (voice coils in series) at the same frequency and sound level. We will further assume that in both cases the driver's EMF-

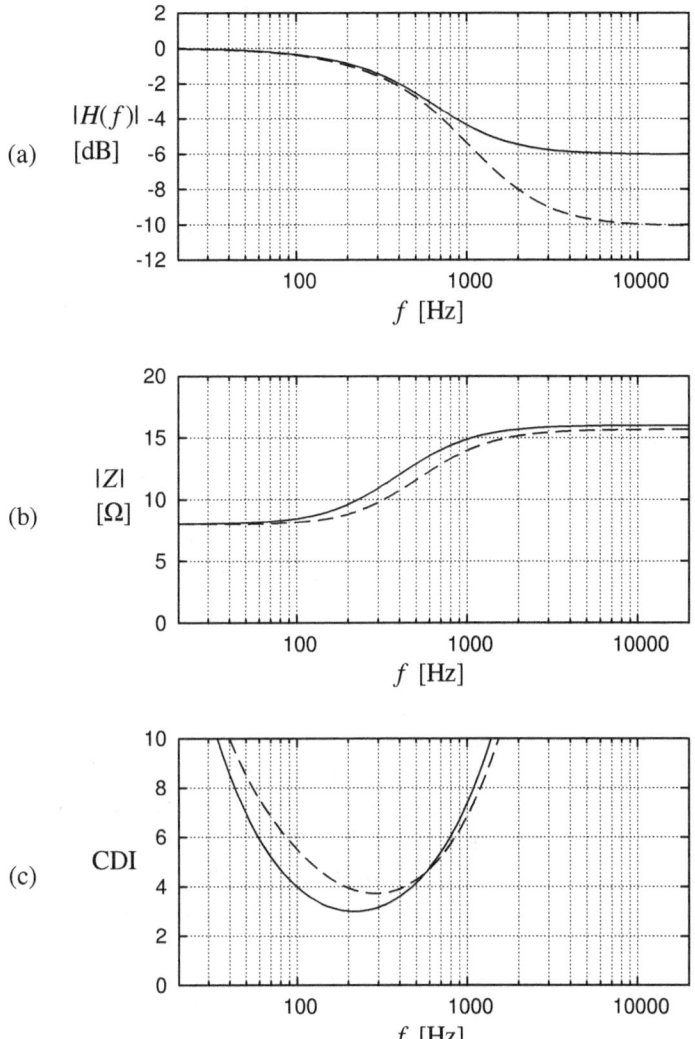

Figure 6.5. Solid lines: behavior of Fig. 6.4a equalizer using example values $R_1 = 12\ \Omega$, $C_1 = 120\ \mu F$, and $L_1 = 4.5$ mH and voice coil impedance of $(4+4)\ \Omega$. Dashed lines: behavior of Fig. 6.4b equalizer using values $R_1 = 21\ \Omega$, $R_2 = 42\ \Omega$, $C_1 = 120\ \mu F$, $C_2 = 30\ \mu F$, and $L_1 = 4$ mH and voice coil impedance of $(4+4)\ \Omega$. a) Attenuation functions (vector sum of voice coil currents). b) Impedance seen from outside. c) Current-drive index of voice coil A.

derived sounds are being heard equally loud, that is, the sound quality is the same.

In the split-coil case, only coil A is able to generate EMF sounds, so the EMF current flowing through it must be 2-fold compared to the unsplit-coil case. In both cases, the velocity of the diaphragm and also the current-generated magnetic flux are the same, so in coil A there is induced half of the motional EMF as well as inductance EMF of the unsplit-coil case. In order that this half EMF would bring about the aforementioned 2-fold current, the impedance opposing it (including voice coil's contribution) must be 1/4 of that in the reference case. Consequently, the Thévenin impedance seen by coil A has to be less than 1/4 of the reference case value.

In terms of current-drive indexes, this means that, when coil B is operating purely current-driven, a sufficient value for the CDI of coil A is slightly less than half of the CDI of the corresponding unsplit coil. Thus, for example, the minimum value of 3, obtained above for coil A, corresponds to a value of about 7 when using the unsplit coil. The improvement over the result of Fig. 6.2c is therefore considerable; and especially at frequencies above 1 kHz, the driving mode of the split voice coil can be deemed even excellent.

Conversely, also coil B can do with half of the index of the corresponding unsplit coil if coil A can be considered fully current-driven.

The dual coil concept also retains its superiority in multiway systems. Figure 6.6a illustrates a similar two-way topology as Fig. 6.3a, except that the attenuator is as shown in Fig. 6.4a.

We will use for the attenuator the same example tuning as in Fig. 6.5 and for the other elements the same values as in Fig. 6.3. The crossover of currents, that is also intended to conform with the crossover of pressures, thus falls also here at 2 kHz (Fig. 6.6b); and differences with respect to Fig. 6.3b are also else very minor.

The current-drive index of coil A (solid line in Fig. c) is only slightly lower than above without the treble arm. The minimum of 2.8 can also here be multiplied by a good two, so the improvement over Fig. 6.3c is goodly. The index of coil B (dashed line) reaches its minimum in the vicinity of the crossover frequency but stays, however, at safe numbers, regarding that the coil A index is at these frequencies already in rapid growth.

A small part of the EMF current introduced by coil A is able to flow via the treble arm but mostly only in the crossover region, where L_2 and C_2 resonate. Because this current part is, however, opposite in coil B with respect to the tweeter, the cancellation effect in conjunction with

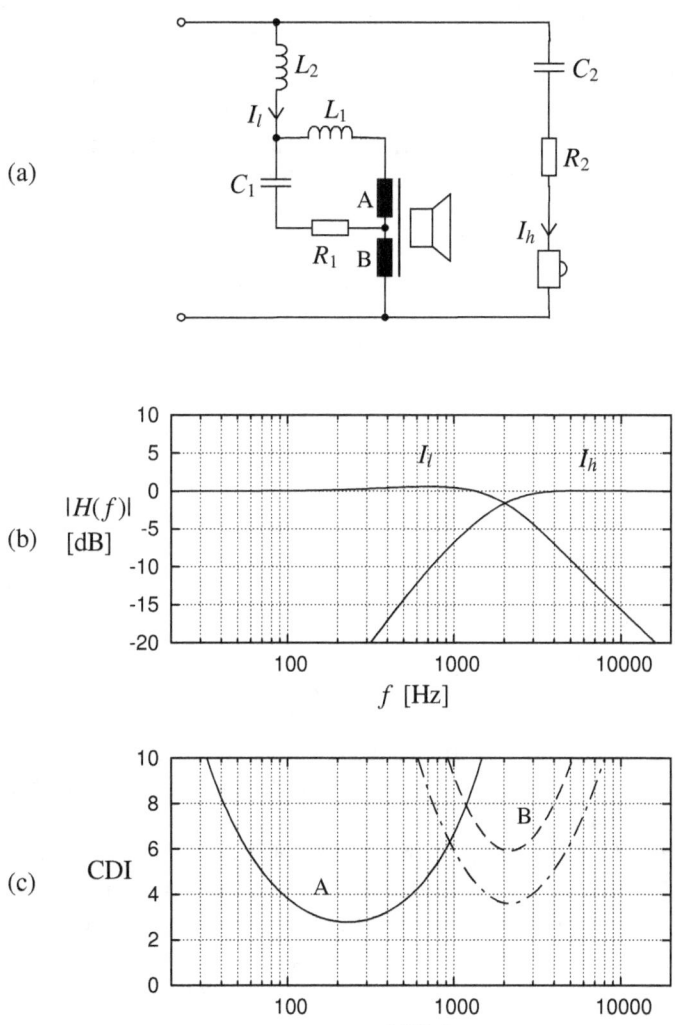

Figure 6.6. a) Two-way system applying the attenuation scheme of Fig. 6.4a. b) Behavior of currents in the Fig. 'a' circuit with an example tuning that yields 2 kHz for the crossover frequency. Settings are the same as in Fig. 6.5; additionally: $L_2 = 1.2$ mH, $C_2 = 4$ μF, and $R_2 = 6$ Ω. Woofer impedance was (4+4) Ω and tweeter impedance 6 Ω. c) Corresponding current-drive indexes for voice coils A (solid line) and B (dashed line) and the tweeter (dash-dot line).

the rather high CDI of coil A keeps the adverse effects well down.

The EMF current introduced by coil B, again, flows in its entirety through the tweeter, so cancellation helps to improve still further the rather good attenuation ratio of coil B. As frequency decreases, greater and greater part of the coil-B-derived EMF current also flows through coil A, though, which, in principle, adds to the interfering effect of this current. However, the CDI of coil B rises at low frequencies so high that any severe harm cannot be introduced even this way.

The CDI of the tweeter (dash-dot line), instead, does not differ just about at all from that shown in Fig. 6.3c since all impedances, including the attenuator block, are the same. The cancellation effect also acts in both cases just in the same way, enhancing the otherwise moderate attenuation ratio of the crossover region. From the viewpoint of the tweeter, it is thus indifferent which compensation scheme is used for the bass-midranger.

In the example case, the phase difference between the currents of the low-frequency and high-frequency arms is at the crossover frequency a good 100°. The peak value of the total impedance will be respectively 20 Ω. By increasing the phase difference, the current-drive indexes can yet be advanced a little if the rise in impedance in the crossover region can be accepted.

Using the option of Fig. 6.4b does not bring any essential changes over the above-described example. The addition of the bypass arm for coil B does establish a new flow route for the EMF currents at high frequencies, but on the other hand, a possibility to use higher values for R_1 makes up for this loss.

Modelling of the impedance of a split voice coil for circuit simulation is a little more complicated than usually because the functioning of the coils is not independent of each other and, for instance, the total impedance of series-connected coils is not equal to the sum of the individual impedances. However, the modelling does not involve any difficulties, as we will find in chapter 7.

A two-part voice coil can be wound in two or four layers without having to take outlets from the coil's bottom rim. The two-layered structure can be constructed so that the wires of both coils tour side by side all the way, from top to bottom and then back again forming the second layer. When using the 4-layer structure, however, it could be of benefit to dispose two of the layers inside the coil former and two outside. This can improve robustness and prevent possible introduction of mechanical hysteresis between wires in the topmost layers and the coil former.

The dual coil compensation loses to the simple RCL compensation mostly only in cost. Besides a more expensive driver, the split-coil method requires more capacitance. However, when sound quality is foremost, the needed capacitance, being of the order of 100 µF, should be realized with plastic dielectric although, when striving for inexpensiveness, electrolytics may seem attractive. Expenses can be minimized by cutting all excessive voltage rating.

6.3 Equalization Using Two Drivers

A similar compensation as above can also be implemented by using two single-coil drivers. Then we have, in fact, a current-drive-suited variation of the presently much used, as if ½-way concept (1.5-way, 2.5-way and so on) where the lack of low frequencies due to the baffle step is leveled with an excessive driver operating only up to a few hundred hertzes. Two drivers, of course, need more space and become more costly than one, but the method also has its virtues.

Figure 6.7 presents circuit solutions, familiar from the previous section, applied for separate drivers. Elements indicated by the dashed lines can be introduced when needed, just like before. In its basic operation, the method does not differ at all from the former, so the same example values are also applicable here.

A significant difference is, however, that as driver A shuts down at high frequencies, its electromotive forces also disappear as well since the coils now function independently. Unlike with the dual voice coil, the impedance seen by driver A thus becomes unimportant at high frequencies. Likewise, when operating in a two-way system, driver A is not capable of effecting any noticeable EMF currents to the loop contai-

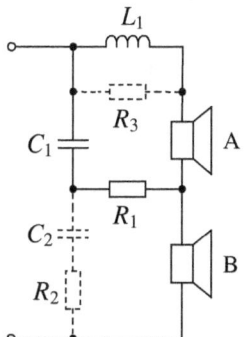

Figure 6.7. Compensation of the rising frequency response using two drivers. C_2 and R_2 are used when the attenuation need is more than 6 dB. R_3 is introduced when the attenuation need is less than 6 dB.

ning the tweeter arm.

Especially with low-sensitivity drivers, the principle of Fig. 6.7 can provide a truly good result. For example, assume that in the circuits of Figures 6.1b and 6.7 one uses otherwise identical drivers but in the latter the numbers of turns of the voice coils and hence their resistances have been halved. (The motional and inductive impedances in fact fall then to a quarter of the original.) Thereby, both methods yield the same sound level at the same current and quite the same input power.

However, at those frequencies where driver A is yet fully active, the displacement required of it is only half of that required of the Fig. 6.1b driver. Because the force factor has also been halved, the motional EMF of driver A is only 1/4 of the reference value. Assuming that driver B is operating without compromise, the EMF current of driver A needs to be, in turn, 2-fold relative to the reference case, in order that interference sounds would remain at the same level. Therefore, the impedance of the current loop containing driver A can be as low as 1/8 of the corresponding Fig. 6.1b impedance to keep sound quality in this respect unchanged. The same result is also obtained in respect of the inductance EMF.

However, the driving mode quickly degrades if the size of the drivers is reduced or the force factor increased. If, in the previous example, the radiation area together with moving mass is halved in drivers A and B, all else being as before, sensitivity and efficiency remain in essence unchanged, but the displacement and motional EMF double. Hence, the impedance of the loop containing driver A also has to be doubled (from 1/8 to 1/4) if the motional EMF detriments are to be kept still the same.

As the radiation emanates from two points, it is well to also note the power response behavior and phase alignment. If the distance between the driver centers is significant with respect to the wavelength, the power response exhibits at these frequencies, in principle, a reduction of 3 dB compared to the condition where the said distance is negligible. This drop in the power response can be prevented or at least minimized by positioning the drivers as adjacently as possible.

Close positioning is indispensable also for the reason that the compensation itself would work as intended. At frequencies where driver A's radiation is yet significant, its phase should remain in correct relationship with the radiation of driver B. In order that phase inaccuracy would not induce too much deviation in the total amplitude, the distance difference of the drivers in the listening direction should not exceed a tenth of the wavelength.

7
MODELLING AND SIMULATION

Computer simulation is generally indispensable in matching speaker drivers for each other or modifying their responses in current-drive systems. With careful modelling and suitable tools, the electrical as well as acoustic behavior of a speaker can be predicted quite reliably and handily. Circuit operation always depends so essentially on the properties of the drivers used that it is better never to rely on any universal calculation formulas or rules of thumb.

The whole design process can be carried out using an ordinary general-purpose circuit simulation software, besides which one may yet need some tool to perform mathematical operations and plotting. Customary, voltage-drive-based box and crossover calculators that understand only specified topologies and only in a specified way are, for our purposes, not of any use.

Circuit simulators, that most often are different SPICE derivatives, are also available in hobbyist-suitable free editions. In some cases, the circuit description can yet be entered in a traditional way as a text form netlist, but many simulators accept today only a circuit diagram created with a schematic editor. Both representations have their own advantages, so the choice is up to one's own needs and preferences. Likewise, there are also differences in whether the results are available as a readable list or merely as a binary file. A text form printout is necessary if the results are to be treated with other programs.

Convergence failures, that have been a major problem in circuit simulators in the past, are nowadays quite rare; but the result displaying interfaces of simulators are often buggy or non-customizable or missing necessary features, so it is recommendable to use a separate plotting software. These are also available both freely and at charge.

7.1 Low-Frequency Modelling

The motional impedance of a driver, being established by means of mechanical motion, can be modelled, by virtue of analogy, by a purely electrical equivalent circuit. By means of a transformer, this model can also be scaled so that its electrical quantities directly correspond to the system's mechanical quantities. By such equivalence modelling, one is able to examine the effects of different mechanical solutions on a speaker's response and impedance.

The motional impedance of a driver in free air or closed enclosure has been expressed in terms of the mechanical parameters in equation (3.8). An equivalent 2nd-order impedance function can also be established by a parallel resonance network (Fig. 2.4a), whose impedance is given by expression (2.18). By equating the expressions, we obtain:

$$\frac{(Bl)^2}{m} \cdot \frac{s}{s^2 + \dfrac{b}{m}s + \dfrac{k}{m}} = \frac{1}{C} \cdot \frac{s}{s^2 + \dfrac{1}{RC}s + \dfrac{1}{LC}} \tag{7.1}$$

For the equation to be satisfied, all coefficients have to be equal on the left and right sides. Thus, we have a set of equations:

$$\frac{1}{C} = \frac{(Bl)^2}{m} \quad , \quad \frac{1}{RC} = \frac{b}{m} \quad , \quad \frac{1}{LC} = \frac{k}{m} \tag{7.2}$$

which yields:

$$C = \frac{m}{(Bl)^2} \quad , \quad R = \frac{(Bl)^2}{b} \quad , \quad L = \frac{(Bl)^2}{k} \tag{7.3}$$

So, the capacitance needed in the equivalent circuit is directly proportional to the moving mass, the resistance inversely proportional to the damping constant, and the inductance inversely proportional to the spring constant.

Figure 7.1 shows the low-frequency model thus obtained. In terms of data sheet parameters, $k = 1/C_{ms}$, $b = R_{ms}$, and $m = M_{ms}$ (in basic units). k and b depend on the enclosure and damping material used, and especially the value of b is often hard to be known in advance; so, in practice, R, C, and L have to be determined by fitting the simulated impedance curve to the measured one. If, however, the impedance curve obtained with the rated parameters essentially differs from the corresponding

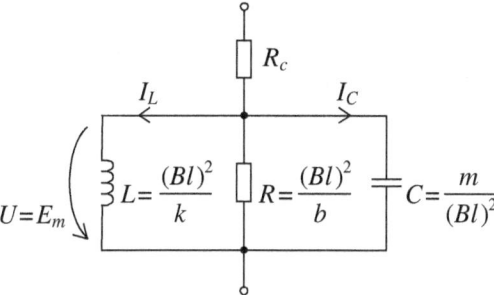

Figure 7.1. Electrical low-frequency equivalent circuit of a driver in free air or closed enclosure. R_c =voice coil's DC resistance. L, R, and C model the motional impedance. Current I_C indicates the acoustic pressure obtained. I_L indicates displacement.

measured curve, something has to be wrong.

The voltage across the parallel resonator, U, corresponds to the voltage drop in the motional impedance, that is, the motional EMF (E_m), that is directly proportional to diaphragm velocity. Current I_C, flowing through capacitor C, is in turn directly proportional to the derivative of U (equation (b9) in appendix B) and consequently reflects the diaphragm's acceleration. This yields a useful result:

Diaphragm acceleration, that indicates the resulting acoustic pressure (excluding the effects of baffle step and other non-idealities), is found out simply by monitoring the current of the capacitor in the resonance network representing motional impedance. This current corresponds to the driver's acoustic behavior as well in frequency as in time domain.

Thus, there is no need to introduce particular modelling circuits for this purpose.

The joint response of two or more drivers can be simulated respectively by summing the currents of the appropriate capacitors (weighted by the sensitivities). In frequency analysis, it must be remembered to perform the summation in phasor form unless the task is already implemented in the circuitry.

If the drivers exhibit delay differences due to mutual positioning, these can be regarded by respective post-processing of the results. If the delay difference is T ($T > 0$), the angle of the phasor representing the rear driver response has to be deducted by $2\pi f T$, or $360° \cdot f T$.

Current I_L, flowing through inductor L, is in turn directly proportional to the time integral of velocity, i.e. displacement. By applying the basic equation (b12), given in appendix B, and the relation $V = j\omega X$, familiar from equation (3.8), we obtain:

$$U = BlV = Blj\omega X = j\omega LI_L \qquad (7.4)$$

thus

$$X = \frac{L}{Bl}I_L \qquad (7.5)$$

With this result, it is easy to determine e.g. when the linear excursion range is exceeded.

If currents through passive components are not directly available in the circuit simulator, they can be measured using voltage sources. Then, to the point from which current is to be measured is inserted a zero-valued voltage source, that does not affect the circuit's operation but whose current can be monitored.

The left-hand side of equation (7.1) represented the motional impedance E_m/I. Because $E_m = BlV$ and $I = F/(Bl)$ (equations (3.7) and (3.1)), we obtain: $E_m/I = (Bl)^2 V/F$. Thus, by dividing the left-hand side of equation (7.1) by the factor $(Bl)^2$, the equivalent circuit is made to model directly the relationship between velocity and drive force, V/F. Then, the solution of the equation comes out to be:

$$C = m \ , \quad R = \frac{1}{b} \ , \quad L = \frac{1}{k} \qquad (7.6)$$

Now, the capacitance corresponds directly to the mass, the resistance to the reciprocal of the damping constant, and the inductance to the reciprocal of the spring constant. The result holds for the numerical values when basic units are used. The units, instead, do not match because the analogy is only a mathematical one.

Figure 7.2a shows the resulting electrical equivalent of the mechanical system. When current I is made to match the drive force F (1A $\leftrightarrow$ 1N), voltage U then matches velocity V (1V $\leftrightarrow$ 1m/s).

In Fig. b, the model of Fig. 'a' has been attached as part of the driver's equivalent circuit. Current-controlled current source I feeds to the resonator current BlI_0, matching the force F, I_0 being the actual current of the driver. Voltage-controlled voltage source E_m, in turn, maps the resonator's voltage U into voltage BlU, representing the motional EMF. The model thus obtained is, looked at the terminals, identical with the model of Fig. 7.1, but the factor Bl is no more needed in the analogies

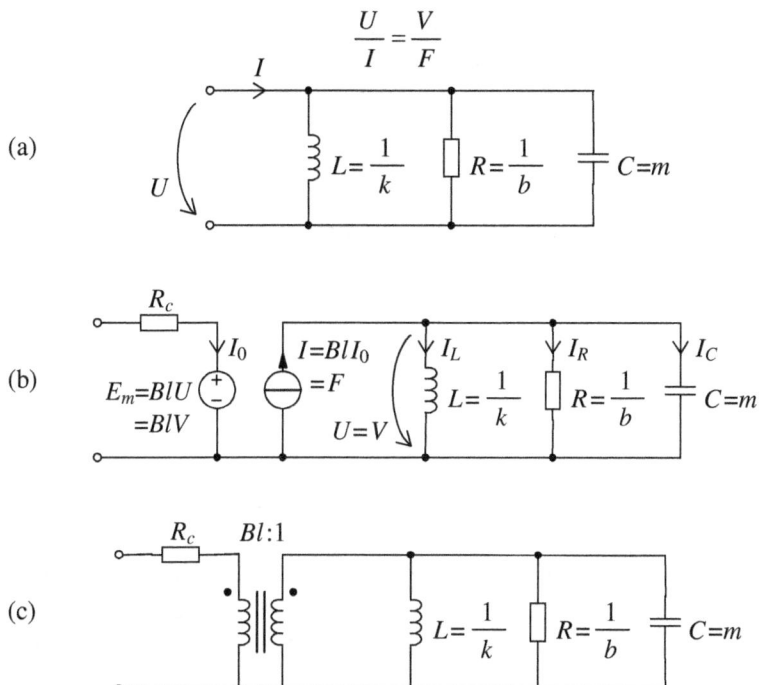

Figure 7.2. a) Electrical model that corresponds to a driver's mechanical system. When current I matches drive force F, voltage U matches velocity V. b) Overall electrical model, using a voltage-controlled voltage source (E_m) and a current-controlled current source (I). R_c, I_0, and E_m represent actual electrical quantities. c) Model of Fig. b presented using an ideal transformer.

between the quantities.

Current I divides into three components: I_C, I_R, and I_L; just as the force F divides into components representing the different counterforces according to equation (3.2). Consequently, the mass is exerted by the force $I_C = mA$, the spring by the force $I_L = kX$, and the mechanical resistance by the force $I_R = bV$. The displacement X is now obtained, instead of equation (7.5), directly as the product of L and I_L.

The controlled sources used in Fig. b are included in the standard elements of circuit simulators, so the modelling circuit is easily constructed, once the positive directions of voltages and currents are correctly chosen.

By examining the connection and operation of the sources, it can further be seen that they in fact make up an ideal transformer since the primary side's voltage is the secondary side's voltage multiplied by the factor Bl and the secondary's current is the primary's current multiplied by the same factor. Thus, the model of Fig. c fully corresponds to that of Fig. b, but the sources have been replaced by an ideal transformer that has a turns ratio of Bl:1 from primary to secondary.

7.2 *Generalization of the Analogy*

The mechano-electrical analogy described above can also be extended to systems that incorporate several moving masses with mechanical connections between them. Although current-drive operation in itself does not require any special enclosure tunings, it is sometimes helpful to gain insight into the functionality of different constructions without having to build the system or master it mathematically. Modelling may also yield new ideas for those who wish to further develop current-drive applications beyond that arrived at in this book.

By comparing the mechanical model of Fig. 3.3b against its electrical equivalent in Fig. 7.2a, one can decide those general principles by which a mechanical model is readily transformable into an electrical one without transfer functions:

- A moving object corresponds to an electrical node, whose voltage with respect to the reference potential (ground) corresponds to the object's velocity with respect to the framework ($V \leftrightarrow U$).

- Mass corresponds to capacitance from the pertinent node to ground ($m \leftrightarrow C$).

- A spring corresponds to inductance between the pertinent nodes ($k \leftrightarrow 1/L$).

- A damper corresponds to conductance between the pertinent nodes ($b \leftrightarrow 1/R$).

- A force exerted on an object corresponds to current going to the node ($F \leftrightarrow I$); and a force that an object exerts on its own mass or a spring or damper corresponds to current leaving the node.

Additionally, the mechanical model may also include levers, whose equivalent in the electrical model is a transformer.

In the following, we will consider a bass reflex speaker, that serves here only as an example of the modelling's usability.

Figure 7.3a explains the notation. m_1 is the driver's moving mass, and m_2 is the mass of the air contained in the reflex port (usually some 0.001 kg). S_1 is the effective area of the diaphragm and S_2 respectively the cross-sectional area of the port. k_1 and b_1 are the spring constant and damping constant of the driver; and b_2 represents the, generally very minor, flow resistance of the port.

Spring constant k_2 represents the air spring acting between the driver and port as seen from the port. k_2 can be calculated in terms of area S_2 and enclosure volume $V \, (\mathrm{m}^3)$ as follows:

$$k_2 = \frac{\gamma p S_2^2}{V} \tag{7.7}$$

where p is the absolute pressure ($p = 100\,000$ Pa) and γ is the adiabatic index, whose value for air is 1.4.

The mechanical model of the system is shown in Fig. b. Because the diaphragm's area is greater than the port's, displacement in the diaphragm tenses the air spring more than an equal displacement in the air column of the port. Hence, the piece representing the diaphragm has been attached to spring k_2 by means of a lever, whose lever ratio corresponds to the areas ratio. Alternatively, the spring could also be placed on the side of m_1 in which case S_2 is replaced by S_1 in expression (7.7).

The positive direction of movement has been defined as the same for both mass pieces, so the sound produced by the vent is in antiphase with the acceleration of the air column in the port.

Mass m_2 and spring k_2 also form a resonator (Helmholz resonator), whose characteristic frequency, i.e. the tuning frequency of the enclosure, can be calculated by formula (3.5).

Figure c shows the resulting electrical model. Circuit elements representing the reflex mechanism have been attached to the driver-related elements by transformer T_2, whose turns ratio corresponds to the mentioned lever ratio. In simulation, both transformers can be represented by a combination of a voltage-controlled voltage source and a current-controlled current source, similarly as in Fig. 7.2. (If the simulator does not accept a loop consisting of voltage sources and inductors, the problem can be circumvented by inserting into the circuit a series resistor of e.g. one micro-ohm.)

Also here, current I_{C1} is directly proportional to the diaphragm's acceleration, thus indicating the pressure produced by the driver. Likewise, I_{C2} is directly proportional to the air column acceleration, whose

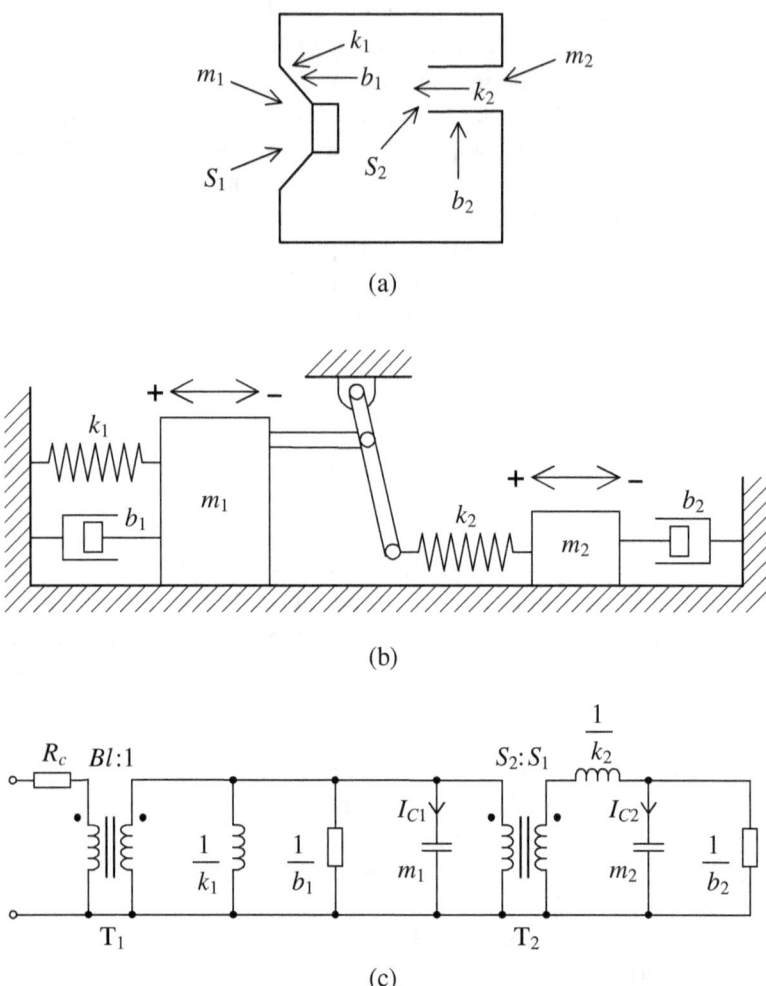

Figure 7.3.   a) Essential structure of a bass reflex speaker, showing the mechanical parameters used in the modelling. Subscript 1 refers to the drive unit and 2 to the reflex port. k_2 represents the enclosure's air spring as seen from the port. m_2 is the mass of the air column in the port. b) Mechanical model of bass reflex operation. The ratio of the lever (assumed to function only laterally) corresponds to the ratio of the areas (S_1 and S_2) of the driver and port. c) Electrical model of a bass reflex speaker in accordance with the mechanical system of Fig. b. Transformer T_2 represents the lever. I_{C1} indicates the acoustic pressure due to the driver, and $-I_{C2}$ indicates the pressure due to the vent.

opposite number indicates the pressure produced by the vent. By scaling I_{C1} and $-I_{C2}$ yet by the pertinent area-to-mass ratio, one obtains commensurate signals, whose sum indicates the system's total response. The model is fully universal and suits as well for time as frequency domain analyses.

The scaling and summing of currents can also be implemented schematically, without need for post-processing. Current-controlled current sources, for example, are practical in this.

The described example model can also be easily extended to systems employing a passive radiator. We then only add a spring between mass piece m_2 and the framework to represent the suspension stiffness of the radiator.

7.3 Modelling of Voice Coil Inductance

In the above studies, the inductance of the voice coil was ignored because we were only interested in low frequencies. In the modelling of the resonance region, the error thus introduced is indeed most often negligible, but in other regions, the results will be way off if the inductance is not included. Also, it will go almost as far from the truth when one uses merely the parameter value specified by the manufacturer. Thus, because of the lossy inductance, it is necessary to create for the inductive impedance a realistic model that works for all frequencies.

Figure 7.4a shows an equivalent circuit scheme that is suited for the purpose. The inductive impedance is modelled by a chain of LR links, whose number can be chosen according to the need. Normally, no more than four links should be needed. Three (as in the Figure) is often sufficient for smallish bass-midrange drivers, and tweeter inductance may be dealt with even by two.

Each LR link introduces a gradually rising step in the total impedance, as shown in Fig. 7.4b. The starting point of the step is set by the respective inductance and the height by the resistance. In practice, the step with the lowest frequency will generally be the thinnest, and vice versa. By optimizing the risings and heights of the steps appropriately, a fairly realistic result can be achieved. There is no need to model the direction angle separately, for it will settle in place automatically once the magnitude curve is in order.

Manual optimization flows best when the measured and simulated impedances are superimposed in the same plot and running a new correction round takes only a few strokes. If there is much of such work,

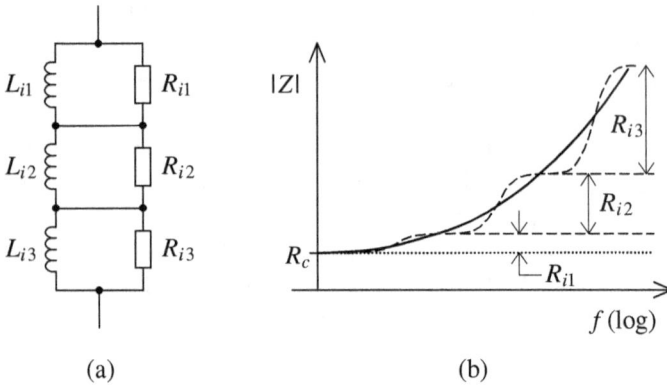

(a)   (b)

Figure 7.4. a) Equivalent circuit modelling the lossy inductance of a voice coil. A needful number of LR links should be included. b) The fitting principle. Dashed lines sketch the impedance increase introduced by each link (R_c =DC resistance). In the result impedance, distinct steps are no more observable.

one may make use of the curve fitting feature that is found in some circuit simulators.

In the measurement, one needs a sine wave source of reasonably low distortion and a suitable AC voltmeter or an oscilloscope. The frequency range of ordinary portable multimeters is, instead, generally insufficient for this purpose.

The measurement can be performed either with constant-current feed or by comparing the object's voltage to the voltage of a known resistor connected in series. The methods are discussed further in chapter 13. The measurement current should not be chosen very low because of the inductance's level dependence, described in section 4.7.

Figure 7.5 shows an example of the impedance modelling of a small bass-midrange driver. The equivalent circuit (Fig. 'a') includes the circuit elements for both the motional impedance and inductance. The element values modelling the inductance are by no means determinate but always depend to some extent on the view of the doer and what regions are to be emphasized.

The accuracy achieved by the model is sufficient for all needs, as can be noticed from Fig. b. Moreover, the sum of inductances L_i is quite close to the driver's rated inductance, which is 0.31 mH. In the Figure is also shown, by dashed line, the impedance obtained with a mere series inductor. The curve rises, at high frequencies, with a steepness that is more than double compared to reality.

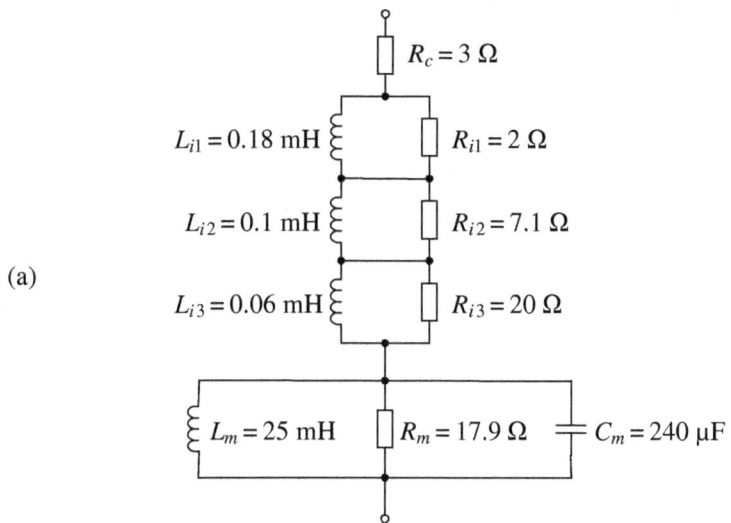

(a)

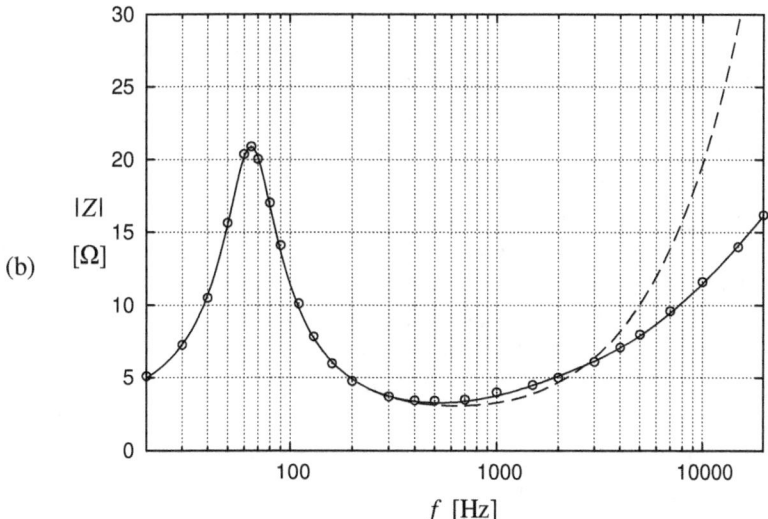

(b)

Figure 7.5. Example of the impedance modelling of a 4½-inch driver (Vifa PL11WH09-04, unenclosed, with 50 mA current) a) Optimized equivalent circuit. Subscript m refers to motional impedance and i to voice coil inductance. b) Fitting result obtained by the model of Fig. 'a'. The circlets represent measurements. Dashed line: inductance modelled by a bare 0.31 mH inductor.

There is reason for the model to cover all frequencies at which the current is yet significant, not just the band assigned for the driver. Also, it won't hurt if the tweeter model worked broadly yet for a couple of octaves above the audio range, for thereby one is also able to examine the current division between the speaker and the high-frequency load necessary in the amplifier.

In lack of measuring equipment, one may also use, for the inductance, the curves published by the manufacturers (so far as they can be deciphered). However, in enclosure testing and low-frequency tuning, own impedance measurements are indispensable. It is also advisable to measure the DC resistance self and not to merely rely on the given number.

7.4 Dual Coil Drivers

When using dual coil drivers in the way presented in section 6.2 and always when the coil currents are unequal, one needs a special simulation model adapted for the dual coil structure. However, the model is not so much more complex than with an ordinary driver.

Currents and the voltage drops occurring in the DC resistances are distinct for both coils. The electromotive forces are, instead, always the same in both coils because their velocities are the same and the inductive flux variation is the same in both. Also, in the generation of electromotive forces, both currents have equal effect, so they may be summed together by current-controlled current sources, as shown in Fig. 7.6; and this sum current may be fed to an equivalent circuit that represents the motional impedance and inductance of a single voice coil. Finally, the EMF voltages produced by the said equivalent circuit are directly copied to the voice coils by voltage-controlled voltage sources (E_1 and E_2).

So, E_m represents here the motional EMF of a single voice coil and E_i the inductance EMF of a single voice coil. The measurements and fitting can be made, however, using either only one or both voice coils. With two coils, the accuracy is a little better because the relative proportions of E_m and E_i are greater.

In Fig. 7.6, the voice coils have been stacked, but nothing prevents from using the model also in other kinds of applications.

The connection indicated by the dashed line has no significance to the functioning of the model itself; but, in order to find the bias point, circuit simulators require that all nodes must be galvanically connected, and this condition can be satisfied e.g. with the link shown.

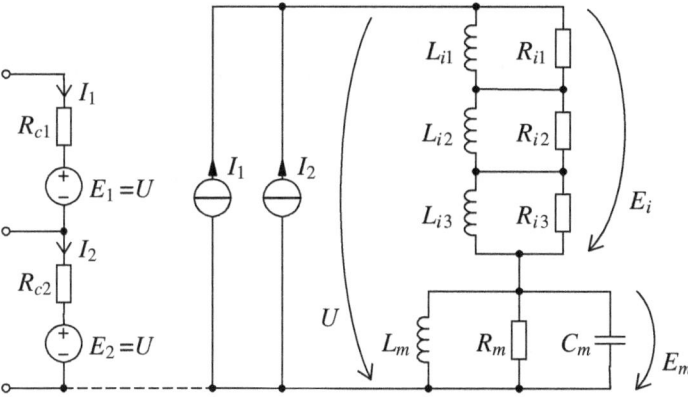

Figure 7.6. Impedance model for a dual coil driver. Voice coil currents I_1 and I_2 are mirrored by current-controlled current sources to feed an equivalent circuit that corresponds to the motional impedance and inductance of a single coil. Voltage U, that represents the total EMF, is in turn mirrored to the voice coils by sources E_1 and E_2.

The current flowing through capacitor C_m still indicates the acoustic pressure and the current through inductor L_m the displacement because E_m is also here directly proportional to the velocity.

L_m, R_m, and C_m can also be scaled to directly correspond to the mechanical parameters, as with the single-coil driver. The said elements are then replaced by the model shown in Fig. 7.2b (without R_c), Bl representing the force factor of a single coil.

7.5 Two Drivers in the Same Space

When two or more drive units share the same enclosure space, it has to be considered whether their currents in the resonance region are similar enough, so that there is no need to regard the mutual mechanical coupling separately. With drivers connected directly in series or parallel, this condition is certainly satisfied, and the air spring seen by each driver is constant, corresponding to the enclosure volume divided by the number of the drivers. If, however, the currents differ in the mentioned region to a noticeable degree in amplitude or phase, as is possible e.g. when applying the scheme of Fig. 6.7, the accuracy of the modelling can be improved by regarding the impact of the other drivers on the air spring seen by each driver.

The coupling between two cones can be modelled, for example, in the manner shown in Fig. 7.7, using the principles presented in section 7.2. Between the nodes representing the cones has been connected an inductance $1/k_a$, that represents the air spring of the enclosure. Spring constant k_a can be evaluated by adapting formula (7.7). k represents then the spring of a bare driver.

The air spring acts, in practice, in the direction that a forward movement in one cone tends to move the other cone backward. This phase opposition has been regarded in Fig. 7.7 by making one of the transformers act invertingly, so that when the cones are moving in phase, the resonator voltages are opposite to each other, effecting current through the air spring inductor.

Equation (7.7) often yields a spring constant that is higher than in actuality because pressure fluctuations have been assumed to occur adiabatically, that is, without heat exchange. When using plenty of damping material, this assumption does not, however, apply very well; and hence it is better to determine the final inductance value by impedance fitting with both drivers in action.

The damping material and inherent losses of the enclosure always act as an extra retarder for cone motion, thus lowering the mechanical Q value. This effect can be modelled in this case also by the resistor drawn by the dashed line.

The interaction between the cones can also be taken into account in another way: by current sources reflecting the spring force. Figure 7.8 illustrates this principle in the case of two drivers.

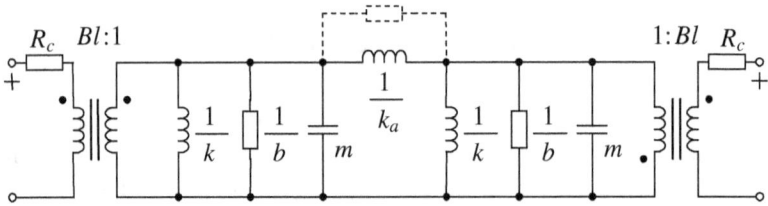

Figure 7.7. Low-frequency model for two identical drivers that share the same closed enclosure space. k_a represents the air spring. One transformer is inverting because positive shift in one cone introduces negative force to the other. By the dash line resistor, one can emulate the retardation effect of the damping material.

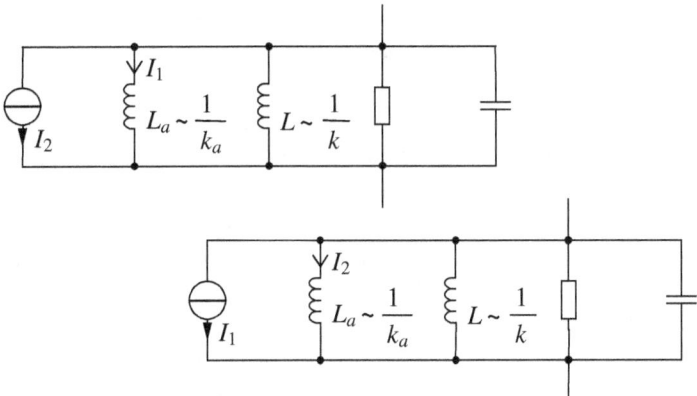

Figure 7.8. Modelling of the common air spring by means of current-controlled current sources. Current through the air spring inductor L_a represents the air spring force due to the driver's own displacement. The force due to the other driver is considered by a parallel current source that follows the displacement of the said other driver.

We know that current through the model's resonator inductor indicates the force exerted on the pertinent spring. In the Figure, inductances representing the suspension and the air spring have been separated. The current flowing through the latter thus indicates the force that the cone exerts on the air spring in the situation where the other cone is in its rest position. Both cones must exert on the air spring also a force deriving from the other cone's displacement, this force being represented by a separate current source. With equal areas, the air spring force is also equal for both cones, so the current source directly mirrors the air spring inductor current of the other driver.

The principle can easily be generalized for any number of drivers. One only has to insert in parallel with every resonator all the current-controlled current sources that represent the air spring forces deriving from all other drivers.

7.6 Filter Coils

Filter inductors used in passive speakers are also always lossy and therefore, in general, need some degree of modelling. It is, nevertheless, not sensible to set about measuring the impedance characteristics of all coils since, compared to voice coils, deviations from ideal inductance are quite small and reasonably well predictable.

In addition to the actual winding resistance, resistive losses are also introduced due to transverse currents occurring in wires in a magnetic field and, when using ferrites or the like, possibly also due to core non-idealities. At low frequencies, where the reactance is low, the dominant loss factor is the winding resistance. As frequency and reactance increase, the resistance loses significance, and other factors rise to dominance.

Figure 7.9 gives some measurement results of the dissipation factors (i.e. ratio of the resistive component to inductive component) of coils that are equal in inductance (0.47 mH) but differ in construction. The measurement method itself has been described in section 13.6.

At low frequencies, the dissipation factor obeys quite well the expression $R/2\pi fL$, where R is the DC-measured resistance. As frequency increases, the extra losses introduced depend, among other things, on the physical size and shape of the winding.

By comparing cases A and B, it can be seen that as the wire becomes thicker (0.71 mm → 1.4 mm), losses at high frequencies increase. In the highest octave, the dissipation factor appears to be even directly proportional to the wire diameter; a noteworthy point when seeking to minimize resistance.

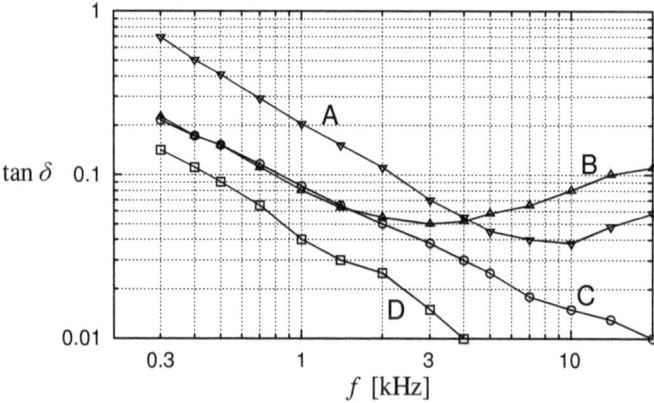

Figure 7.9. Dissipation factors, i.e. tangent of the loss angle (δ), vs. frequency in four different coils having 0.47 mH inductance. A: air coil with 0.71 mm wire, winding inner diameter 14 mm, height 21 mm, and resistance 0.59 Ω. B: air coil with 1.4 mm wire, winding inner diameter 28 mm, height 36 mm, and resistance 0.19 Ω. C: copper foil coil, foil width 30 mm, inner diameter 24 mm, resistance 0.16 Ω. D: ferrite bobbin coil with 0.95 mm wire, winding inner diameter 24 mm, height 19 mm, and resistance 0.10 Ω.

In the devices examined, the shape of the windings was, however, not very optimal. Also in general, the inner diameter of the coil is usually made too small and the height too great. The high-frequency losses namely substantially decrease when using a wider and shallower structure. The wider coil does not require more wire than a narrower one of the same inductance because increase in the loop area makes up for the lessening of the turns.

In the air-cored copper foil roll (curve C), the high-frequency losses are even surprisingly low, so the structure seems to work. The least losses are, however, achieved with the ferrite coil (curve D) because it can do with less wire, and the magnetic flux penetrating the wire is lesser. The ferrite's own losses do not show up at least at low currents.

In cases C and D, it is enough to model the losses by a mere series resistance, whose value can be set slightly higher than the DC resistance, to regard the high frequencies a little. In cases resembling curve B, a parallel resistance may also be included, as shown in Fig. 7.10a.

Figure 7.10b shows the modelled dissipation factor fitted to the data of curve B in Fig. 7.9. When the parallel resistance is used, the agreement is good for all frequencies, whereas with the mere series resistance, the result is as given by the dashed line.

The excess losses occurring at high frequencies are, in practice, quite independent of the inductance value, so in this regard the results of Fig. 7.9 are quite universal. The modelled high-frequency losses are made

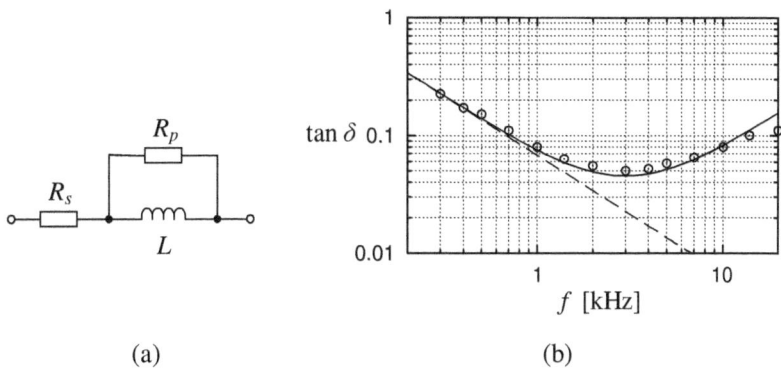

(a)   (b)

Figure 7.10. a) Equivalent circuit suitable for a filter coil. R_s represents mostly the winding resistance and R_p the excess losses at high frequencies. b) Example fitting of the dissipation factor with the model of Fig. 'a'. The measurement data used is from curve B in Fig. 7.9. Solid line: $L = 0.47$ mH, $R_s = 0.2$ Ω, and $R_p = 390$ Ω. Dashed line: $R_p = \infty$.

independent of inductance by keeping the ratio of R_p and inductance constant.

7.7 Operational Amplifiers

Operational amplifiers are basic elements in analogue electronics, and essential especially in the implementation of active filter circuits. In some simulators, the operational amplifier is included as a built-in device, that can be used like ordinary circuit symbols, by only giving the necessary parameter values. If this feature is not provided, one has to use an own macro model, to which all necessary characteristics may be incorporated.

Figure 7.11a shows an operational amplifier without feedback, the input terminals being exposed to a differential voltage U_d. The differential gain, or raw gain, of an operational amplifier, U_o/U_d, that is somewhat misleadingly also termed open-loop gain (Factually, an electrical loop is either closed or nonexistent.), is typically as shown in Fig. b.

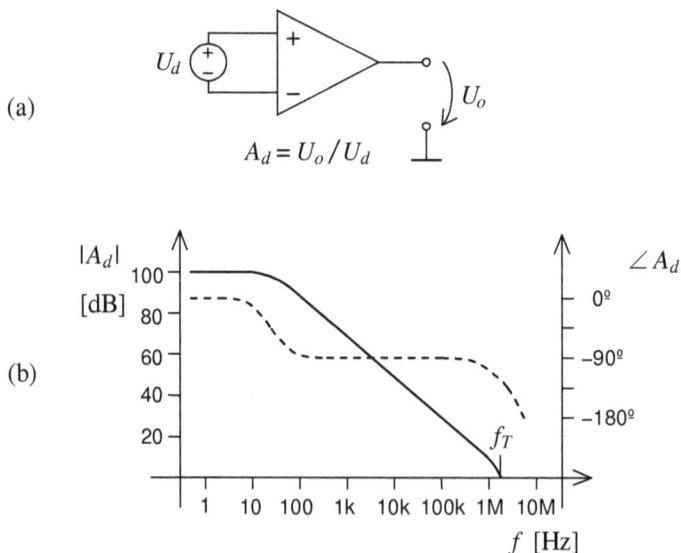

Figure 7.11. a) Operational amplifier connected to exhibit the differential gain (A_d). Generally, the supply voltage is assumed to be two-sided, so that U_o can be of either polarity, like also the input terminal voltages with respect to ground. b) Typical operational amplifier Bode plot. The first pole is located in the region of 20 Hz, the other poles being found only in the megahertz range.

At the lowest frequencies, the gain is very high, generally something around 100 000, i.e. 100 dB; but rolls off after about 20 Hz in inverse proportion to frequency so that the limit of 0 dB is crossed at a few megahertzes (f_T). The second pole must lie at a frequency high enough that the phase lag at frequency f_T is still below 180°. Then, the amplifier is stable with all resistive feedbacks.

Figure 7.12 shows a simple operational amplifier equivalent circuit that models the frequency behavior of Fig. 7.11b. Other non-idealities don't need to be considered when doing basic simulations. For instance, the slew rate limit is not yet encountered at audio frequencies.

The low-pass stage formed by R_1 and C_1 establishes the first pole, which at the values given occurs at 20 Hz. By R_2 and C_2 is yet created another pole, whose frequency is here set to 2 MHz. The latter stage does not load the former noticeably because C_2 is only a hundredth of C_1. R_1, in turn, must be set so high that the low-pass network does not load the external connection in any circumstances.

Voltage source E, that follows the voltage across C_2, provides the necessary gain and isolates the output from the input circuit. Operational amplifiers don't have a specific ground connection, but in the model, such is, however, needed because the other end of the source must be connected somewhere.

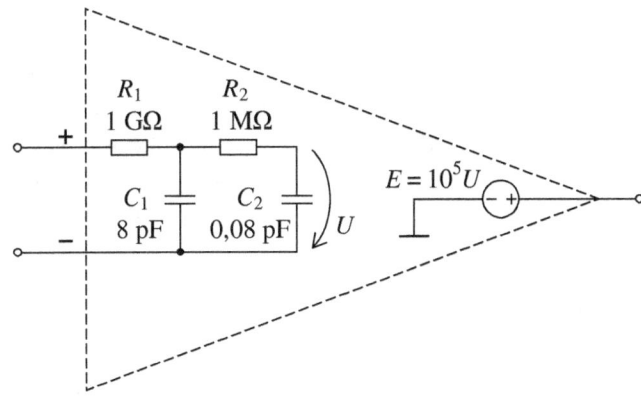

Figure 7.12. Operational amplifier model sufficient for basic simulations. R_2 and C_2 are not necessary for audio frequencies but improve the model's veracity near the unity-gain frequency (f_T).

8

RESONANCE COMPENSATION

As was noticed in section 5.5, drive units with suitable mechanical Q values for direct current-drive are at present not available; so to eliminate the bass peaking, the resonance has to be damped either by passive or active circuits.

Passive equalization has, of course, the advantage that the amplifier doesn't have to be adjusted separately for each loudspeaker model and the user doesn't have to be aware of the resonance values of his speaker. On the other hand, the low-frequency limit cannot be determined freely, and the source impedance also suffers somewhat.

Active equalization, again, is best suited for systems where the amplifier and speaker are firmly connected to each other. It is, however, entirely possible to also build an equalizer that is easily tunable to correspond to the parameters of diverse speakers. In addition to high source impedance, the benefits include better efficiency in the resonance region because any power-consuming damping network is not needed.

8.1 Passive Equalization

Passive bass response leveling can be implemented on the principle shown in Fig. 8.1. A series resonance network, connected in parallel with the driver, passes part of the current at those frequencies which the driver in its enclosure emphasizes. The inclusion of the resonator raises the order of the system in fact to four, but the attenuation slope of the lowest frequencies remains unchanged, that is, at 12 dB per octave.

The resonator has to be tuned close to the driver's resonant frequency, so the needed inductance and capacitance are rather high. However,

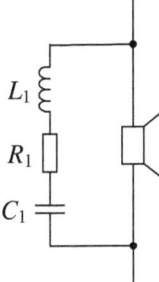

Figure 8.1. Passive compensation of a driver's resonance peaking. The series resonance network is tuned to the driver's response peak.

the size of the coil is not a problem since the winding resistance, that merges with R_1, may also be high. In fact, by choosing the thickness of the wire appropriately, the actual resistor can be omitted altogether.

Plastic dielectrics should be preferred in the capacitor since the dissipation factor of electrolytics is so high as to already affect the tuning. The usual 10% accuracy of bipolar electrolytics is also not quite sufficient for the purpose.

The impedance of the resonator equals R_1 at minimum and increases with a 1st-order slope when moving away from the resonance region, as shown in Fig. 2.5. In the bass region, such a current path appearing in parallel with the driver is, however, acceptable, as was reasoned in section 4.10.

The inductance should be high enough that the resonator does not decrease too much the impedance seen by the driver (or drivers) at the lower mid-frequencies. On the other hand, one should not increase the resonator Q too much, to avoid a dip in the response curve. Suitable element values are found by simulation, after the resonance of the enclosed driver has been modelled.

The method is best suited for rather low Q_m values. With a suitable driver and copious use of damping material, values lower than 2 can be achieved which keeps the attenuation need and accuracy requirements yet moderate.

It is not necessarily worthwhile to seek to level all the accentuation introduced by the resonance because, due to the baffle step effect, the bass frequencies in any case attenuate with respect to other regions. By leaving a mild prominence in the vicinity of 100 Hz, part of the baffle step can be compensated already by this, so that the attenuation need at high frequencies is a little relieved and sensitivity correspondingly enhanced.

Figure 8.2 shows simulated examples of the application of passive equalization, using typical values for smallish 8-ohm hifi drivers. The resonant frequency has been chosen to be 60 Hz, which is a fully possible value with a properly filled enclosure of about 10-15 litres. In Fig. 'a', $Q_m = 1.2$, in Fig. b, 1.6, and in Fig. c, 2.0. At least the two last values are easily achievable even with existing drivers when effective damping material is employed.

In all these cases, the response can be made rather neat without having to compromise the driving mode in the midrange considerably. Even in the case of Fig. c, where L_1 is at the lowest, the impedance of the damping arm reaches 70 Ω at 200 Hz, an 8.8-fold value with respect to the driver's nominal impedance.

As the response peak is leveled, the load impedance also becomes leveled accordingly. A gentle impedance peak means a nearly resistive load, which is easy to drive for the amplifier, unlike vented speakers, for instance, whose impedance curve exhibits two high spikes.

The −6-dB cut-off frequency (referenced to the 100-Hz level) occurs in all examples a little below 40 Hz. Only few voltage-driven loudspeakers of the same size class reach as low. An advantage over the bass reflex principle is also that the roll-off in the lowest octave occurs more gradually and so the variation in phase response is also lesser.

In Figures b and c, a prominence of a scant 2 dB has been left in the 100 Hz region, this emphasis being thus available in the compensation of the rising response and to also improve the current-drive index a bit. The circuits used for resonance and rising response compensation interact also otherwise a little, so, in practice, these two have to be optimized together.

8.2 Voltage Divider Equalizer

Shaping of the resonance region in the above fashion can also be performed by an equalizer stage incorporated in the amplifier, one such scheme being shown in Fig. 8.3. The circuit is in effect a passive band-stop filter, whose practical realization though also requires active components.

Resistor R_1 forms with the series resonance network a voltage divider, whose division ratio reaches its minimum $(R_2/(R_1+R_2))$ at the resonator's characteristic frequency, that is obtained by formula (2.19). Far from this frequency, the ratio approaches unity.

Assuming negligible loading, the same current flows through all the

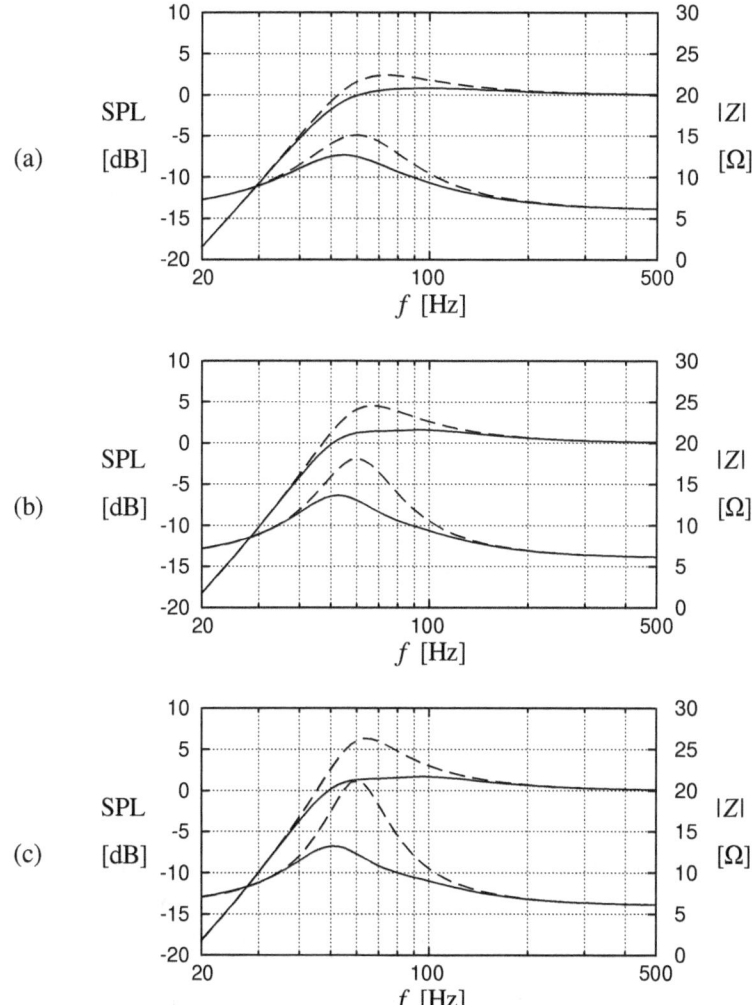

Figure 8.2. Examples of frequency responses shaped with the circuit of Fig. 8.1, and corresponding load impedances at three different Q_m values, the driver's resonant frequency being 60 Hz. The equivalent circuit of Fig. 7.1 was used with values $R_c = 6\ \Omega$, $C = 350\ \mu F$, and $L = 20.1$ mH, R being varied. The relative sound pressure level (SPL) corresponds to the ratio of current I_c to the applied current. The original responses and impedances are indicated by dashed lines and the modified ones by solid lines. a) $Q_m = 1.2$ ($R = 9.1\ \Omega$), $L_1 = 100$ mH, $R_1 = 60\ \Omega$, and $C_1 = 40\ \mu F$. b) $Q_m = 1.6$ ($R = 12.1\ \Omega$), $L_1 = 90$ mH, $R_1 = 39$ Ω, and $C_1 = 50\ \mu F$. c) $Q_m = 2$ ($R = 15.2\ \Omega$), $L_1 = 60$ mH, $R_1 = 26\ \Omega$, and $C_1 = 80$ μF.

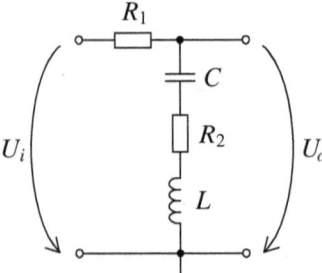

Figure 8.3.   Passive 2nd-order band-stop filter for bass resonance compensation before power amplification. U_i =input voltage, U_o =output voltage.

elements, so voltages correspond directly to the impedances. The capacitor impedance being $1/sC$ and the inductor impedance correspondingly sL, we can thus write as the network transfer function:

$$\frac{U_o}{U_i} = \frac{\dfrac{1}{sC} + R_2 + sL}{R_1 + \dfrac{1}{sC} + R_2 + sL} \tag{8.1}$$

from which we obtain by multiplying and arranging:

$$\frac{U_o}{U_i} = \frac{s^2 + \dfrac{R_2}{L}s + \dfrac{1}{CL}}{s^2 + \dfrac{R_1 + R_2}{L}s + \dfrac{1}{CL}} \tag{8.2}$$

By comparing to the general form (2.17), one can see that the poles and zeroes have here the same characteristic frequency. The Q of the zeroes is, instead, always greater than that of the poles.

The order of the overall system will be 4, as also in the method of Fig. 8.1. However, the filter can also be tuned to retain the order at 2. Then, instead of trial-and-error optimization, the element values are obtained by easy calculations, but the cut-off frequency is left somewhat higher.

The transfer functions of series-acting linear systems are multiplied together, so we can write for the overall system:

$$H(s) = \frac{s^2 + \dfrac{R_2}{L}s + \dfrac{1}{CL}}{s^2 + \dfrac{R_1 + R_2}{L}s + \dfrac{1}{CL}} \cdot \frac{s^2}{s^2 + \dfrac{\omega_0}{Q}s + \omega_0^2} \tag{8.3}$$

where the latter expression represents the speaker high-pass function.

By making the characteristic frequency and Q of the zeroes of the equalizer function equal to the characteristic frequency and Q of the poles of the speaker function, that is, by setting $1/(CL) = \omega_0^2$ and $R_2/L = \omega_0/Q$, the respective polynomials are canceled out, and we are left with a second-order high-pass function that has the same characteristic frequency as the speaker but whose Q can be set to the desired value with R_1.

Labeling the desired pole Q value by Q_p, we then have:

$$\frac{R_1 + R_2}{L} = \frac{\omega_0}{Q_p}$$

$$\Leftrightarrow R_1 = \frac{\omega_0 L}{Q_p} - R_2 \tag{8.4}$$

In practice, the inductance L must be at least several henries in order that the current taken by the equalizer would stay modest. If constructed of wire, such a coil would take a very large number of turns and could produce interference. Instead, using virtual implementation, also high inductances are fully possible.

A virtual inductor topology that serves the purpose is shown in Fig. 8.4. In addition to the operational amplifier, that functions as a voltage follower, only a capacitor and two resistors are needed. The series resistance introduced by this circuit is relatively high, so, in practice, it is best to also incorporate equalizer resistor R_2 in the impedance produced by the circuit.

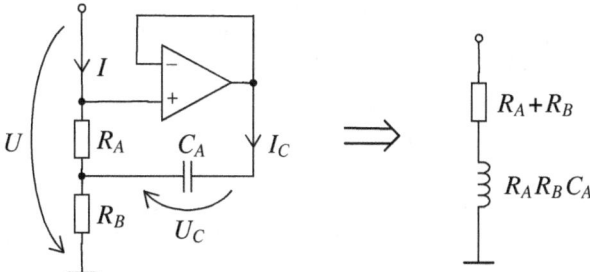

Figure 8.4. Circuit for generating ground-connected inductance. R_A and R_B are chosen for the desired series resistance ($R_A > R_B$) and C_A for the desired inductance. A floating inductance is established by connecting two identical circuits face-to-face, merging then resistors R_B.

Assuming the operational amplifier to function ideally, so that voltage across its inputs is zero, we can write using the notation given:

$$\begin{cases} U = U_C + R_B\,(I + I_C) \\ U_C = R_A I \end{cases}$$

(8.5)

By using yet the current-voltage relationship of capacitors (equation (b10) in appendix B), we obtain:

$$U = R_A I + R_B I + R_B\,j\omega\,C_A U_C$$

(8.6)

Dividing yet by I yields:

$$\frac{U}{I} = R_A + R_B + j\omega\,R_A R_B C_A$$

(8.7)

The constant part represents resistance, while the coefficient of $j\omega$ represents inductance, as seen from equation (b12). So, the inductance generated by the circuit is $R_A R_B C_A$, the series resistance being $R_A + R_B$.

In practical operational amplifiers, the raw gain rolls off with increasing frequency (Fig. 7.11), correspondingly increasing the voltage difference between the inputs. Therefore, virtual inductors are not able to function at high frequencies, but after a certain limit, the impedance turns capacitive. The impedance of the presented circuit stays inductive at least up to a few kilohertzes which suffices for this purpose. The limit is also affected by the difference between R_A and R_B so that it is advisable to keep R_A greater than R_B.

Figure 8.5 displays bass responses obtained with the equalizer, using a driver whose resonant frequency is 60 Hz (as in the examples of Fig. 8.2) and Q value 2.5. The dashed line represents a tuning where the goal has been low cut-off frequency without significant prominence. The dotted line, in turn, indicates the ordinary 2nd-order Butterworth response, that is obtained by the above-described pole replacement method by setting the Q value to 0.71.

In the first-mentioned fashion, the response extends a little lower, even though at the lowest frequencies the difference becomes diminutive. In some cases, and especially if the speaker is used near wall surfaces, the latter option may nevertheless be more appropriate.

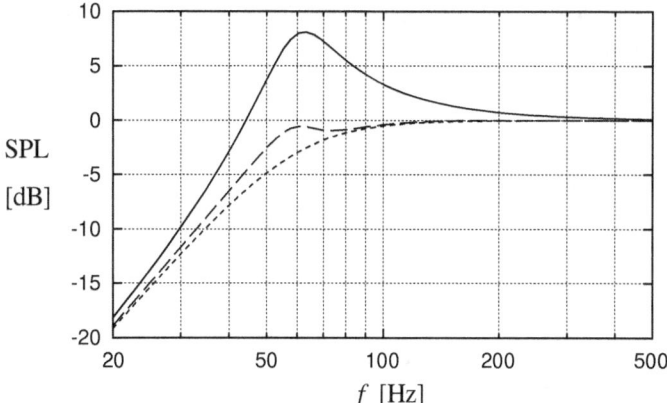

Figure 8.5. Simulated examples of the Fig. 8.3 equalizer performance. Solid line: original response, the resonant frequency being 60 Hz and the Q value 2.5. Dashed line: equalized response using values $R_1 = 100$ kΩ, $R_2 = 58$ kΩ, $C = 21$ nF, and $L = 300$ H. Dotted line: equalized response using values $R_1 = 114$ kΩ, $R_2 = 45.2$ kΩ, $C = 23.5$ nF, and $L = 300$ H; in which case the order of the overall system reduces to two with a Q value of 0.71.

8.3 Pole Shifting

Above, we already found how, by canceling the poles of the speaker function, a corrected 2nd-order overall response can be obtained. However, in the circuit used (Fig. 8.3), the constant terms of the numerator and denominator were equal, so equalization could be performed only for the Q. By using an equalizer in which the characteristic frequency of the poles is lower than that of the zeroes, also the cut-off frequency can be tailored while the order is still retained at two.

In general, the product of a second-order equalizer function and the speaker transfer function can be written as

$$H(s) = \frac{s^2 + \dfrac{\omega_z}{Q_z} s + \omega_z^2}{s^2 + \dfrac{\omega_p}{Q_p} s + \omega_p^2} \cdot \frac{s^2}{s^2 + \dfrac{\omega_0}{Q_0} s + \omega_0^2} \qquad (8.8)$$

By setting $\omega_z = \omega_0$ and $Q_z = Q_0$ (like before), the equalizer function numerator cancels the speaker function denominator, whose both parameters can thus be replaced by the desired values, ω_p and Q_p. In practice however, due to circuit-related restrictions, these values cannot be chosen all freely.

Figure 8.6 illustrates the replacement of the poles when extending the operation range. Figure 'a' shows the pole-zero diagram of the driver high-pass function, consisting of a complex pole pair and two zeroes located at the origin. In Fig. b has been marked, in addition to these, the zeroes of the equalizer function, that coincide with the said poles, and the poles introduced, which in this case are real. Figure c shows the net result where the new poles lie closer to the origin than the former ones and are lower in Q.

Figure 8.7 depicts the effect of pole shifting on the frequency response. The dip established in the equalizer response (solid line) completely compensates the prominence occurring in the speaker response (dashed line); and as frequency decreases toward ω_p, the increase in the equalizer gain compensates the decline in the speaker response for this region. The equalized response (dash-dot line) turns down essentially at ω_p, below which the equalizer gain levels off. The Figure also shows the asymptotes that the curves approach far from the corner frequencies.

As the equalizer function is minimum-phase, the overall system also remains minimum-phase, so the phase response is automatically modified to correspond to the equalized amplitude response.

When expanding the operation range, one always has to regard the sufficiency of the driver's linear excursion. When frequency is halved while keeping the sound pressure unchanged, the required excursion namely becomes 4-fold; so when using program material with plenty of low frequencies, it is better not to take ω_p very far from ω_0.

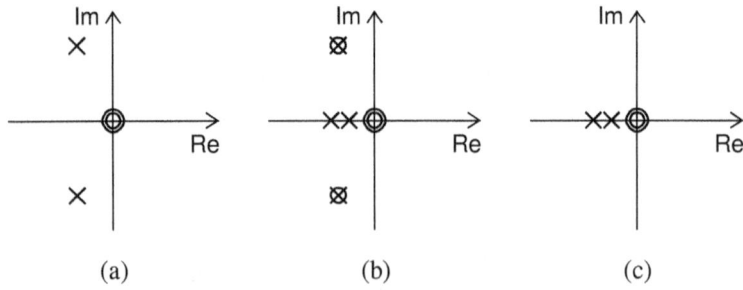

Figure 8.6. Principle of pole shifting. a) Pole-zero diagam of the original speaker function. b) Diagram of the equalized system, the equalizer zeroes coinciding with the speaker poles. c) Result after cancellation.

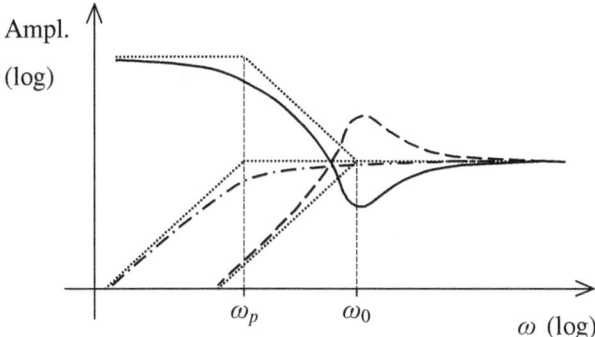

Figure 8.7. Frequency response equalization by pole shifting. Dashed line: original high-pass response of the speaker. Solid line: response of the active equalizer. Dash-dot line: resulting, expanded response, in which the original resonance has been canceled. The asymptotes of the curves intersect at frequencies ω_p and ω_0.

On the other hand, the practical realization of the equalizer may restrict ω_p to be at a certain minimum distance from ω_0. Then, it is expedient to set ω_p way below the desired corner frequency, and the actual high-pass filtering can be implemented by separate circuits.

Even active compensation methods are unsuitable for very high Q values, for the narrower the resonance peak is, the easier it is for different parameter variations to effect discrepancy between the response to be equalized and the equalizer function. Fortunately though, there is generally no need to use high Q values since when damping the enclosure so as to suppress the inner sound level, also the Q can be made moderate. As a rule of thumb, unless the Q value stays below 3, it is better to use another driver.

If the same equalizer is to be used with different speakers, at least ω_z and Q_z should by some means be adjustable. The realization schemes presented in the following differ e.g. in this respect much from each other.

8.4 The Linkwitz Equalizer

The pole-shifting-based operation range expansion works similarly also in voltage-driven loudspeakers. In this, it has become customary to employ the circuit published by S. Linkwitz in 1978, shown in Fig. 8.8. With certain cautions, the circuit is also usable on current-drive.

The number of reactive elements (capacitors) is 4, so the order of the topology is actually also 4. The order is, however, reduced to 2 by matching certain elements, as indicated, and by keeping the following condition: $R_1C_1 = R_3C_3$.

The transfer function will then be [1]

$$\frac{U_o}{U_i} = -\frac{s^2 + \left(\dfrac{R_2}{R_1^2 C_1} + \dfrac{2}{R_1 C_1}\right)s + \dfrac{1}{R_1^2 C_1 C_2}}{s^2 + \left(\dfrac{R_2}{R_3^2 C_3} + \dfrac{2}{R_3 C_3}\right)s + \dfrac{1}{R_3^2 C_2 C_3}} \tag{8.9}$$

Using the symbology familiar from equation (8.8), we obtain as the analysis equations:

$$\omega_z = \frac{1}{R_1 \sqrt{C_1 C_2}} \tag{8.10}$$

$$\omega_p = \frac{1}{R_3 \sqrt{C_2 C_3}} \tag{8.11}$$

$$Q_z = \frac{R_1}{2R_1 + R_2} \sqrt{\frac{C_1}{C_2}} \tag{8.12}$$

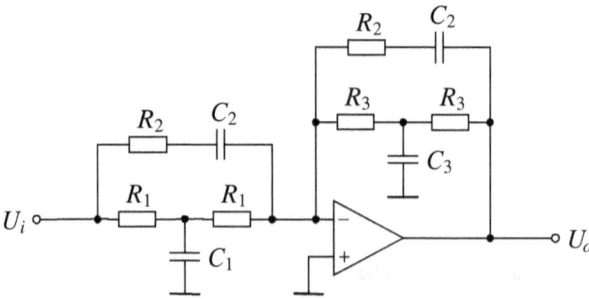

Figure 8.8.   The commonly known, inverting pole shifting circuit named after Linkwitz. The amount of elements affecting the tuning is quite high.

$$Q_p = \frac{R_3}{2R_3 + R_2} \sqrt{\frac{C_3}{C_2}} \qquad (8.13)$$

For the choice of element values, there exists different procedures, that are, however, not repeated here.

The parameter relationships are confined by the inequation pair

$$\frac{\omega_p}{\omega_z} < \frac{Q_z}{Q_p} < \frac{\omega_z}{\omega_p} \qquad (8.14)$$

in which especially the latter restriction is easily encountered. Because Q_p should be considerably lower than Q_z, it may often be necessary to set ω_p lower than what is appropriate.

The circuit is hard to be made adjustable, so it is mostly applicable in active loudspeakers and other integrated appliances. Also, the rather large amount of critical elements and their mutual matching requirements introduce unnecessary inaccuracy factors, that may cause the response to deviate from the 2nd-order response.

The transfer function has a minus sign, so the circuit changes the signal's polarity. This can be an unwanted property unless the change of sign is canceled in some other stage. An inverting operational amplifier circuit also loads the input, so the feeding source has to be a buffered one.

It is also not so stylish to use a 4th-order circuit topology to realize a 2nd-order transfer function. There should be no more than two capacitors. Thus, there is a need to find more practical alternatives for the circuit solution, especially in terms of adjustment possibilities.

8.5 Equalizer with Active Feedback

One way to establish the needed equalizer circuits is based on the remark that the transfer function of a non-inverting operational amplifier circuit is in fact the reciprocal of the transfer function of the feedback block used. Consequently, if we are able to realize the reciprocal of the desired equalizer function, $1/H(s)$, which is sometimes easier, the function $H(s)$ is obtained by placing a system that realizes the reciprocal function as the feedback of an operational amplifier.

The case can be verified with formula (b26), given in appendix B, and the associated Figure, B14a; once the summing block is replaced by a difference block, as in operational amplifiers, so that the minus sign appearing in formula (b26) is changed to a plus. Then, H_1 corresponds to the amplifier's differential gain and H_2 the feedback network. As the product H_1H_2 is very large compared to unity (because H_1 is very large), we obtain the result $H \approx 1/H_2$, which was to be proved.

From Fig. 8.7, it can be concluded that mirroring the equalizer response about a horizontal axis results in a high-boost type response, that is obtained from an ordinary high-pass response by moving the zeroes from the origin to a suitable frequency. There are several possibilities to realize a corresponding circuit.

We will consider the circuit shown in Fig. 8.9a, which is a variation of the common Sallen-Key type high-pass topology. Without resistors R_1, the circuit produces the usual 2nd-order high-pass function, but with the resistors, the response levels off at low frequencies to a constant, thus providing the needed high-boost operation (Fig. 8.9b).

The transfer function of the circuit is worked out by applying Kirchhoff's current law to nodes 'a' and b and assuming the operational amplifier inputs to be at the same potential (U_y). The admittance (reciprocal of impedance) of a capacitor is, in general, Cs, so we can write as the current-equations:

$$\begin{cases} \dfrac{U_y}{R_3} + (U_y - U_a)\left(\dfrac{1}{R_1} + Cs\right) = 0 \\[2mm] (U_a - U_x)\left(\dfrac{1}{R_1} + Cs\right) + (U_a - U_y)\left(\dfrac{1}{R_1} + Cs + \dfrac{1}{R_2}\right) = 0 \end{cases} \tag{8.15}$$

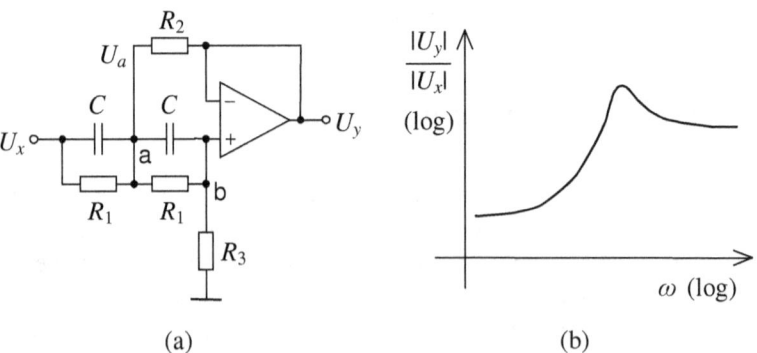

(a)   (b)

Figure 8.9. a) Circuit topology to produce the high-boost response useful in equalizer feedback. b) Obtained amplitude response, that is the inverse of the desired equalizer response.

Solving for voltage U_a in the upper equation gives:

$$U_a = \frac{U_y\,(1/R_3 + 1/R_1 + Cs)}{1/R_1 + Cs} \qquad (8.16)$$

By substituting this expression in the lower equation, one obtains, after a few algebraic steps, as the reciprocal of the transfer function (U_x/U_y) and thus as the transfer function of the actual equalizer (labeled U_o/U_i)

$$\frac{U_x}{U_y} = \frac{U_o}{U_i} = \frac{s^2 + \left(\dfrac{2}{R_1 C} + \dfrac{2}{R_3 C}\right)s + \dfrac{1}{C^2}\left(\dfrac{1}{R_1^2} + \dfrac{2}{R_1 R_3} + \dfrac{1}{R_2 R_3}\right)}{s^2 + \dfrac{2}{R_1 C}s + \dfrac{1}{R_1^2 C^2}}$$

$$(8.17)$$

The equalizer itself is shown in Fig. 8.10. The feedback signal of A_2, generated by the circuit of Fig. 8.9a, has been attenuated by the voltage divider R_4-R_5. Stimulus is applied to the non-inverting input of A_2.

The feedback has to be attenuated somewhat to ensure stability since at high frequencies the phase lag arising in A_1 can, in conjunction with the lag of A_2, make the circuit oscillate. However, based on testing with a TL-072 amplifier, the attenuation need is not great. With equal values for R_4 and R_5, the system was already fully stable and also free from high-frequency peaking. The attenuator though increases the gain factor of A_2, but the increase is easily compensated elsewhere.

From the transfer function (8.17), the following analysis equations

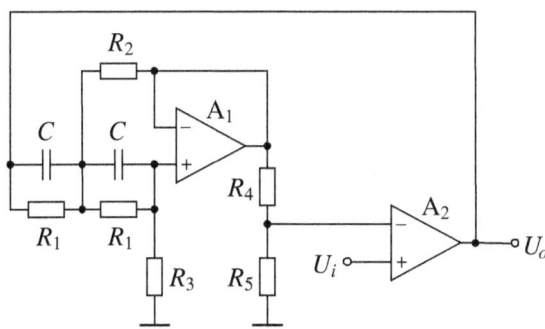

Figure 8.10. Equalizer using active feedback. Voltage divider R_4-R_5 is fitted so that stability is ensured.

can be derived:

$$\omega_z = \frac{1}{C}\sqrt{\frac{1}{R_1^2} + \frac{2}{R_1 R_3} + \frac{1}{R_2 R_3}} \qquad (8.18)$$

$$\omega_p = \frac{1}{R_1 C} \qquad (8.19)$$

$$Q_z = \frac{\sqrt{R_3(R_3 + 2R_1 + R_1^2/R_2)}}{2(R_1 + R_3)} \qquad (8.20)$$

$$Q_p = \frac{1}{2} \qquad (8.21)$$

Q_p is thus always 0.5, but this is not of great significance since also here (as in the Linkwitz equalizer) one often has to set ω_p very low and perform the filtering of the lowest frequencies by other means.

Design can be carried out by the following procedure:

1. Choose ω_p so that

$$\omega_p < \frac{\omega_z}{2Q_z} \qquad (8.22)$$

2. Choose a practical value for C, e.g. 0.1 µF.

3. Solve R_1 from equation (8.19).

4. Determine R_3 from the equation

$$R_3 = \frac{2}{C\left(\dfrac{\omega_z}{Q_z} - \dfrac{2}{R_1 C}\right)} \qquad (8.23)$$

which yields a positive value if condition (8.22) is satisfied.

5. Finally, determine R_2 from the equation

$$R_2 = \frac{1}{R_3 C^2 \omega_z^2 - \dfrac{R_3}{R_1^2} - \dfrac{2}{R_1}} \qquad (8.24)$$

If some resistance is impractically low or high, a normal impedance scaling can be performed, in which C is multiplied by a number and R_1, R_2, and R_3 are divided by the said number.

The source that feeds the equalizer should have relatively low impedance, so that parasitic feedback introduced by the stray capacitance between the output and plus input of A_2 would not be able to make the circuit oscillate. If the input voltage is allowed to fluctuate, the circuit may namely enter into a square wave mode, but it is also generally well to avoid using high source impedances, just because of the coupling that occurs through stray capacitances.

Tuning can be performed by mere resistors, so the large gap between successive standard values of capacitors (generally about +50%) doesn't matter. Also, the signal is not inverted, and the preceding stage is not loaded. However, the adjustments are not very linear nor independent of each other, so this realization is also best suited for stand-alone systems.

The DC gain of the equalizer is ω_z^2/ω_p^2 multiplied by the extra gain introduced by the voltage divider R_4-R_5, in practice even of the order of 100; so in the choice of operational amplifiers, one has to pay attention to their offset voltages and input currents. At least the latter can virtually be done away with by using amplifiers with JFET (Junction Field Effect Transistor) inputs. Without any precautions, there may accrue in the output, in an extreme case, a DC voltage of the order of volts. Due to the accumulation of DC voltage, that eats signal headroom, it is better not to set ω_p very much lower than condition (8.22) necessitates.

8.6 Equalizer with RCL Feedback

The equalizer topology presented next is also based on the reciprocality of the transfer functions of a non-inverting amplifier circuit and its feedback network. Here, the feedback block consists of a passive RCL network in accordance with Fig. 8.11, including one large inductance. This inductance is, however, easily implemented by an active circuit, as was done in section 8.2.

The transfer function of the equalizer is found quite conveniently by regarding that, as the operational amplifier inputs are currentless, the feedback network functions only as a complex voltage divider, whose division point potential is U_i. The impedance of two parallel-connected impedances always equals the product of the impedances divided by their sum, so we can write for the voltage divider, using the notation of Fig. 8.11:

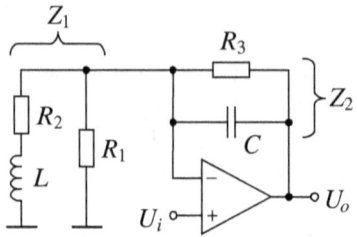

Figure 8.11. Equalizer topology using RCL feedback. Due to its size, inductance L requires virtual implementation.

$$\frac{U_o}{U_i} = \frac{Z_1 + Z_2}{Z_1} = \frac{\dfrac{R_1(R_2 + sL)}{R_1 + R_2 + sL} + \dfrac{R_3/(sC)}{R_3 + 1/(sC)}}{\dfrac{R_1(R_2 + sL)}{R_1 + R_2 + sL}} \tag{8.25}$$

By first converting the numerator terms into a common denominator, one obtains, after some reduction:

$$\frac{U_o}{U_i} = \frac{s^2 + \left(\dfrac{R_2}{L} + \dfrac{1}{R_1 C} + \dfrac{1}{R_3 C}\right)s + \dfrac{1}{CL}\left(\dfrac{R_2}{R_3} + \dfrac{R_2}{R_1} + 1\right)}{s^2 + \left(\dfrac{R_2}{L} + \dfrac{1}{R_3 C}\right)s + \dfrac{R_2}{CLR_3}} \tag{8.26}$$

The denominator is in factored form $(s + R_2/L)(s + 1/(R_3 C))$, from which it can be seen that the poles are always real, that is, $Q_p \leq \frac{1}{2}$ where the equality is achieved when the pole frequencies R_2/L (labeled ω_L) and $1/(R_3 C)$ (labeled ω_C) are equal.

For the 1st-order coefficient of the numerator, we may now write:

$$\omega_L + \frac{1}{R_1 C} + \omega_C = \frac{\omega_z}{Q_z} \tag{8.27}$$

so Q_z is best adjusted by R_1.

If R_1 is to be kept positive, one has to obey the condition

$$\omega_L + \omega_C < \frac{\omega_z}{Q_z} \tag{8.28}$$

which approximately corresponds to the earlier restrictions (8.14) and (8.22).

However, condition (8.28) can be circumvented by making R_1 negative which is accomplished by replacing the resistor by the converter circuit shown in Fig. 8.12. The negative resistance established between the driving node and ground is $-R_A R_C / R_B$.

Function (8.26) yields as the operation parameters:

$$\omega_z = \sqrt{\frac{1}{CL}\left(\frac{R_2}{R_3} + \frac{R_2}{R_1} + 1\right)} \tag{8.29}$$

$$\omega_p = \sqrt{\frac{R_2}{CLR_3}} \tag{8.30}$$

$$Q_z = \frac{\sqrt{CL\left(\dfrac{R_2}{R_3} + \dfrac{R_2}{R_1} + 1\right)}}{R_2 C + \left(\dfrac{1}{R_1} + \dfrac{1}{R_3}\right)L} \tag{8.31}$$

$$Q_p = \frac{\sqrt{R_2 R_3 CL}}{R_2 R_3 C + L} \tag{8.32}$$

It is best to fix the value of C and use L to set ω_z. It must be noted, however, that L affects similarly both frequencies, so when L is changed, ratio ω_z/ω_p and hence also the DC gain remain unchanged.

In this case, the inductance is best realized by the circuit of Fig. 8.13, using two voltage follower amplifiers.

The circuit's operation can be described by the equation pair:

$$R = \frac{U}{I} = -\frac{R_A}{R_B} R_C$$

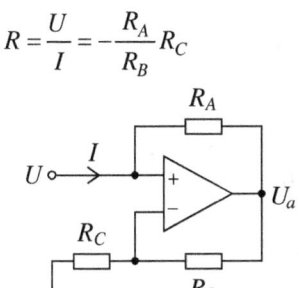

Figure 8.12. Creating negative resistance to ground. When determining the ratio of R_B and R_C, one must consider the headroom of voltage U_a.

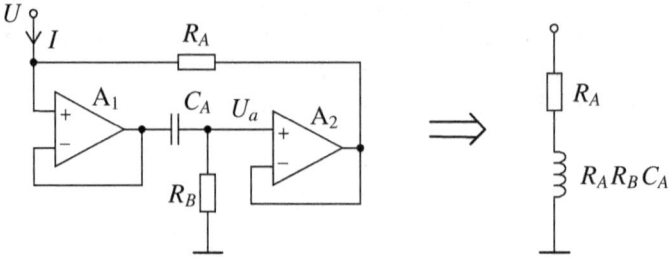

Figure 8.13. Virtual inductor topology suitable for the inductance appearing in Fig. 8.11. The input current of A_2 must be low to avoid generating too much DC voltage in R_B.

$$\frac{U_a}{U} = \frac{R_B}{\dfrac{1}{sC_A} + R_B} \quad ; \quad I = \frac{U - U_a}{R_A} \qquad (8.33)$$

Solving for the impedance between the driving node and ground gives:

$$\frac{U}{I} = R_A + R_A R_B C_A \, s \qquad (8.34)$$

Consequently, the inductance generated by the circuit is $R_A R_B C_A$ and the series resistance R_A. C_A may be chosen quite freely, after which the inductance can be adjusted by R_B, R_A acting also as a coefficient.

The topology of Fig. 8.13 differs from Fig. 8.4 in fact only in that a buffer amplifier is now used to drive R_A, making the series resistance lower and dependent only on R_A. Also in this case, the impedance stays inductive, in practice, only up to a few kilohertzes and turns then capacitive; but at these frequencies, the matter is already irrelevant.

In the numerator of the transfer function (8.26), all five variables are found both in the constant term and in the coefficient of s, so writing a straightforward design procedure is not so simple in this case. The element values can be found, however, by iteration with equations (8.29) – (8.32). It is also possible to construct contour plots from which the correct values are obtained by interpolation, after the point representing the required operation parameters has been determined. Such charts also provide insight into how sensitive the parameters are to changes in the appropriate resistor values.

Figure 8.14a shows the resulting equalizer, in which the inductor and

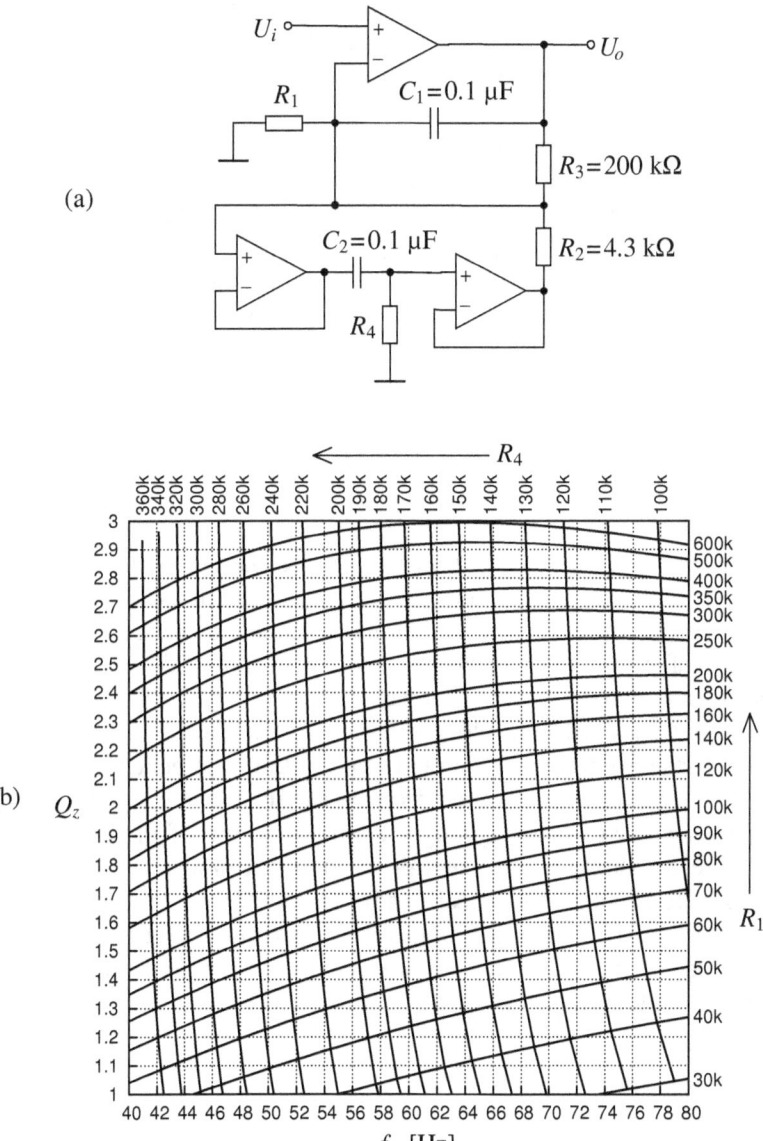

Figure 8.14. a) Practical example of the equalizer shown in Fig. 8.11. R_1 is mainly determined by the required Q_z value and R_4 mainly by the required f_z. b) Tuning map for the Fig. 'a' circuit. The curves have been obtained by keeping the other resistance constant while varying the other. Every f_z-Q_z pair is corresponded by a specific R_1-R_4 pair obtained by interpolation.

the accompanying series resistor have been replaced by the equivalent shown in Fig. 8.13. Elements regarded as fixed have been given example values, that have been used to construct the tuning map of Fig. 8.14b. With these values, the ratio ω_z/ω_p is about 7 and the DC gain correspondingly about 50.

The map helps to determine resistances R_1 and R_4 in the circuit of Fig. 'a' when f_z ($=\omega_z/2\pi$) and Q_z are known. For example, if the needed operation parameters are $f_z = 70$ Hz and $Q_z = 2.5$, we obtain, by reading the contours, $R_1 = 220$ kΩ and $R_4 = 125$ kΩ.

From the map one can also see that R_1 affects quite little the characteristic frequency and R_4, in turn, affects quite little the Q value; so, in this case, the adjustments do not interfere very much with each other.

8.7 Double-Integrator Method

A good way to generate biquad responses without restrictions is to use a feedbacked double-integrator structure, that produces at the same time both high-pass, band-pass, and low-pass functions. By comparing expressions (2.5), (2.12), and (2.16), it can namely be seen that any 2nd-order transfer function can be realized by summing these three basic responses in appropriate proportions.

Figure 8.15 shows the block diagram of such a system. The high-pass response is obtained from the output of the difference block S_1, the band-pass response is obtained from between the integrators and the low-pass response correspondingly from the output of the latter integrator. The weighting coefficients c_1, c_2, and c_3 can also be made interdependent to accomplish a specific adjustment effect.

Let the transfer function of the first integrator be $1/(\tau_1 s)$ and that of the second, $1/(\tau_2 s)$ (as in the Figure). By following the flow of signals through the integrators and S_1, we obtain the equations:

$$U_a = \frac{U_i - U_a - U_b}{\tau_1 s} \quad ; \quad U_b = \frac{U_a}{\tau_2 s} \tag{8.35}$$

For the output signal, it holds:

$$U_o = c_1(U_i - U_a - U_b) + c_2 U_a + c_3 U_b \tag{8.36}$$

By solving voltages U_a and U_b from equations (8.35) and substituting the results in equation (8.36), we obtain, after a few steps, as the overall

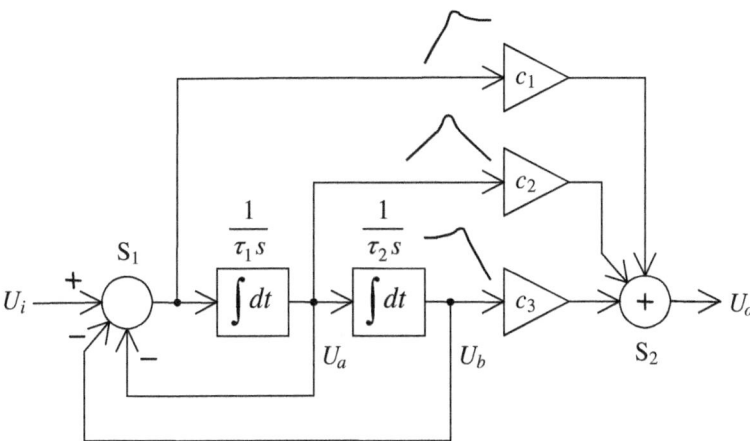

Figure 8.15. System capable of realizing any 2nd-order filtering function. The feedbacked double-integrator structure provides three basic response functions that have common poles (common denominator) and that can be summed to obtain the desired response.

transfer function

$$\frac{U_o}{U_i} = \frac{c_1 s^2 + \dfrac{c_2}{\tau_1} s + \dfrac{c_3}{\tau_1 \tau_2}}{s^2 + \dfrac{1}{\tau_1} s + \dfrac{1}{\tau_1 \tau_2}} \tag{8.37}$$

The poles can be determined easily and without limitations by the time constants τ_1 and τ_2, after which we still have free hands to set the zeroes using coefficients c_2 and c_3. c_1 has to be kept at unity if the general signal level is not to be changed. When applying formula (2.17), it must also be noted that the coefficients of the 2nd-order terms must be unities.

The analysis equations come out to be:

$$\omega_z = \sqrt{\frac{c_3}{c_1 \tau_1 \tau_2}} \tag{8.38}$$

$$\omega_p = \frac{1}{\sqrt{\tau_1 \tau_2}} \tag{8.39}$$

$$Q_z = \frac{1}{c_2} \sqrt{\frac{c_1 c_3 \tau_1}{\tau_2}} \qquad (8.40)$$

$$Q_p = \sqrt{\frac{\tau_1}{\tau_2}} \qquad (8.41)$$

For an equalizer that is fixed or tunable by replaceable resistors, one can use the circuit shown in Fig. 8.16. A_1, R_1, and C_1 make up the first integrator and A_2, R_2, and C_2 the second. A_3 with its resistors forms the addition and subtraction block for the feedback signals, and A_4 with its resistors performs the final summation.

The integrators are inverting, so the output voltage of A_1 is added to the input signal instead of subtracting. In the final summation, the voltage of A_1 is correspondingly taken into account as inverted.

Design can be performed in the following way, assuming that the level of high frequencies is not changed:

1. Choose values for R_3 and R_6, e.g. 10 kΩ. (The impedance of the feeding source must be negligible compared to resistance R_3, or then the value of the source resistance must be deducted from the value of the input resistor.)

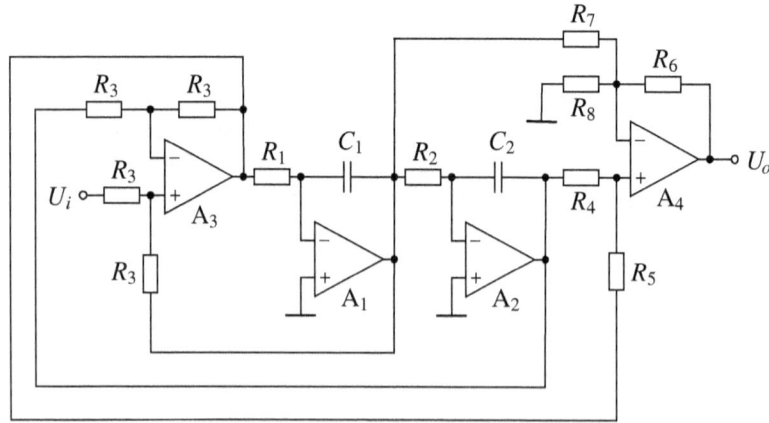

Figure 8.16. Equalizer circuit working on the double-integrator principle of Fig. 8.15, using fixed tuning resistors.

2. Calculate τ_1 from the formula $\tau_1 = Q_p/\omega_p$ and choose R_1 and C_1 so that $R_1 C_1 = \tau_1$.

3. Calculate τ_2 from the formula $\tau_2 = \tau_1/Q_p^2$ and choose R_2 and C_2 so that $R_2 C_2 = \tau_2$.

4. Choose resistors R_4 and R_5 so that $R_5/R_4 = \omega_z^2/\omega_p^2$. For practical reasons, it is best to keep the sum $R_4 + R_5$ in the range 5-50 kΩ.

5. Determine R_7 from the formula

$$R_7 = \frac{\omega_p Q_z}{\omega_z Q_p} R_6 \qquad (8.42)$$

6. Finally, determine R_8 from the formula

$$R_8 = \frac{R_4 R_6 R_7}{R_5 R_7 - R_4 R_6} \qquad (8.43)$$

which yields a positive value with normally used Q_z/Q_p ratios.

If R_7 and R_8 don't come out close to standard values, R_6, R_7, and R_8 can be scaled anew by multiplying them all by the same number.

8.8 Adjustable Equalizer with Fixed Poles

If the same pole shifting equalizer is to be used with different loud-speakers, it should be easily re-adjustable to conform to each particular need. Thus, in an equalizer intended to be adjustable by an ordinary cus-tomer, at least the f_z and Q_z settings should function independently and preferably in a linear fashion.

The double-integrator scheme is well suited just for this purpose. Ei-ther ω_p and Q_p can be held constant, so that the cut-off frequency of reproduction becomes independent of speaker parameters, or then ratio ω_z/ω_p can be held constant, so that the cut-off frequency follows ω_z at a specified distance. The solution presented next operates in the first-mentioned fashion.

When seeking to adjust the characteristic frequency and Q value of the zeroes independently of each other and the poles, two problems are to be solved: c_3 appears in equation (8.38) under a square root, so in order to have a linear ω_z adjustment, c_3 should be made proportional to the square of the desired frequency value. Secondly, Q_z is dependent on c_3, that is used for the frequency adjustment (equation (8.40)), and this effect should be canceled somehow.

Figure 8.17 shows a scheme for arranging the coefficients in such a way that both ω_z and Q_z are adjustable with potentiometers linearly and independently of each other. The coefficient of the low-pass function, c_3, is implemented by two identical amplifier stages, so that the coefficient of both phases will be $\sqrt{c_3}$, or ω_z/ω_p. The adjustment devices of the amplifiers have been ganged mechanically. The band-pass signal, in turn, is led after a Q_z adjustment amplifier through the second ω_z adjustment amplifier, so that coefficient c_2 becomes proportional to $\sqrt{c_3}$ and hence Q_z becomes independent of c_3.

Q_z is inversely proportional to its adjustment coefficient, c_2, but this does not form a problem in practice.

Figure 8.18 presents an example of an adjustable equalizer circuit based on the above-described method. With the values given, frequency f_z is adjustable from 40 to 80 Hz and Q_z from 0.5 to 3. f_p $(=\omega_p/2\pi)$ has been set to 30 Hz while Q_p is 0.5.

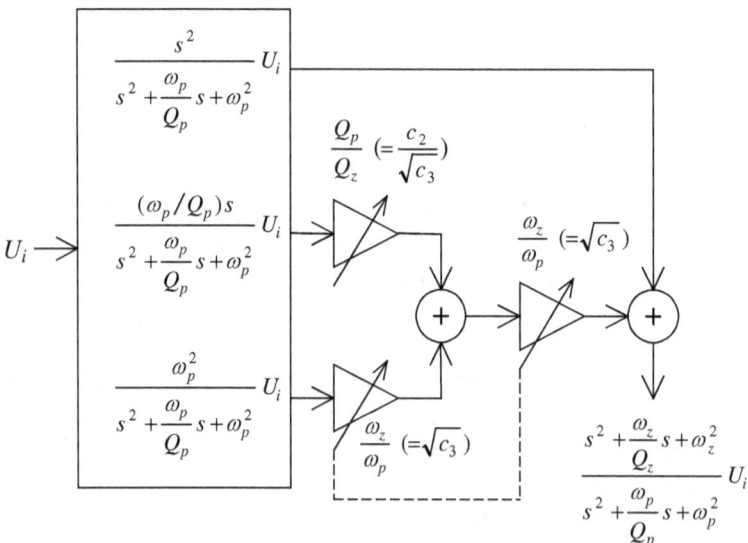

Figure 8.17. System that enables linear and orthogonal adjustment of the characteristic frequency and Q of the zeroes while keeping the poles fixed. The basic responses presented in the box can be implemented as in Fig. 8.15.

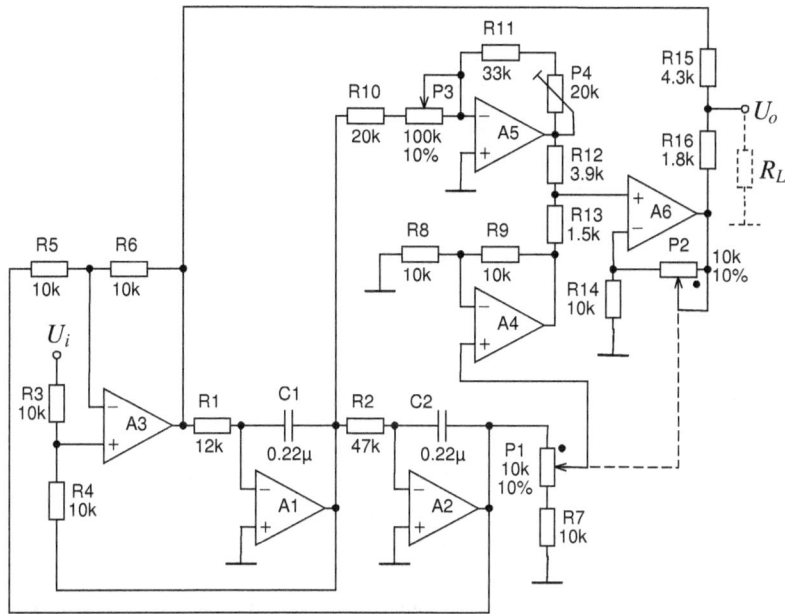

Figure 8.18. Linearly and orthogonally adjustable equalizer circuit, in which f_p = 30 Hz and Q_p = ½. f_z can be set by dual potentiometer P_1-P_2 within range 40-80 Hz and Q_z by potentiometer P_3 within range 0.5-3. For the operational amplifiers, e.g. TL074 (quad) or TL072 (dual) are suitable.

The integrators and the feedback adder operate similarly as in Fig. 8.16.

Operational amplifier A_4, potentiometer P_1, and resistors R_7, R_8, and R_9 make up the first frequency adjustment amplifier and A_6 with feedback resistors R_{14} and P_2 the second. Potentiometers P_1 and P_2 must be on the same shaft.

A_5 with its resistors forms an inverting amplifier for the band-pass signal. The gain of the stage is inversely proportional to resistance R_{10} +P_3, so Q_z is directly proportional to this resistance.

Resistors R_{12} and R_{13} form a summing stage by which the band-pass and low-pass signals are mixed in correct proportions before the common gain stage. Likewise, with R_{15} and R_{16}, the high-pass signal is combined with the other signals.

Output buffering is not presented because such is not necessarily always needed. The output may, however, be loaded by any resistive load R_L without affecting the response shape. R_L can be, for instance, a volume adjustment potentiometer.

c_1 will be, due to the voltage division, $R_{16}/(R_{15}+R_{16}) = 0.295$, so the circuit produces an attenuation of 10.6 dB. Some amount of overall attenuation is even necessary to prevent possible clipping at the lowest frequencies, that are boosted many times over with high f_z values. Also, the ratio of the applied signal to the supply voltage must be such that the output of A_6 will not become saturated in any normal conditions.

The resistance tolerance of the potentiometers must be 10% to keep gain factor errors sufficiently small. For the same reason, the frequency adjustment amplifiers have been realized differently although both have the same gain. Thereby, the resistance error of P_1 has mostly effect at the low end of the f_z region and the resistance error of P_2, in turn, at the high end of that region. In this case, a tolerance of 10% introduces a gain error of 5% at most; and because ω_z is proportional to $\sqrt{c_3}$, the frequency error due to the potentiometers stays at 2.5%.

The inaccuracy effects of P_3, instead, cannot be neglected. Hence, the error in Q_z can be minimized by trimmer P_4. The gain of the stage (from the output of A_1 to that of A_5) is intended to be exactly $1/Q_z$. If the gain is adjusted in place e.g. with a Q_z setting of 2, the error with values that are of interest on current-drive is left almost negligible.

All the other passive components have also some effect on the tuning accuracy (except the series connection R_{11}-P_4, whose resistance is adjustable), so it is recommendable to use precision components for the resistors and especially for the capacitors.

The rather high amount of operational amplifiers may raise some doubts in those who appreciate as short as possible signal path and simplicity. To this, it can be noted, however, that outside the bass region mostly only A_3 is operative, the other operational amplifiers remaining virtually passive.

In many listening rooms, there are problems with standing-wave-induced narrow-band peaks or dips at frequencies below 100 Hz. With the presented equalizer, it may be possible to help this somewhat by trying parameter settings that differ a little from the speaker values.

8.9 Adjustable Equalizer with Tracking Poles

If the operation range is extended substantially and the material to be played is rich in the lowest frequencies, there is the risk of mechanical overloading, that effects heavy distortion and ultimately even voice coil damage. Hence, it may be of advantage to keep the ratio ω_z/ω_p constant and relatively low, thus making the equalizer better suited for speakers of different sizes.

By examining equations (8.38) - (8.41), it can be seen that such an operation can be accomplished by adjusting the frequency by the time constants τ_1 and τ_2, instead of c_3. In addition, by keeping the ratio of τ_1 and τ_2 constant, the Q values remain unchanged.

The adjustment could be realized directly by varying the resistance of both integrators, but then the frequency's dependence on potentiometer position will be nonlinear. Thus, a more practical way is to add an adjustable gain stage for both integrators, whereupon the time constant becomes inversely proportional to this extra gain (see Fig. B3b in appendix B).

Figure 8.19 shows an example circuit that operates on the described principle. Potentiometer P_1 affects time constant τ_1, and potentiometer P_2, that is mechanically coupled with P_1, affects time constant τ_2. Also here, the adjustment amplifiers have been made different, for the same reason as before.

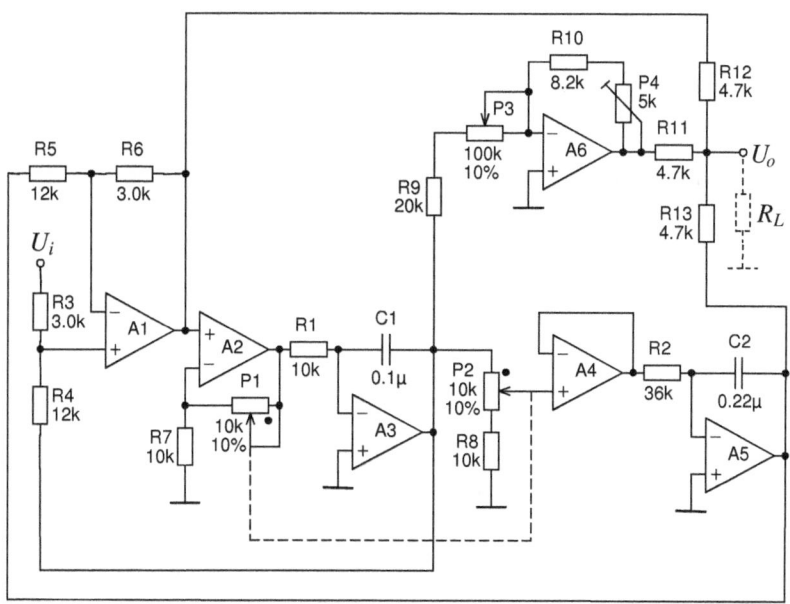

Figure 8.19. Linearly and orthogonally adjustable equalizer circuit, in which $f_z/f_p = 2$ and $Q_p = \frac{1}{2}$. f_z can be set by dual potentiometer P_1-P_2 within range 40-80 Hz and Q_z by potentiometer P_3 within range 0.5-3.

Also here, f_z has been confined to the range 40-80 Hz and Q_z to the range 0.5-3. The ratio f_z/f_p has been fixed to 2, i.e. these frequencies always stay one octave apart from each other. Q_p is solidly 0.5, so that the response descends softly and the risk of overdrive remains minor.

The double-integrator structure differs from the previous ones also in that the feedback signals coming from the integrators are attenuated in the summing stage by 1:4. Thereby, the low-pass and band-pass signals are magnified 4-fold which reduces the amplification need afterwards.

The Q value is adjusted by potentiometer P_3 just like in Fig. 8.18. The resistance inaccuracy of P_3 is compensated by trimmer P_4, which in this case must be set so that gain from the output of A_3 to the output of A_6 is $1/(4Q_z)$ in the region where the best accuracy is needed. Of course, the potentiometer scale must be properly aligned before undertaking the tuning.

The level of the low-pass signal obtained from A_5 is fourfold relative to the high-pass signal obtained from A_1, the ratio being thus already equal to ω_z^2/ω_p^2, so these signals can be combined as 1:1. After A_6, the level of the band-pass signal is also such that all the summing resistors (R_{11}, R_{12}, and R_{13}) can be made equal. Again, any resistive load R_L may be attached to the output without affecting the response shape.

---

[1]   R. A. Greiner and M. Schoessow, "Electronic Equalization of Closed-Box Loudspeakers", *Journal of the Audio Engineering Society*, vol. 31, March 1983, p. 125-134.

9
RESONANCE VARIATIONS

The successful use of the compensation methods presented in the previous chapter necessitates that the resonant frequency and Q of the enclosed drive unit are known with sufficient accuracy and the values stay sufficiently unchanged in different circumstances. However, experiments show that there can be large variation in fulfilling this requirement, mostly due to the properties of the surround material used.

One should not rely on the specified resonance parameters as such, for they often contain systematic error too. Namely, the author of this has yet never come across such a bass driver whose resonant frequency would have been essentially lower than specified. Instead, excesses of 10-20 Hz over the nominal value are very common. The mechanical Q value also exhibits remarkable spread.

For the reasons mentioned, even a hobbyist designer should have the facilities to determine resonance parameters. Methods for this have been presented in chapter 13. The following measurements have been performed mainly by making use of the EMF extractor appliance described in section 13.2.

The stability of resonance was tested in six divergent bass drivers, as introduced in Table 3. The devices differ from each other above all by their outer suspension but also in size. When writing this, some of them are already out of production.

The properties under inspection were the dependency of the resonant frequency and mechanical Q value on current, temperature, and stress. Additionally, the drivers were exposed to UV light. The results are even a little surprising.

Table 3.   Drive units used in the tests. f_0 and Q_m are the specified res. frequency (Hz) and mech. quality factor.

	Type	Size	f_0	Q_m	Surround
A	Seas P17REX	6½"	34	1.21	rubber (high loss)
B	Vifa M13SG-09-08	5"	54	1.5	rubber
C	Alpine SXS-1357 (bass)	5"	-	-	butyl rubber
D	Vifa PL18WO09-04	6½"	37	2.53	NR rubber
E	Peerless 833429 (WF165)	6½"	47.5	1.63	foam
F	P. Audio HP-10W	10"	48	5.9	textile

9.1 Current-Dependence

It is customary to consider the resonant frequency and Q values of drivers so constant that the signal level used in their determination is generally not an issue. However, as it appears from the current-dependences shown in Fig. 9.1, in many cases there would be need for such a clarification.

The measurement current was varied from 3 to 200 mA, except for sample C, which, being 4-ohmic, tolerated yet 300 mA. Higher currents cannot really be used without exceeding the driver's linear excursion.

In all samples, f_0 decreases as current and cone excursion increase. The same applies, generally speaking, also to the Q_m value. The effects are nonlinear by nature since the properties of linear systems can never depend on signal level.

The current-dependences are least in samples D and E, where the surround is of NR rubber and foam, respectively. With these drivers, the deviation of the resonant frequency from the specified value is also the least, whereas in others, the provided frequencies are hardly even rough estimates.

The sample representing butyl rubber, C, and the one representing coated cloth, F, also behave satisfactorily in terms of f_0; but the change in Q_m is, within the measurement range, over 20%. Even this variation is not necessarily a problem if the final Q value is determined largely by the damping material.

In samples A and B, that represent rubber grades commonly used in bygone years, f_0 descends substantially with increasing current, in A, even about 30%, Q_m variation being again of the order of 20%. Such a

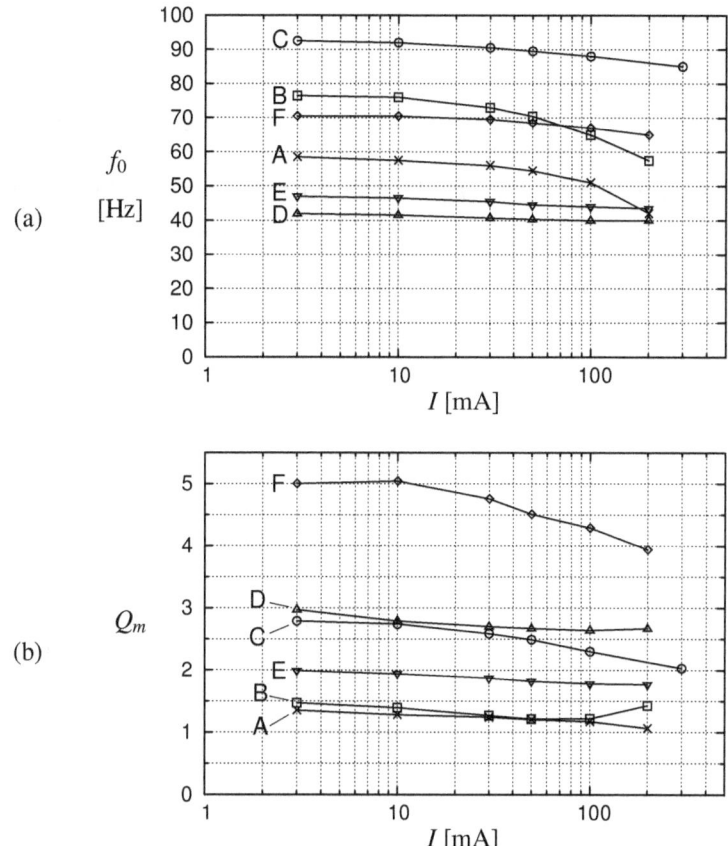

Figure 9.1. a) Resonant frequency vs. measurement current in the drivers of Table 3. b) Corresponding results for the mechanical Q value.

driver calls for a relatively small enclosure if the spring force is to be linearized and distortion kept low.

Reasons for the dependences are not easy to figure out offhand, but the differences are anyway related to the deformation properties of the outer suspension since the inner suspension is in all samples very similar.

The test is best coped with by sample D, whose resonant frequency changes only 5% and whose Q value also levels off as excursion increases. Thus, rubber materials with good enough dynamic performance already exist, but the issue would deserve more general attention, even without any resonance compensation being used.

9.2 Temperature Dependence

The temperature dependence of resonance also varies much between different surround materials, as can be seen from Fig. 9.2. However, the direction is the same in all, i.e. when temperature rises, f_0 decreases and Q_m increases.

The measurements were made with 50 mA current, which already generates moderate excursion, avoiding yet the non-idealities of high signal levels.

The dependences of samples C, D, and E are comparably mild in the examined 8-degree range. In all these, the change in f_0 is less than 5%

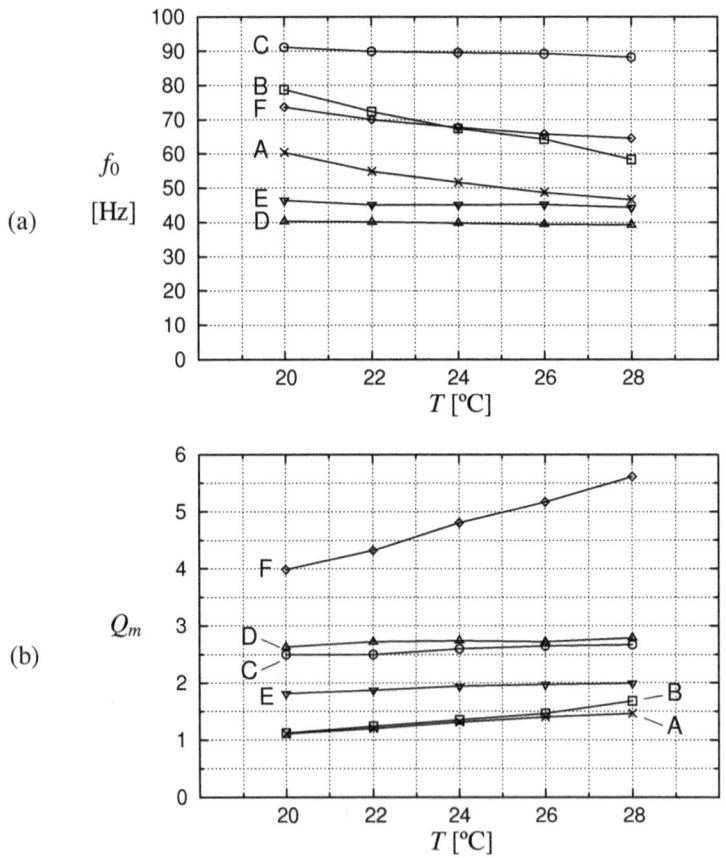

Figure 9.2. a) Resonant frequency vs. ambient temperature in the drivers of Table 3. b) Corresponding results for the mechanical Q value.

and the change in Q_m less than 10%. The best-ranking sample is again D, which, at least when writing this, belongs to a series still in production.

The temperature dependences of sample F are already of significant level. Similar surrounds are mostly used in PA speakers, and especially in outdoor use, the bass reproduction of such a driver may vary a little according to weather conditions, even though sealed enclosure structure and good damping relieve these variations also.

Samples A and B, that exhibited the greatest current-dependences, are also the most sensitive to temperature. In A, the sensitivity of the resonant frequency is about 1.7 Hz per Celsius degree and in B about 2.6 Hz per degree. The Q value rises, within the observation range, 32% in A and 50% in B. With such sensitivities, resonance compensation may already be hampered too much if the usage cannot be restricted to a rather narrow temperature range.

9.3 "Burn-in"

According to a common view, the suspension of a new bass driver becomes a little more flexible during the first exercises, due to which e.g. the resonant frequency should accordingly lower somewhat. The process is generally termed burn-in or break-in, during which driver properties are supposed to settle in place. The performed stress tests indicate, however, that the prevailing picture of the subject is lacking.

The drivers of Table 3, familiar from the previous tests, were each stressed for 12 hours by a sinusoidal 30 Hz current whose amplitude was set so as to effect a peak displacement of ±3 mm or a little over. The trial can be considered even comparably demanding, for in typical use the linear excursion limit is exceeded only seldom, if ever. After this burn-in, the resonance parameters were monitored for 40 days, at first more frequently and finally in 8-day intervals. All measurements were performed with 50 mA current and at the same temperature.

The results are depicted in Fig. 9.3. Excluding sample D, both the resonant frequency and Q value fall distinctly, but generally embark on a rise, however, immediately after the stress.

In sample D, f_0 decreases only a hertz and also recovers quite rapidly close to its original value. Q_m, again, does not change in the stress virtually at all.

Changes occurring in sample E, that represents plastic foam, are also quite minor. Both f_0 and Q_m decrease 4% and recover completely in a

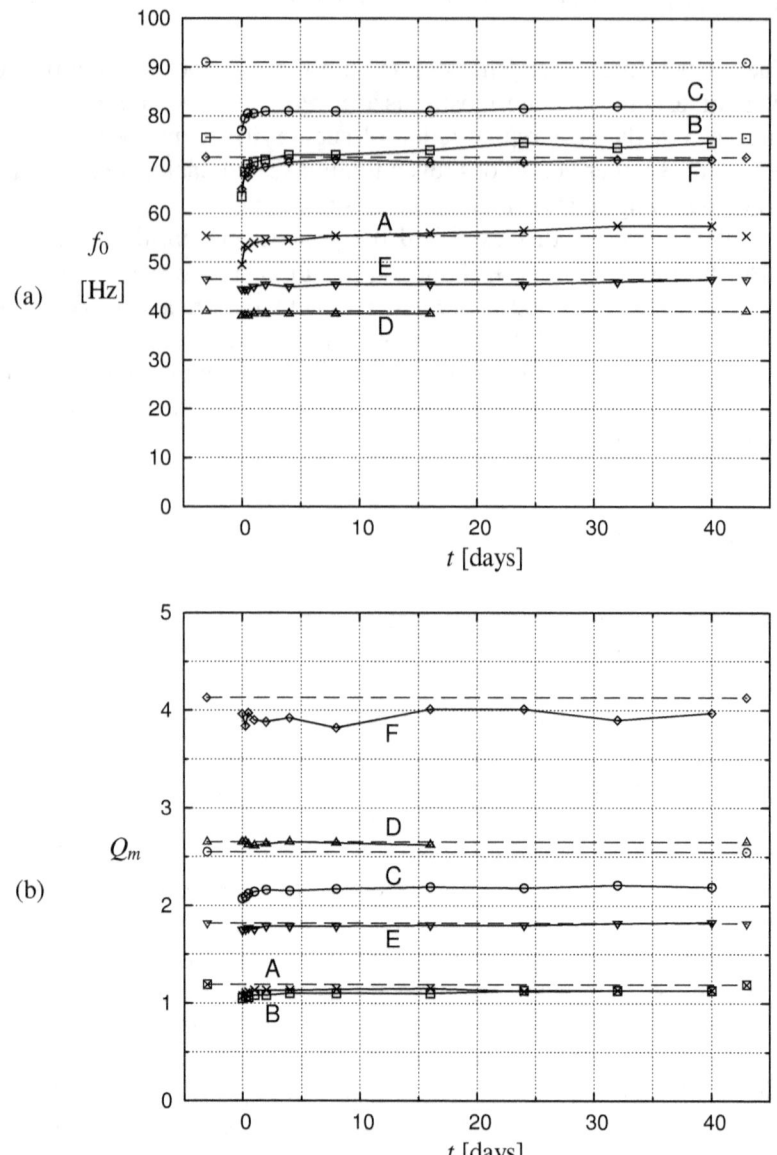

Figure 9.3. a) Resonant frequency behavior during 40 days after a stress test of 12 hours in the drivers of Table 3. Dashed lines represent values measured before the stress. b) Corresponding behavior of the mechanical Q. (Results A and B partly coincide.)

few weeks.

Also here, samples A and B behave similarly. The parameter decreases are in the range 10-16%, but Q_m is not able to yet recover fully in 40 days. For some reason, the resonant frequency of A increases, however, even a little higher than the pre-stress value.

The resonant frequency of the PA driver F decreases in the stress by 9%, recovering, however, already in a week very close to the original value. Q_m, instead, seems to permanently end up to a lower level, but only slightly.

The only significant seemingly permanent changes occur in the car kit unit C, whose parameters recovered in the observation period only partially. However, the resonant frequency of this driver was measured yet a couple of months after the last recorded measurement, whereupon the difference from the original value had reduced to 5 Hz. Thus, even this sample would seem to recover from the stress, although extremely slowly.

It is naturally partly case-dependent how long-lasting changes can be regarded as "permanent" since the suspensions may experience hard use even quite frequently. Minor excursions, however, don't have any influence on the parameters.

Summing up:

The loosening of the suspensions that occurs in mechanical stressing of a driver is, in general, temporary by nature, and the resonant frequency recovers to or near its original value within a few weeks or months after the stress. The mechanical Q also often returns close to its pre-stress value.

9.4 UV Sensitivity

Hardening of the rubber surrounds is a subject that every now and then emerges in discussion forums. Of course, the surround properties should noway change in the course of time, but if the rubber hardens to such degree that the change is observable by hand probing, the problem is already evident, for the resonance parameters of such a driver have already drifted far from the original. The hardening may impair reproduction also at higher frequencies since matching between the mechanical impedances of the cone and surround may suffer.

The phenomenon is due to normal ultraviolet radiation of daylight, and unfortunately, such UV sensitivity has been quite common in drivers manufactured in earlier years. Nowadays, it seems, however, that we would be getting rid of the use of such rubbers.

The light sensitivity of the driver set was examined by keeping them in a closet at a half-metre distance from a 25 W UV lamp. Before the experiment, all samples had been kept, from the time of their purchase, in darkness.

Samples C, D, E, and F passed the test cleanly, that is, no light-induced changes were detectable in the resonant frequency.

A and B, instead, exhibited strong increase in their resonant frequency. During the 8-week long exposure, A's frequency rose 17.5 Hz (51 → 68.5) and B's even 37 Hz (66 → 103), and the rise would still have continued.

It was also noteworthy that the rise was not steady but could stay in some periods in both drivers almost at zero. From this, it can be concluded that, besides UV radiation, some other climatic factors are also playing a role.

The rate of rise achieved is roughly only about 10-fold compared with what it is in ordinary room lighting. The author has experience, for instance, of two 5-inch drivers from a well-known manufacturer, both of which exhibited a rise in the resonant frequency up to 140 Hz (specified value 35 Hz) in little more than a year, merely by keeping them near a window, even so that they didn't see the sun.

A 4-fold resonant frequency relative to the rated value respectively denotes a 16-fold suspension stiffness, so the changes occurring in the rubber can be even quite dramatic.

Even light-sensitive drivers can, however, be used if they are kept from every side in sufficient darkness. For example, a black foam mask of 3 cm thickness already blocks 99% of the radiation, slowing the hardening by a corresponding amount.

10
AMPLIFIER REALIZATIONS

When talking about current-output amplifiers, it is still justified to use the term "amplifier", even though the output quantity (current) is other than the input quantity (voltage), and thus any actual gain factor does not exist. In practice, power and current are, however, at the output always greater than at the input, so amplification is anyway occurring.

However, a more precise designation would be a *transconductance amplifier* or a *transconductor*, that, in general, refers to a means that converts input voltage to corresponding output current. The established symbol for transconductance is g_m and the unit, siemens ($1S = 1A/1V$).

10.1 The Series Resistor Method

For lack of better, a conventional voltage-output amplifier can be used as a substitute for a real transconductor by inserting in the speaker connections appropriate series resistors. The operation then corresponds to Fig. 5.1a, and the current-drive index obtained depends directly on the applied series resistance. The downside is, of course, that the available current magnitude is only a fraction of what the amplifier is normally capable of producing.

For the amplifier, the load is easy because it is almost resistive, and no high currents or powers are taken. Essential is the amplifier's voltage capacity, and if the nominal impedance of the output is selectable (as in some valve appliances), it is best to use the highest value.

The value of the resistor should be preferably about 10-fold relative to the speaker's nominal impedance. Then, the highest obtainable sound level is about 20 dB lower than in the normal voltage use. With music signals, the power rating of the resistors doesn't have to be very high, but they should be placed so that possible heating does not pose a danger.

By inserting a potentiometer of say 100 Ω as the resistor, one is able to examine how the sound quality changes with the driving mode. With speakers intended for voltage drive, increasing the series resistance may bring about undesirable changes too, as frequencies at or near the impedance peaks are boosted.

10.2 Current-Feedback

Realizing a current-output amplifier is not anything more complicated or tricky compared with voltage-output ones. The normal way is to connect in series with the load a low-valued resistor, whose voltage indicates the current flowing through the load. By using this voltage as the feedback signal, the result is *current-feedback*, due to which the output current seeks to follow the input signal.

Power amplifiers have traditionally been implemented by the basic scheme shown in Fig. 10.1a, that is, using *voltage feedback*. The signal to be amplified, U_i, is applied to the non-inverting input of a differential amplifier (triangle), and a specified proportion of the output voltage U_o, as determined by resistors R_1 and R_2, is brought back to the inverting input. Ideally, the circuit sets its output voltage always so that the difference voltage, U_d, remains at zero. Voltage at the common point of the resistors is then U_i, and because no significant current goes to the input terminals, the voltage division causes U_o to follow U_i, amplified by the factor $(R_1+R_2)/R_1$.

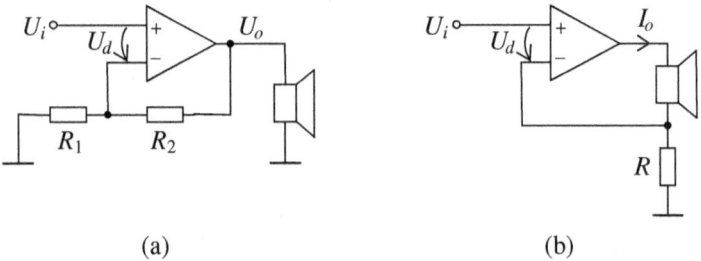

(a)   (b)

Figure 10.1. a) General principle of voltage-output amplifiers. b) Basic topology for current-output amplifiers.

Figure 10.1b shows, correspondingly, the use of current-feedback in its simplest form. Again, the non-inverting input terminal acts as the input, but now the feedback signal applied to the inverting terminal reflects the current flowing through the load (I_o), instead of the load voltage.

The operation of the loop can be conceived as follows: Suppose the difference voltage U_d is, for some reason, growing positive. The differential amplifier responds to this by raising its output voltage, making thus current I_o and the voltage drop in R change in the positive direction. This increase continues until the feedback signal reaches the input signal and U_d is nulled. Likewise, when U_d is becoming negative, I_o and the voltage of R decrease until the deviation in U_d has been eliminated. Because the resistor's voltage thus follows the input voltage, we obtain as the circuit's transconductance $I_o/U_i = 1/R$.

The differential amplifier itself can be of similar construction in both cases. The differential amplifier's own output impedance does not have any appreciable significance in either circuit because the impedance seen by the speaker is determined by the feedback used. Even ordinary power operational amplifiers can be used for the purpose if crossover distortion is not excessive and other properties are suitable. However, it is more advisable to use actual audio amplifier chips, which are also available without built-in feedback.

Amplifier designs out of discrete components are not presented here. Existing topologies may well be used as a basis, and skilled designers are able to further refine them to conform to the needs of the current-drive scheme.

A suitable value for the current-monitoring resistor is about half an ohm. With such a value, the power consumed in the resistor generally amounts to less than a tenth of the power taken by the speaker, so the loss is not very significant. It is better not to increase the resistance very much, for, besides the power loss, there may arise a risk of instability due to the high feedback factor. Too low a resistance is neither for good because, as the circuit gain becomes high, the raw gain of the differential amplifier is no more sufficient to keep the circuit's output impedance high enough for high frequencies. Additionally, the direct-current component at the output, caused by the offset voltage of the input terminals, increases with the transconductance.

If high transconductance is desired but without R becoming unreasonably low, the feedback signal can be attenuated as shown in Fig. 10.2. R_2 and R_3 or at least one of them should be preferably of the order of a kilo-ohm at most, so that the impedance feeding the inverting input ter-

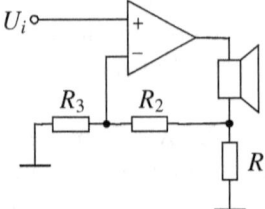

Figure 10.2. Current-feedback in which the transconductance has been raised with a voltage divider (R_2-R_3).

minal would not become overly high. Namely, from the standpoint of sound quality and also stability, it is the safer, the less various wiring-related stray capacitances are able, when charging and discharging, to vary the voltages of the input terminals.

Above, we found the transconductance of the Fig. 10.1b circuit to be $1/R$, assuming that the difference voltage U_d stays at zero which requires an ideal differential amplifier. A real differential amplifier has, however, only finite gain (designated A_d), that rolls off with increasing frequency in the manner shown in Fig. 7.11b. Also, the output always has some internal impedance (designated Z_o) even though its value is generally not directly specified.

Next, we will determine the transconductance and output impedance of the current-feedbacked amplifier, using the notation of Fig. 10.3a and the corresponding signal diagram shown in Fig. 10.3b.

The transfer function from the difference voltage to output current equals the voltage gain A_d divided by the total impedance of the current path, $Z_o + Z_L + R$ where Z_L is the impedance of the load. As for the feed-

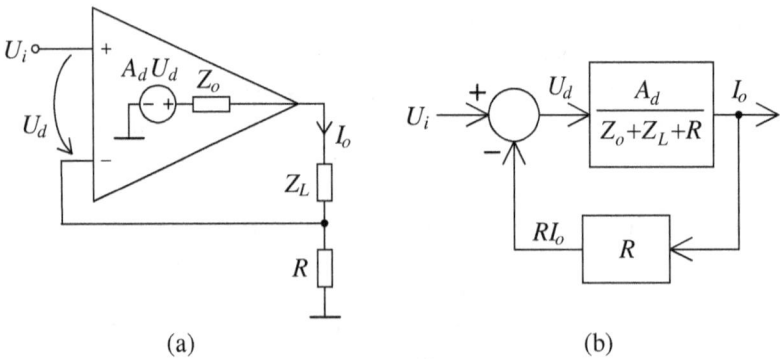

(a)   (b)

Figure 10.3. a) Current-feedbacked amplifier showing the parameters used in the analysis. b) Block diagram representation of the same.

back, the transfer function is mere R. The difference block is an inherent part of the differential amplifier.

By applying, to the feedbacked system, formula (b26), derived in appendix B, we obtain as the transconductance

yielding:

$$g_m = \frac{I_o}{U_i} = \frac{A_d/(Z_o + Z_L + R)}{1 + A_d R/(Z_o + Z_L + R)} \qquad (10.1)$$

$$g_m = \frac{A_d}{(A_d + 1)R + Z_o + Z_L} \qquad (10.2)$$

As A_d is high in absolute value, Z_o and Z_L become negligible, and we are left with $g_m \approx 1/R$, as it should. It is most important that the transconductance's dependence on the load impedance Z_L is made small, and this is achieved when $|A_d|$ and R are high enough.

The output impedance can be solved on the basis of the Thévenin method by dividing the source voltage of the output terminals by the short-circuit current. From Fig. 10.3a, it can be reasoned that with the load disconnected the output terminals exhibit voltage $A_d U_i$. (The feedback signal being then zero.) Further, by marking Z_L to zero in equation (10.2), we obtain as the short-circuit current $A_d U_i/[(A_d+1)R+Z_o]$. By performing the above-mentioned division, we obtain as the output impedance, i.e. the total impedance seen by the load

$$Z_t = (A_d + 1)R + Z_o \qquad (10.3)$$

Thus, Z_o being small, the output impedance is about directly proportional to the differential gain. If R is 0.5 Ω and $|A_d|$ is at 20 kHz yet 500, as common, the minimum value of $|Z_t|$ will be a good 250 Ω, which is, in practice, very satisfactory.

The final output impedance will though often come out lower than equation (10.3) gives, for it may be necessary to add in parallel with the speaker a compensation branch to reduce the reactivity of the load at very high frequencies (section 10.3). Even so, the impedance can be retained moderate, minding yet that the hearing range of an adult person averagely extends up to some 15 kHz.

It is also possible to construct differential amplifiers with high Z_o, thus reducing the importance of A_d and R.

The direction angle of A_d is negative, so Z_t is, correspondingly, capacitive. This reactivity does not, however, introduce any harm as long as

$|Z_t| \gg |Z_L|$ since, for the driving mode of a speaker, only the magnitude of impedance is essential.

The topology of Fig. 10.1b has been regarded as unsatisfactory in performance [1],[2] on grounds that are, however, virtually unfounded. It has been raised as a problem that the output impedance depends on frequency and that the transconductance is, as given by equation (10.2), dependent on the load impedance and its nonlinearities.

However, the output impedance's dependence on the raw gain and hence on frequency is not any detriment in itself since the only property required of the output impedance is its sufficiency. In practice, it is only important that the amplifier does not unduly decrease the current-drive indexes of the drivers at audible frequencies.

As for the transconductance's load dependence (which is the same thing as the output impedance but expressed differently), it is in the case of Fig. 10.1b not any greater than in any other circuit topology producing corresponding output impedance.

For voltage amplifiers, it is generally considered a good achievement if the output impedance is less than a hundredth of the nominal load impedance. Then, the effect of load impedance variations on the load voltage is, in practice, already negligible, assuming the cable is resistanceless. Correspondingly, for a current-output amplifier, it ought to be considered fully sufficient that the output impedance is 100-fold relative to the nominal load. With the topology in question, this requirement can be fulfilled at least for the major part of audible frequencies.

However, in the highest octaves the capacitive nature of Z_t comes yet to aid. There, the load impedance is generally nearly resistive or lightly inductive. The direction angle of the term $(A_d+1)R$ in expression (10.2) is, in turn, nearly $-90°$, so the changing of $|Z_L|$ as a function of frequency affects the transconductance's magnitude in fact much less than one might conclude by considering the mere absolute values of the terms.

So, despite its simplicity, the described transconductor principle is fully fit even for demanding use, nor is it inferior in terms of power losses to more elaborate designs.

The reactive nature of a typical loudspeaker load causes that, with practical program signals, the ratio of the peak values of voltage and current may differ even much from the speaker's nominal impedance and even from the minimum or maximum impedance. Under voltage drive, this means that the amplifier should be capable of feeding many times higher peak currents than would be needed with resistive load.

On current-drive, however, things get reversed; and, instead of extra current, the amplifier should correspondingly be able to provide extra voltage. The requirement is, however, somewhat relieved because bass reflex enclosures, that constitute heavily reactive loads, are not needed in current-drive use.

10.3 Stability

Many power operational amplifiers and other chips of interest require from the circuit a specified minimum gain (e.g. 20 dB) in order to function stably. This condition has to be satisfied especially at those frequencies where oscillation is possible, in practice at about one megahertz or so. The specified gain limit applies, however, only with resistive feedbacks, such as e.g. in Fig. 10.1a. Instead, when using current-feedback, the speaker forming the load is part of the feedback network, that hence is no more resistive.

At high frequencies, the inductivity of the speaker and its cable causes in the case of Fig. 10.1b that the feedback signal exhibits certain phase lag with respect to the output voltage. This lag adds to the total lag of the loop and can hence jeopardize the circuit's stability although the gain requirement would be fulfilled. In order to make the feedback more resistive at high frequencies, an RC network can be attached in parallel with the speaker, as shown in Fig. 10.4; this arm acting like an extra "way" in the crossover filter and establishing the principal current path for the highest frequencies. It is better to connect the arm in the amplifier rather than in the speaker, so that cabling inductance will also be eliminated.

Also in conventional voltage amplifiers, it is common to use an RC branch from the output to ground in order to reduce load inductivity.

A suitable value for resistor R_2 is some twenty ohms. In normal use, the current and power dissipation in R_2 are negligible, but if the amplifier, for instance during testing, gets into an oscillatory mode, the power assumed by R_2 may become considerable, and it is well to prepare for

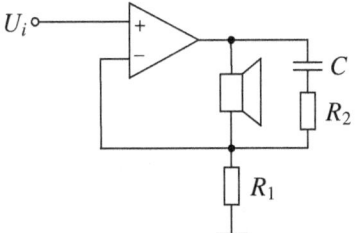

Figure 10.4. Current-feedback stabilized with an RC arm in parallel with the load.

this, at least to some extent.

C should be chosen low enough that the impedance seen by the speaker does not suffer too much at the operating frequencies. On the other hand, *C* should be so large that the total impedance of the speaker and RC arm is no more inductive at the frequencies where the risk of oscillation exists. A suitable capacitance is generally of the order of tens of nanofarads, depending a little on the load's high-frequency impedance and the required output impedance. When testing the amplifier with a resistive dummy load, it may yet be necessary to connect a small inductance in series with the load resistor, so that the minimum gain would be retained.

The impedance of a dynamic drive unit is in the region of one megahertz generally hundreds of ohms but can drastically vary with frequency in both magnitude and angle. The reactivity of the speaker cable and especially its capacitance, that is typically of the order of a nanofarad for a length of 10 m, also strongly affects the load seen by the amplifier.

A strongly capacitive load may also make some amplifier circuits oscillate, and the presented RC arm also helps in such a case. Additionally, the arm constrains possible high-frequency noises, whose conveyance to the speaker is not meaningful.

If an amplifier is oscillating at say 2 MHz, the amplitude of the oscillation may remain quite low because the slew rate value of the amplifier limits the rise and fall time of the voltage. Therefore, the amplifier may still work almost normally despite the oscillation. Sound quality is, however, degraded since the distorted high-frequency vibration produces reflection effects also in the audible range. In a borderline case, the oscillation can also be only occasional, appearing, for instance, only at certain output voltage values but not necessarily when the signal is zero.

Similar phenomena are, however, also possible in voltage-output amplifiers; and in general, the only means to be assured of the reliable operation of a given amplifier circuit is to use an oscilloscope and suitable signal generator.

10.4 Ground-Connecting the Load

For the above-described basic transconductor topology, it is characteristic that the load is not attached to the ground rail, even though the resistance between them is very minor. This separation from the ground does not, however, introduce any actual harm, and speaker cables do not

generally need any screening. Nevertheless, some may prefer a grounded speaker output; and in some situations, it is possible that the user
may, due to an old habit or ignorance, make wrong connections.

The load can be attached to the ground by using the variation shown
in Fig. 10.5 in which the load and series resistor have been interchanged. Because this resistor is now separated from the ground, its voltage
drop has to be replicated into a ground-referenced feedback signal with
a distinct difference amplifier, consisting of A_2 and four resistors.

From the standpoint of the power amplifier, nothing else has changed, except that at high frequencies the difference amplifier introduces
in the feedback signal some extra phase lag, that may have effect on the
circuit's stability. To prevent the danger of oscillation, A_2 should thus
have quite a large bandwidth (like e.g. NE5534).

The offset voltage of A_1 appears as such in the voltage across R_1; the
offset voltage of A_2 appearing, respectively, 2-fold. Therefore, to minimize the direct current flowing to the load, both amplifiers have to be
regarded. It would be best that either one would have a nulling possibility, by which both DC components can be canceled.

It is advisable to keep resistance R_2 relatively low, so that capacitive-
ly coupled interferences at the inputs of A_2 are well suppressed. A suitable value is around 2-3 kΩ.

10.5 Using One-Sided Power Supply

In the circuits presented thus far, the supply voltage has been assumed to be two-sided, so that all signal potentials can be of either polarity with respect to ground. However, many times, e.g. for cost reasons,

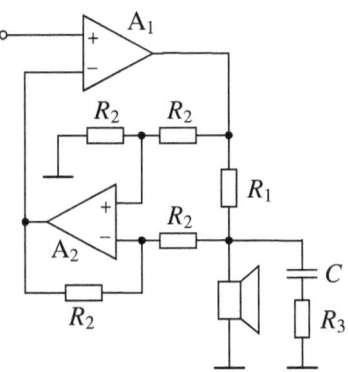

Figure 10.5. Current-feedback for a
ground-connected load.

only a one-sided supply is available in which case it is necessary to use
a series capacitor in the load circuit to block direct current.

Figure 10.6 shows an amplifier suitable for one-sided voltage. At the
operating frequencies, the circuit employs current-feedback but changes
to voltage feedback below the operating band, so that the output would
not be saturated by the direct voltage accumulating in C_3.

The voltage feedback established with resistors R_2 and R_3 keeps the
output bias voltage in the middle of the supply voltage. At DC and the
lowest frequencies, gain from the non-inverting input terminal to the
output is $1+2R_3/R_2$ and can be set to roughly correspond to the output
voltage developed at the operating frequencies. Resistors R_1 and capacitor C_1 are needed only in the case that the bias voltage provided by the
preceding stage is not half of the supply voltage.

The voltage alterations of the current-sensing resistor R_5 are shifted
by C_2 to the inverting input to form the current-feedback. C_2 as well as
C_3 and C_4 assume half of the supply voltage. C_2 has to be so large as to
exhibit negligible impedance, at the operating frequencies, compared to
resistances R_2 and R_3. Suitable values amount to microfarads, but it is
better not to use electrolytics for the purpose, neither for C_1.

C_3, in turn, should be so large that its impedance remains lower than
the speaker impedance for all signal frequencies. However, the import
of C_3 essentially differs from what it is in a conventional amplifier:

In a current-output amplifier operating on a one-sided power

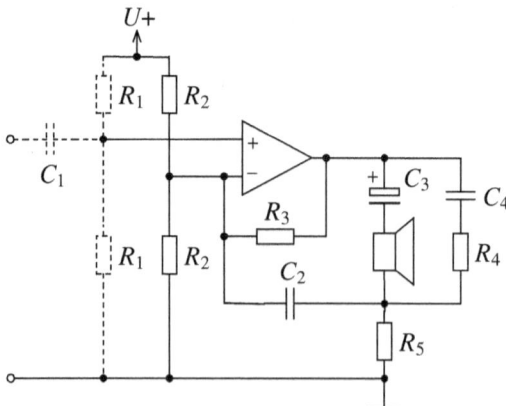

Figure 10.6. Transconductor topology suitable for one-sided supply voltage.
C_2 and C_3 separate the load from the direct voltages of the amplifier. The level
shifter stage indicated by dashed line is used if needed.

supply, the electrolytic capacitor used in series with the load does not degrade sound quality, as can happen in a voltage-output amplifier. In current-use, the non-idealities of the electrolytic are reflected only in its own voltage and are not able to affect the load current, as happens under voltage drive.

Also, the current-driven capacitor does not introduce a pole in the low-frequency response but, instead, assumes more voltage down to those frequencies where the voltage feedback takes control.

In terms of stability requirements, the circuit does not differ from the former two-sided supply case. The reactivity-reducing RC arm is preferably connected over the load and the series capacitor, as in Fig. 10.6.

10.6 Bridge Connection

The concept of bridging, much employed in voltage amplifiers, can also be applied to current-drive systems. In bridge connection, the speaker is fed as though from both ends which enables the load current and voltage to be 2-fold with respect to single-ended feeding with the same supply voltage.

Figure 10.7 shows a bridge-connected transconductor topology, in which one power amplifier (A_1) controls the current flowing through the load while an other (A_2) mirrors the output voltage of the former to op-

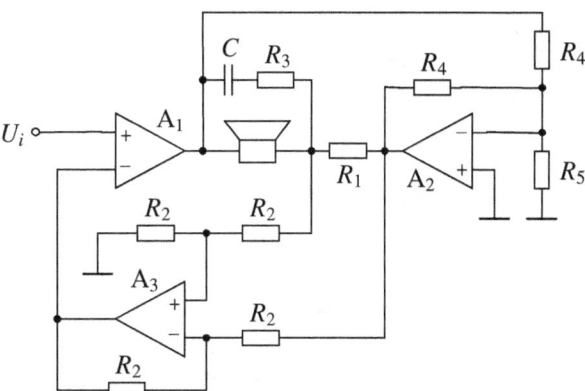

Figure 10.7. Transconductance amplifier with bridge connection, by which the supply voltage can be utilized maximally. With 8-Ω load, R_1 can be e.g. 0.27 Ω. A_1 and A_2 can be similar, and R_4 can be equal to R_2 (2-3 kΩ).

posite polarity. A_3 and resistors labeled R_2 make up a difference amplifier that provides a load current dependent feedback signal on the principle familiar from Fig. 10.5.

The current taken from the output of A_1 is now 2-fold compared to a
corresponding unbridged circuit. If the gain of the aforementioned difference amplifier is yet set to unity, as is reasonable, the value of the
current-sensing resistor (R_1) can be half of that one would use without
bridging; the feedback factor remaining then the same and the transconductance being 2-fold. Further, by a corresponding study that was used
in the derivation of equations (10.2) and (10.3), it can be shown that the
output impedance remains then about the same.

In terms of stability, the bridged circuit is somewhat more demanding than the usual. Of course, the difference amplifier and the inverting
amplifier must in themselves be stable. In addition, A_1 has to keep stable despite the phase lags developing in the aforementioned amplifiers
at high frequencies. Anyway, in experiments the circuit has been found
to be fully steady and well functioning at least with the used R_1 value of
0.33 Ω.

As for stability analysis, the circuit used for the difference amplifier
corresponds to a gain factor of 2. (Here, we consider the non-inverting
gain, as if the signal were applied to the plus input of the operational
amplifier.) Generally, small-signal operational amplifiers are stable or at
least can be compensated to be stable at this gain.

Without resistor R_5, the circuit of the inverting amplifier also corresponds in the stability sense to a gain factor of 2. If A_2 is stable at this
gain, R_5 can be left out. Else, R_5 should be chosen so low (often about
$1/10$ of R_4) that the stability of A_2 is secured. R_5 affects similarly as if
the raw gain of A_2 were decreased, that is, achieving stability is facili-
tated but at the expense of obtained bandwidth.

10.7 The Modified Howland Transconductor

Another usable way to realize transconductance amplifiers is the so-
called improved Howland current pump, that is represented in Fig. 10.8.
(The basic Howland circuit incorporates only 4 resistors so that R_4 is
replaced by a short.) Here, we also have a kind of current-feedback, but
now the difference voltage is established by means of resistor divisions.

The load can be ground-connected without additional arrangements,
but, as a small inconvenience, resistors need to be precisely matched to
achieve the best output impedance.

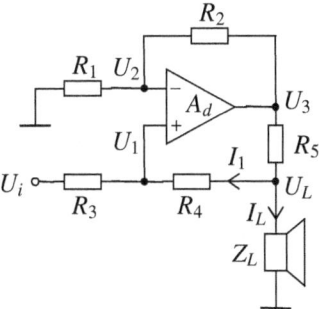

Figure 10.8. The improved Howland circuit, that is suited for currents required by loudspeakers. The circuit can also be made inverting by applying the signal via resistor R_1 and respectively connecting R_3 to ground.

Using the notation given, all information about the circuit's operation is contained in the following equation set:

$$\begin{cases} U_2 = \dfrac{R_1}{R_1+R_2} U_3 \\ U_1 = U_i + R_3 I_1 \\ U_L = Z_L I_L \\ U_L = U_1 + R_4 I_1 \\ U_3 = U_L + R_5 (I_L + I_1) \\ U_3 = A_d (U_1 - U_2) \end{cases} \tag{10.4}$$

The equations are six in number, as also the unknowns (U_1, U_2, U_3, U_L, I_1, and I_L).

After a bit laborious but routine solving, we obtain as the circuit's transconductance

$$g_m = \frac{I_L}{U_i} = \frac{A_d (R_4 (R_1+R_2) + R_1 R_5) + R_5 (R_1+R_2)}{((A_d+1) R_1 R_5 + R_2 R_5)(R_3+R_4) \dots} \tag{10.5}$$

$$\dots + Z_L [(R_1+R_2)(R_3+R_4+R_5) + A_d (R_1 R_4 - R_2 R_3 + R_1 R_5)]$$

The transconductance should, of course, be as independent as possible of the load impedance Z_L, so the absolute value of the square bracket expression in the denominator should be made as low as possible. This is achieved by marking $R_1 R_4 - R_2 R_3 + R_1 R_5 = 0$ which yields the condition:

$$\frac{R_1}{R_3} = \frac{R_2}{R_4 + R_5} \tag{10.6}$$

R_5 acts as the current-sensing resistor and has thus very low value compared to the other resistors. Hence, equation (10.6) is, in practice, satisfied fairly well even by choosing equal values for all resistances from R_1 to R_4.

However, the required matching precision is so high that the usual 1% resistor tolerance is not sufficient. With 0.1% tolerance instead, the load dependence of the transconductance already begins to remain at an acceptable level.

R_5 being very low, $|A_d|$ being very high, and condition (10.6) being satisfied, expression (10.5) is simplified to the form

$$g_m \approx \frac{R_2}{R_1 R_5} \tag{10.7}$$

from which we can see that when R_1 and R_2 are equal (implying R_3 and R_4 are also equal), the transconductance will be the reciprocal of the sensing resistor's resistance, as also in the basic topology of Fig. 10.1b.

By determining the idling voltage and short-circuit current of the output, we can yet solve the circuit's output impedance, obtaining

$$Z_t = \frac{A_d R_1 R_5}{R_1 + R_2} \tag{10.8}$$

However, the prerequisite for this is that condition (10.6) is satisfied accurately.

Thus, also here, the output impedance is directly proportional to the differential gain. However, due to the factor $R_1/(R_1+R_2)$, $|Z_t|$ always remains a little lower than in the earlier-discussed basic transconductor (equation (10.3)) with the same sensing resistor value. For simplicity, Z_o has however been ignored in this analysis.

The improved Howland circuit does not differ much from the simple current-feedback in stability properties either because the differential amplifier's load is the same in both and with zero input the difference voltage is in both the sensing resistor's voltage or a definite portion of it. Thus, using the RC compensation branch in parallel with the speaker is in the Howland circuit as recommendable as elsewhere.

In the Howland case, the stability margin is improved and the transconductance increases when ratio R_2/R_1 increases, as also happens in the circuit of Fig. 10.2 when ratio R_2/R_3 increases. However, in both, the output impedance decreases when so doing.

The requirement of accurate resistance matching also connotes that stray capacitances appearing in parallel with the resistors must not disturb the achieved balance at least in the operating frequency range. This is worth minding in circuit board design, but it is as well important that the resistor values are scaled low enough.

The stage preceding the Howland circuit must have very low output impedance because it directly adds to resistance R_3. Therefore, the feed must be taken directly from an operational amplifier output. Also, the Howland circuit loads the preceding stage by negative input impedance (positive U_i gives rise to positive I_1), so the feeding amplifier stage must be able to handle currents as though of wrong direction.

10.8 A Do-It-Yourself Project

In the following are presented complete building instructions for a current-output stereo amplifier, by which the friends of natural sound may partake in a novel enjoyment without having to await the awakening and coming forth of industrial practitioners.

The project is suitable for crafters having some experience. However, because the appliance operates on mains power, nobody should set about assembling the apparatus who is not absolutely sure about his/her competence in dealing with mains voltage parts.

The power amplifier circuit used is TDA2040, whose properties are suitable for the purpose and that worked in testing reliably at all signal levels, unlike certain examined operational amplifier chips, that exhibited weird low-frequency oscillation at high negative output currents.

Figure 10.9 shows the schematic of the amplifier with a suitable voltage supply, the corresponding parts list being given in Table 4. In addition to the transconductance stage, the amplifier includes the resonance compensation circuit presented in section 8.6 and an adjustable active baffle step compensation for loudspeakers into which this equalization has not been built.

The intaken signal is attenuated somewhat at the beginning by voltage divider R_1-P_1-R_2, so that voltage at the outputs of A_1 and A_4 would not become clipped in any situation despite the fact that the resonance compensation stage (A_1 with its feedback network) amplifies the lowest frequencies strongly. Overdrive should not occur if the peak values of the input signal are 3 V at most and the resonant frequency used (f_z) is

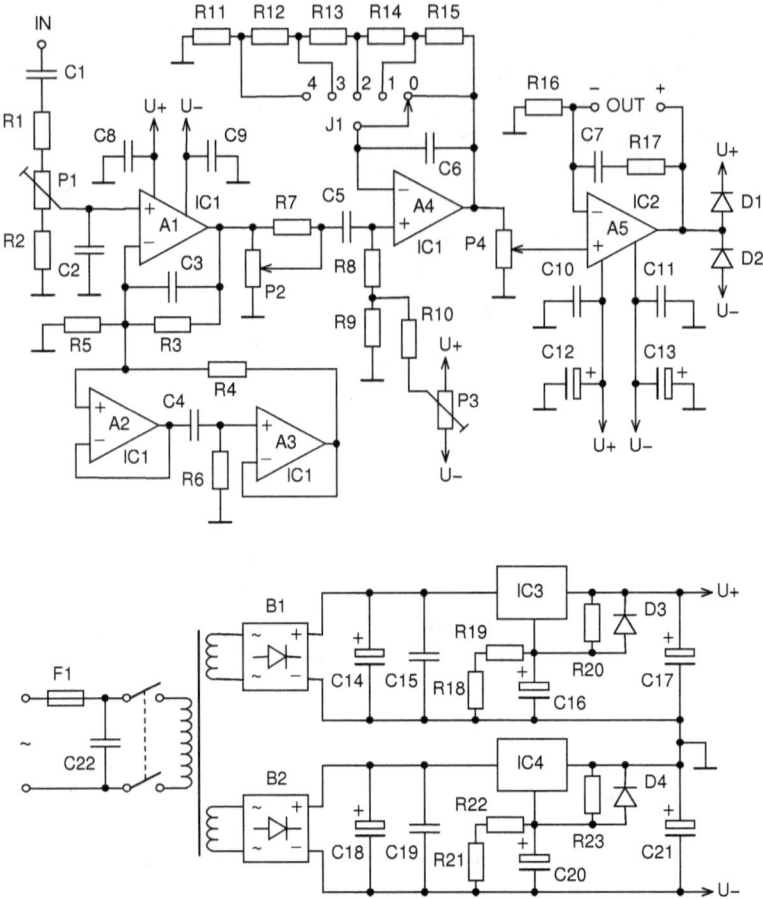

Figure 10.9. Circuit diagram for one channel of a transconductance amplifier equipped with a pole shifting equalizer and optional baffle step compensation, also showing the voltage supply common for both channels.

80 Hz at most. Trimmer P_1 is used for fine-tuning the channel sensitivities to equal levels.

Capacitor C_2 is for the filtering of possible high-frequency interferences. With a value of 68 pF, the cut-off frequency will be about 180 kHz.

The pole shifting circuit itself is the same as shown in Fig. 8.14a, so the setting of the resonance parameters can be performed using the map of Fig. 8.14b, only regarding that the adjusted resistances are here la-

Table 4. Parts list for the Fig. 10.9 circuit diagram. Resistor values are from the E24 series and have 1% tolerance unless otherwise stated.

R1 .... 18kΩ	C1 .... 0.1μF; 5%
R2 .... 30kΩ	C2 .... 68pF
R3 .... 200kΩ	C3 .... 0.1μF; 5%
R4 .... 4.3kΩ	C4 .... 0.1μF; 5%
R5 .... see text	C5 .... 0.1μF; 5%
R6 .... see text	C6 .... 0.1μF; 5%
R7 .... 3.0kΩ	C7 .... 22nF
R8 .... 51kΩ	C8, C9, C10, C11 .... 1μF
R9 .... 1kΩ	C12, C13 .... 100μF; 50V
R10 .... 750kΩ	C14, C18 .... 10 000μF; 50V
R11 .... 7.5kΩ	C15, C19 .... 0.1μF
R12 .... 1.5kΩ	C16, C20 .... 1000μF; 25V
R13 .... 1.6kΩ	C17, C21 .... 22μF; 50V
R14 .... 2.0kΩ	C22 .... 0.1μF; see text
R15 .... 2.4kΩ	D1, D2 .... 1N4001
R16 .... 0.56Ω; 2W; 5%	D3, D4 .... 1N4001
R17 .... 18Ω; 1W; 5%	IC1 .... TL074ACN, TL074IN
R18, R21 ...... 1.5kΩ; 0.25W	IC2 .... TDA2040
R19, R22 .... 56Ω	IC3, IC4 .... LM338K
R20, R23 .... 120Ω	B1, B2 .... rect.bridge; 2A
P1 .... 5kΩ	F1 .... 1A; slow
P2 .... 10kΩ; lin	Transformer 2x18V; 100VA
P3 .... 100kΩ	
P4 .... 5kΩ; log	

beled R_5 and R_6. The map has been constructed for an f_z range of 40-80 Hz, but higher resonant frequencies can also be used. R_5 and R_6 can be attached to the circuit board by means of screw terminals, so that the values are quite easily alterable when the speaker is changed.

The equalizer sets the characteristic frequency of the poles to about one-seventh of the original resonant frequency, as was discussed in the context of Fig. 8.14. A corresponding drop in the system's cut-off frequency would too easily cause the woofer to run out of excursion, so the signal yet needs to be high-pass-filtered to keep the frequency range moderate.

This confinement is performed by two passive 1st-order filters, one of which is formed by C_1 with the resistor chain R_1-P_1-R_2 and the other by C_5 with resistors R_8 and R_9. The corner frequency of both filters has been set to 30 Hz, at which point the total attenuation will thus be 6 dB. So, the final bass response always extends to 30 Hz, the Q value being 0.5. If desired, this frequency can be changed by scaling the capacitors C_1 and C_5.

When using speakers with built-in resonance compensation, active compensation is no more needed. In this case, the amplifier's frequency response can be leveled so that the zeroes introduced by the equalizer are used to cancel out the above-described 30-Hz poles. This canceling is simply achieved by setting $f_z = 30$ Hz and $Q_z = 0.5$ which makes the transconductance at audible frequencies next to flat. The point is not seen in the map of Fig. 8.14b, but the needed resistances are $R_5 = 32$ kΩ and $R_6 = 750$ kΩ.

Potentiometer P_2 acts as the balance control. Each channel has its own potentiometer on a common axis, connected so that while one is turned open, the other turns shut. Thereby, one avoids the interchannel crosstalk, possible in the usual balance control circuit, stemming from current partially drifting past the grounded slide. The effect of R_7 is that attenuation at the potentiometer's middle position is about 3 dB when otherwise the attenuation would be 6 dB.

A_4 with its feedback forms the baffle step compensation stage and also acts as a buffer, feeding the volume potentiometer P_4. With jumper (or switch) J_1, one can select either an unmodified response (0) or one of the four compensation levels that are shown in Fig. 10.10.

At position 4, low frequencies are boosted a full 6 dB, the different options being arranged at 1.5-dB spacing. Because, due to power response aberration, it is generally not worthwhile to eliminate the baffle step wholly, option 3 or 2 is often the most appropriate for frequency balance.

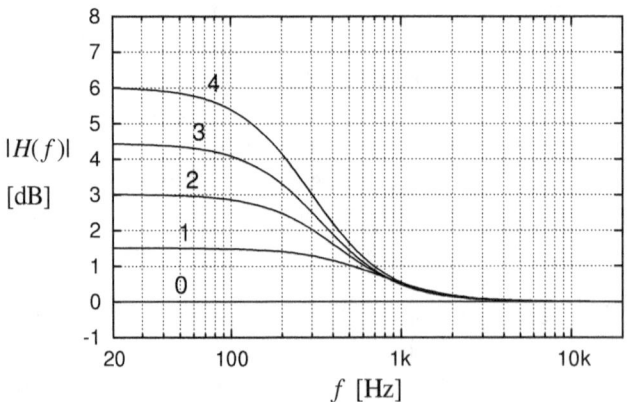

Figure 10.10. Amplitude response options available in the baffle step equalizer used in the Fig. 10.9 amplifier. At position 0, A_4 acts as a mere voltage follower. Full 6-dB compensation is achieved at position 4.

The center frequency of the step has been set to approximately correspond to a speaker width of a good 20 cm. If the size differs much from this, the frequency can be scaled by changing the value of C_6.

The value of the volume potentiometer P_4 is only 5 kΩ, so that the impedance feeding the power amplifier would stay low enough at all slide positions. A potentiometer of 3-4 kΩ would be, in this respect, yet more ideal, but such ones are not available.

By trimmer P_3, one can reduce the direct current flowing to the load, stemming from the offset voltages of A_4 and A_5. However, due to the potentiometer between these stages, the direct current cannot be entirely nulled except at some specific position of P_4.

The offset voltages of A_1 and A_3 have also significance although C_5 blocks them from affecting the output. Namely, both voltages appear at the output of A_1 about 50-fold; so, for example, with 10 mV offsets, the direct voltage accruing at the output of A_1 can be 1 V. Therefore, it is advisable to use the given circuit types, whose offset voltage should be at least less than 6 mV. The operational amplifiers have to have JFET inputs in order that the voltage drops due to the input currents in various resistors (like R_6) would not introduce in the signal large DC components either.

The purpose of diodes D_1 and D_2 is to protect the power stage from possible overvoltage peaks caused by an inductive load. Rising of the output voltage over the supply voltage can be possible if the connection to the load for some reason becomes loose or the amplifier is heavily overdriven.

C_8 ... C_{13} are normal supply voltage line stabilization capacitors, that are always best to be situated near the circuit whose voltages are to be steadied. The electrolytics C_{12} and C_{13} mostly have effect in the region below 100 kHz, whereas the 1 µF capacitors are for higher frequencies.

The best choice of dielectric material for the signal path capacitors C_1, C_3, C_4, C_5, and C_6 is polypropylene, whose non-idealities are considerably lesser than those of polyester. Polypropylene capacitors are generally a little bigger, but the matter is irrelevant here. Polycarbonate would also be a decent material for audio use, but for some reason is no more manufactured.

The power supply has been implemented using the adjustable regulator LM338, whose current supplying capability is more than adequate for the purpose. Because negative regulator chips are not available for high currents, the negative side has also been realized with a positive regulator. The arrangement works anyway all well.

Using a regulated power supply also for the power amplifier is not so common today but is, however, well justified here because, besides avoiding hum problems, with a regulated supply the operating voltage range of the amplifier chip can be utilized better than with an unregulated one.

The maximum allowed supply voltage for the TDA2040 is ±20 V. Without regulators, establishing a suitable and safe voltage would be troublesome because the secondary voltage ratings of transformers vary in steps of 3 V (9 V, 12 V, 15 V, 18 V) and the peak values correspondingly in steps of over 4 V and because the actual secondary voltage can be under free-running condition significantly higher than the nominal value and because the mains voltage may vary in both directions many percents and because the voltage dropped in the rectifying diodes essentially depends on the current flowing through them. If, in view of these considerations, we left an adequate safety margin against the 20 V limit and reserved yet a couple of volts room for the ripple, the obtained output power would be left unnecessarily low.

It is vain to suspect anything such that a regulator circuit would not be able to deliver current fast enough in all possible transient situations. To prove this, it is enough that the voltage variations remain practically negligible with the highest load current and the highest operating frequency. The use of regulators does not endanger the amplifier's stability either, as long as the necessary bypass capacitors have been appropriately connected.

Moreover, regulation suits for a current-output amplifier better than for a voltage-output one because in the first-mentioned, the required peak currents are at the same power level much lower due to the fact that the reactivity of the load no more affects the instantaneous values of current.

The supply voltage has been set with resistors R_{18}, R_{19}, R_{21}, and R_{22} to ±17.5 V, so that the same source can also be used for the operational amplifier chip, that can withstand ±18 V. Amplitude peaks are then able to reach ±15 V, which yields an output power of 12 W into 8 Ω load. One could obtain a little more power by raising the supply voltage of IC_2 to some ±19.5 V, but then IC_1 would require an own supply.

The values of C_{16} and C_{20} are much higher than usually in order to curb the power-on snap. The output terminals of the regulator develop a voltage of a good volt instantaneously after power-on, but due to the slow charging of these electrolytes via R_{20} and R_{23}, the supply voltage reaches its full value only after a couple of seconds.

Typically, regulators make a feeble whistling sound when loaded in a suitable frequency range. The phenomenon occurs, however, also in other power semiconductors, and the sound is not so loud as to have practical significance.

The purpose of diode D_3 is to protect the regulator in case of a short circuit by preventing the discharge current of C_{16} from flowing via the regulator's adjustment terminal.

C_{22} must be a so-called class X capacitor, intended for mains voltage interference suppression. Ordinary capacitors should not be used in the place.

The voltage rating of electrolytics C_{12}, C_{13}, C_{14}, C_{17}, C_{18}, and C_{21} has been set to 50 V although the voltage they actually assume is much lower. The reason for this is that 50-100 V electrolytics generally have a better dissipation factor than types of lower or higher voltages. There is nothing harmful in such undervoltage use.

Figure 10.11 shows the top layout view for a two-channel amplifier, using the circuit board that has been shown in actual size in appendix F. The layout view for the power supply is in Fig. 10.12 and the artwork respectively in appendix G.

All external connections to the boards are purposed to be made using screw terminals except the inputs, for which are suggested gold-plated pin headers with corresponding female crimp terminals. To improve mechanical firmness, 3-way terminals have also been used for many 2-way connections.

For resistances R_5 and R_6, there are used 4-way screw terminals to enable series connection since it is unlikely that the needed resistances would coincide with commercial standard values. When mounting the resistors, one must remember to remove from their leads all adhesive residues coming from the tape.

It is better not to attach to the screw terminals very stiff wire but nor very thin. Also, the wire strands should be preferably tin coated since plain copper oxidizes quickly degrading contact reliability. It is well to check the tightness of the screws yet the next day after mounting, for they have a tendency to loosen a little.

The supply voltage wires can be drawn for both channels separately from the power supply board, the ground wire being common. All five should be bundled together to reduce inductances; and also in general, all wires should be kept as short as possible. From the GND terminal of the amplifier board is also drawn contact to the chassis.

The input signals are brought to the possible selector switch and the amplifier board by screened cables. To avoid ground loops, the bodies of the RCA connectors are isolated from the appliance chassis.

When using separate heat sinks for both power amplifiers, the ther-

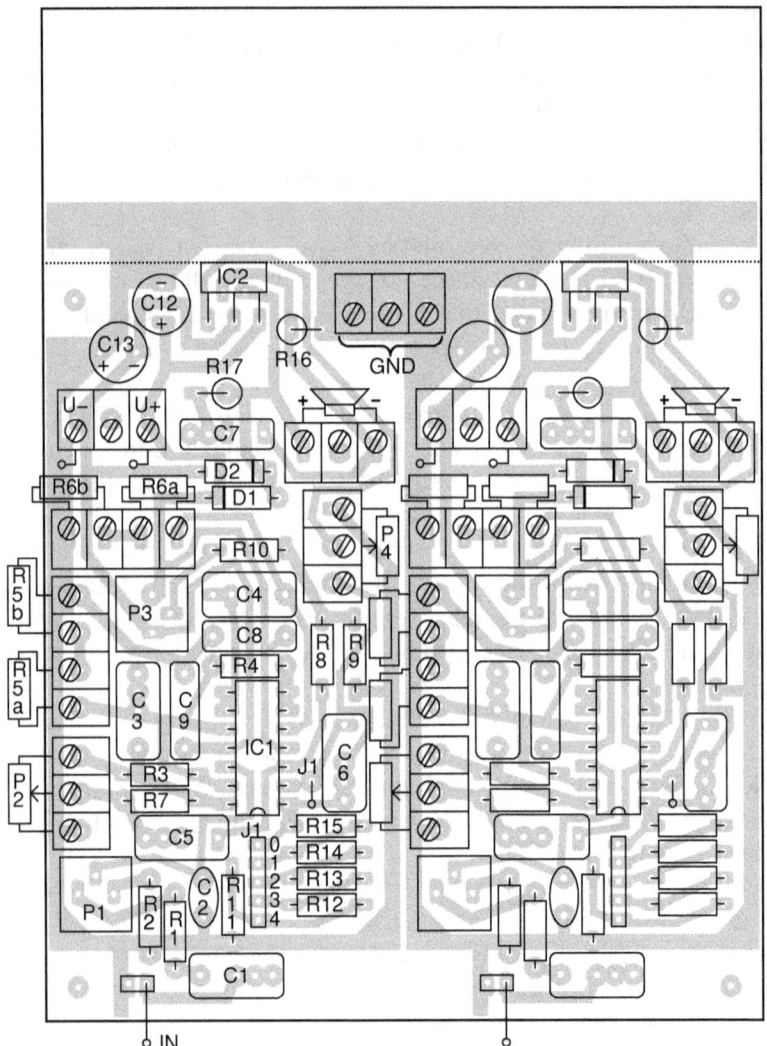

Figure 10.11. Layout design for the transconductance amplifier shown in Fig. 10.9. The components have been named only for one channel, but the other is furnished the same way. The rectangles marked with a screw head represent screw terminal cells. The inputs and jumpers J_1 may be connected by means of pin headers. For some capacitors, there has been provided different pitch options. C_{10} and C_{11} are not seen in the picture but are situated on the copper side directly from the supply voltage pins of IC_2 to the ground foil.

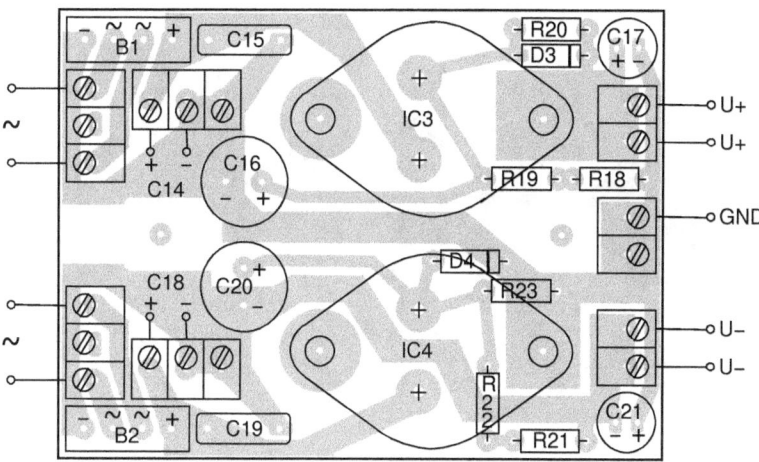

Figure 10.12. Layout view for the power supply shown in Fig. 10.9. The regulators are purposed to be fixed with their heat sinks on top of spacers above the board.

mal resistance should be no more than about 5 °C/W and with a common sink, correspondingly, half of that. This suffices well for music listening even though not necessarily for the severest testing. The lid of the apparatus must naturally have adequate ventilation holes.

When using a common heat sink, it is advisable to isolate the amplifiers electrically from the sink, so that it would not convey operation currents of the amplifiers. The heat sink is then connected to ground potential.

The thermal resistance of the regulator heat sinks should be at least less than 10 °C/W. If directly suitable ones are not available, several smaller ones may be attached together.

In the installation of the regulators, one has to see to it that the case, which forms the output terminal, has reliable contact with the pertinent copper foil, minding that the blackened heat sink surface or an oxidized copper surface may form an insulating layer. If needed, a separate short wire should be used.

Figure 10.13 shows the interior of a completed amplifier. The box could be somewhat larger, so that the power supply board would also fit horizontally.

From three accommodated input sources, the selected one has own output connectors, e.g. for a recorder. However, it is important that to

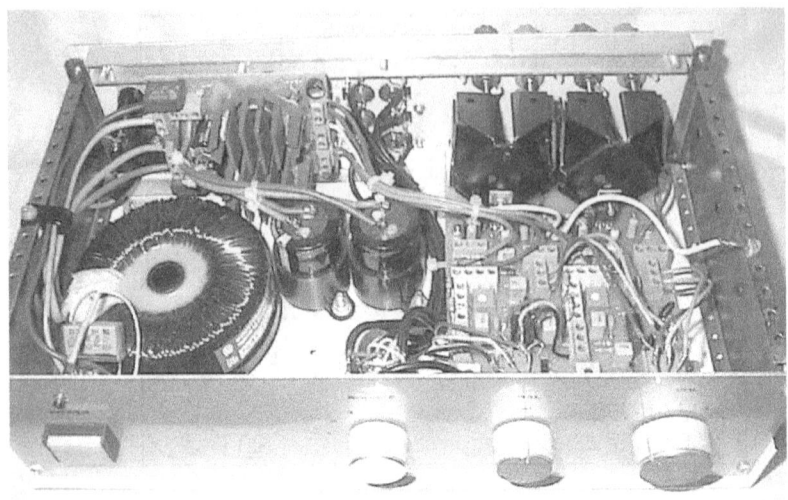

Figure 10.13. Embodiment of the project amplifier.

such parallel interfaces are not attached any unpowered devices because the load presented by them can be indefinite.

The selector switch and the potentiometers should be of higher than ordinary quality if they are desired to preserve their contacts still after years.

10.9 Using Headphones

Current-drive can also be applied for electro-dynamic headphones though the achieved benefits are usually very minor compared with the improvement brought about in loudspeaker operation. This is mainly because in the impedance of headphone transducers the relative proportion of the DC resistance is generally much higher than in speaker drivers, so the current components produced by the electromotive forces are intrinsically left small.

In headphones, a greater problem is generally constituted by the unevenness of frequency reproduction and its dependence on the ear canal shape. A feasible solution to this might involve a personal equalizer that would be tuned by ear with a sine wave source.

As with loudspeakers, the frequency response of headphones also exhibits certain changes when moving to current-drive. These changes may, of course, also result in undesirable impressions.

If only low listening levels are needed and when the current-sensitivity of the headphones is sufficient, the current can also be taken directly from the line outputs of the program source. This enables listening from appliances that don't have a headphone output. To limit current, a series resistor is needed; a suitable value being 2 kΩ or more. Lower values may burden the output stage excessively.

In principle, the amplifier introduced in the previous section can also be used to drive headphones. It is, however, as such too powerful for the purpose, thus endangering hearing and hampering the adjustment of volume. Moreover, the usual stereo headphone connection, where one pole is common for both channels, could not be used.

With the auxiliary arrangement shown in Fig. 10.14, these problems can, however, be set aside. In headphone listening, the loudspeaker is replaced by a dummy load, R_2, so that the amplifier operates as a usual voltage amplifier, and the headphone is fed in the traditional fashion via a series resistor (R_4). R_2 can be chosen to be e.g. $100 \cdot R_1$, also minding power dissipation. A suitable value for R_4 is around 1 kΩ.

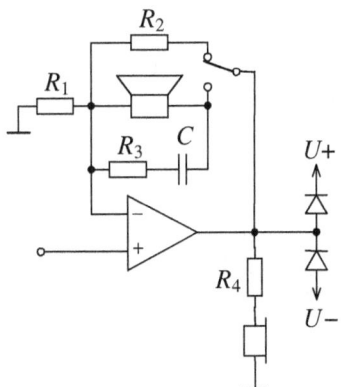

Figure 10.14. Providing a headphone output in a transconductance amplifier. With the switch is selected either speaker operation or headphone operation, in which R_2 acts as a dummy load.

[1] P. G. L. Mills and M. O. J. Hawksford, "Distortion Reduction in Moving-Coil Loudspeaker Systems Using Current-Drive Technology", *Journal of the Audio Engineering Society*, vol. 37, March 1989, p. 129-148.

[2] P. G. L. Mills and M. O. J. Hawksford, "Transconductance Power Amplifier Systems for Current-Driven Loudspeakers", *Journal of the Audio Engineering Society*, vol. 37, October 1989, p. 809-822.

11

SPEAKER REALIZATIONS

11.1 Principles of Enclosing

In current-drive systems, we are mostly interested in sealed enclosure structures. The absence of the bass reflex port has many advantages, of which all are not even commonly recognized.

One significant detriment concerning reflex vents is that they form an open channel for all internal cabinet noise to get out and mix with the actual sound signal. Further, it is customary to shape the tube ends horn-like to reduce turbulences and linearize operation which, for its part, yet enhances the leak in the mid-frequency region. The cross-sectional area of the tube is though generally much smaller than the cone area of the driver; but on the other hand, the mass acting in the tube is negligible with respect to the cone mass, so the sound leakage occurring through the tube can be practically of the same order of magnitude than the cone leakage, which was shown, in section 4.2, to be very significant.

Reflex tubes also have a tendency to rattle themselves by the effect of transverse forces. The joint between the tube and panel namely easily resonates unless the glueing has been done very carefully. The rattle may only show up at one low frequency, so in music playing the effect is easily left unidentified.

In terms of cabinet firmness and tightness, at least the same requirement level should be followed for which it has been strived in comparable class voltage feed speakers, since the giving in and resonation of the walls naturally rise more easily as significant error sources, once having managed to remove or reduce other aberrations. On the other hand, by ample stuffing, the interior sound pressure can be considerably decreased, especially in the mid-range (Fig. 4.3), which reduces the resonation ills.

The most suitable material especially for hobbyist use is MDF board, but particle board, plywood, and joint panel are also decent. A cabinet can also be constructed out of metal or even stone if skills and tooling are found. Instead, out of plastic materials cannot be made sufficiently rigid structure for any seriously taken hifi purpose although they are regrettably common in supermarket products.

A recommendable thickness for the MDF board is 19 mm for interior volumes up to some ten litres. In larger enclosures it is justified to use 22 mm board. Additionally, the walls can be strengthened with supportive braces and panels, as needed.

It is worthwhile to also pay attention to the screw threads. A screw with too steep a thread may loosen in the course of time due to vibration, and a too gradual one easily breaks the hole, losing its hold.

In major part of loudspeakers on the market and building instructions for hobbyists, the cabinet is so narrow that the driver hardly fits it without hitting the side walls. Many times, the flange has yet been truncated to nip off the last millimetres. The practice is, however, somewhat disadvantageous from the standpoint of both front and back radiation.

In a narrow cabinet, the edge of the front panel comes close to the cone (Fig. 11.1); a thing which is apt to reinforce the diffraction experienced by the sideward radiation. Also, there is established a relatively sharp angle between the cone surface and the side wall although in general, round edge shapes should be preferred.

In a narrow cabinet, the backward radiation of the driver is also a little stronger because the side walls act like a horn at certain mid-frequencies. The most disadvantageous case in this respect is naturally that

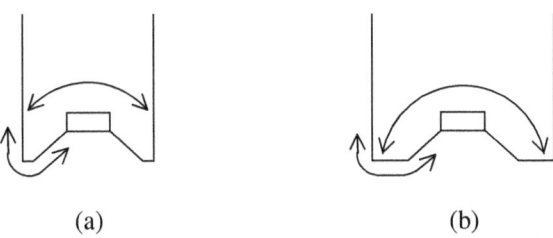

(a)   (b)

Figure 11.1.   Effect of enclosure width on the acoustic environment of the driver. a) In a narrow enclosure, the sideward radiation encounters a sharp angle, and the solid angle of the back radiation is left small. b) In a wide enclosure, no sharp angle is formed, and the solid angle of the back radiation is greater.

the bottom and top panels also come close to the driver. A cramped or tubular back space is not for good at low frequencies either because it can develop unwanted reflex effects, that colour reproduction.

In sealed speakers and especially in current-driven ones, the enclosure volume is not a very critical factor, and increasing it only expands the frequency range, so there can be left moderately margin on the sides of the drivers, and there is also room for rounding the cabinet, if desired. For laymen, it should naturally also be explained that the cabinet is not merely for keeping the drivers fixed, but the shape and size have an essential import to the quality of reproduction.

In sections 5.4 and 5.5, it already became clear that, in multiway systems, it is important to control the distance difference of the drivers' acoustic source points in the listening direction at the frequencies where the drivers work together. This distance difference is, however, not easy to be determined from the drivers beforehand very accurately, for the phase behavior is also affected by the enclosure shape. If it is afterwards found that the difference did not quite fall in place, the matter can nevertheless be helped by changing the recommended listening altitude relative to the speaker.

The phase of the upper-frequency driver can be lagged with respect to the lower-frequency driver by stepping the front panel or inclining it backward, as shown in Fig. 11.2. Inclination is the better alternative out of these, for the step in itself effects changes notably in the upper-frequency driver's response, thus hampering the matching. Also, because the step has to be gradual, the drivers end up quite far from each other.

The inclination angle is mostly limited by the fact that the tweeter should not be directed very much aside from the listener. A small aside orientation can be even beneficial if the highest frequencies need atte-

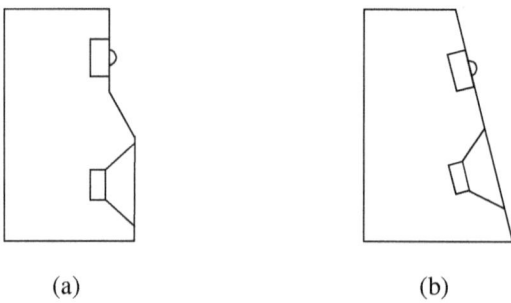

(a)   (b)

Figure 11.2. Two ways to align the acoustic source points of the drivers. a) Stepped front panel. b) Inclined front panel.

nuation. Inclining the bass-midrange driver can also be advantageous as the floor reflections weaken a little.

Enclosing increases the resonant frequency and Q value of a driver the more, the smaller is the volume since the spring constant of the air spring, being inversely proportional to the volume, is added to the driver's own spring constant. When the volume equals the equivalent volume of the driver, these springs should be equally stiff, so the spring constant of the air can be expressed as $k_a = k_d V_e/V$ where k_d is the driver's spring constant, V_e is the equivalent volume (of the driver), and V is the enclosure volume.

Thus, we obtain as the resonant frequency of an enclosed driver, on the grounds of equation (3.5),

$$\omega_0' = \sqrt{\frac{k_d + k_a}{m}} = \sqrt{\frac{k_d(1 + V_e/V)}{m}}$$

$$= \omega_0 \sqrt{1 + \frac{V_e}{V}} \tag{11.1}$$

where ω_0 is the resonant frequency in free air.

Consequently, when using, for example, an enclosure of the size of V_e, the resonant frequency becomes, in principle, $\sqrt{2}$-fold and with a size of ½V_e, respectively, $\sqrt{3}$-fold.

The equivalent volume is proportional to the square of the effective area, so the relative increase in the resonant frequency is the least with small drivers. On the other hand, ω_0 increases as the moving mass decreases, so the final resonant frequency does not, in practice, depend very much on the driver size.

The Q value is, according to equation (3.6), also directly proportional to the square root of the spring constant, so we obtain as the Q value of the enclosed driver, correspondingly,

$$Q' = Q \sqrt{1 + \frac{V_e}{V}} \tag{11.2}$$

In practice, formulas (11.1) and (11.2) are only estimative because the enclosure walls are not wholly ideal, nor are its dimensions totally negligible relative to the wavelength.

By increasing the enclosure, both parameters can thus be decreased, but on the other hand, at large displacements the air spring functions much more linearly than the driver's own suspension which speaks for keeping the enclosure smaller than the equivalent volume.

11.2 Damping Materials

Besides effectively relieving the interior sound pressure, damping material can also make the Q value much lower because the stuff, when impeding air motion, increases the damping constant b. By effective stuffing, b can be increased even so that the final Q ends up lower than in a free driver.

Damping material also lowers the resonant frequency itself because the air compression process changes from adiabatic to isothermal and the adiabatic index, that affects the spring constant, is left out (see equation (7.7)).

In the following are presented some results of the effect of various materials as modifiers of resonance properties. The experiments have been made in a firm 10-litre cabinet, using a 15-cm driver manufactured by Seas, CA15RLY, for which the specified resonant frequency is 44 Hz and the mechanical Q value 1.88; the measured values being, at 50 mA current, 54 Hz and 2.27.

The impedance curves in Fig. 11.3 have been measured using, as the damping material, sheet cloth (Fig. 'a'), polyester fibre (Fig. b), plastic foam (Fig. c), and household tissue paper (Fig. d) at different filling degrees. The curves indicate illustratively how the adding of damping material changes the resonance parameters.

In all cases, there has been left some empty space around the driver. If this is not done, low mid-frequencies may exhibit response unevenness that shows up in the impedance curve as small bumps.

With an empty cabinet, the resonant frequency rises to 76 Hz, from which it decreases with both the cloth, fibre, and foam, at full stuffing, down to 71 Hz, or about 7%, which corresponds quite well to the theoretical maximum drop.

With the household tissue, the peak frequency can be made distinctly lower; here, down to about 65 Hz. However, the curve becomes then asymmetric which denotes that the system is no more purely of second order. Soft tissue is, nevertheless, a usable and, above all, inexpensive alternative for enclosure stuffing.

The Q value rises in the empty cabinet to 3.37, that is, a little more, in proportion, than the resonant frequency. The ability of the damping materials to decrease the Q varies, however, a lot.

The most effective Q value lowerer is here the cloth while the se-

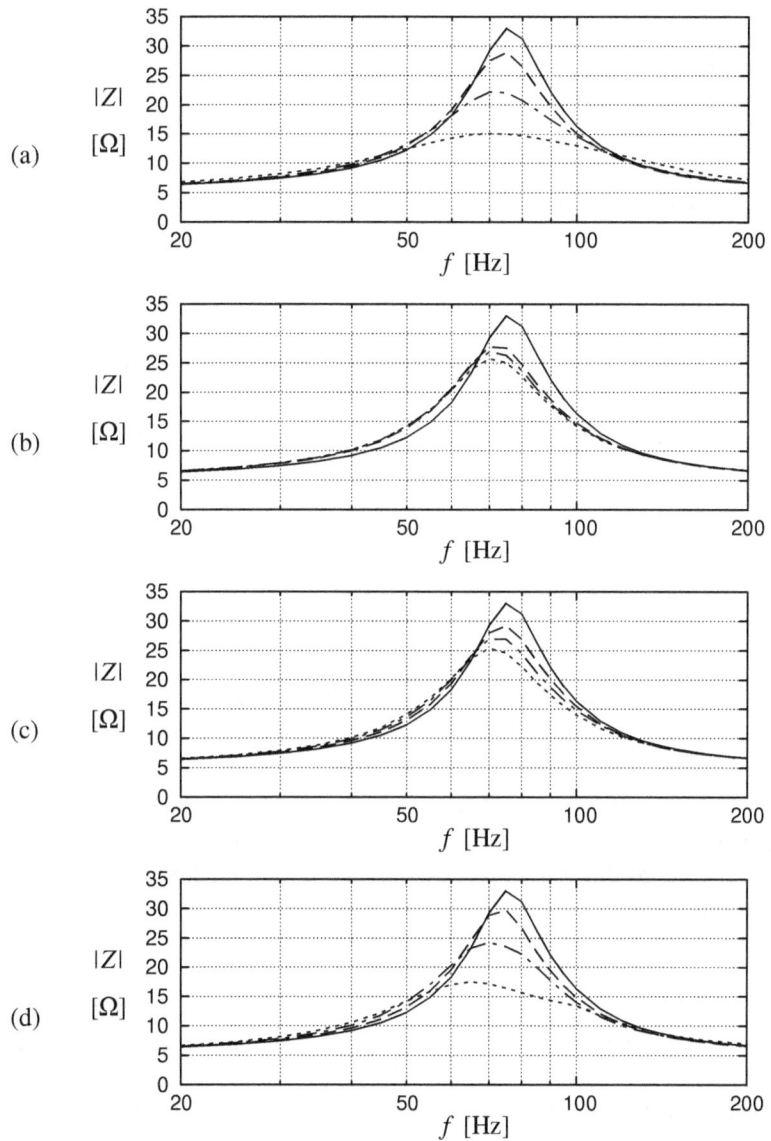

Figure 11.3. Effect of different amounts of damping materials on the impedance curve of a 5½-inch example driver in a 10-litre cabinet. a) Cotton cloth. b) Polyester fibre. c) Plastic foam 20 kg/m³. d) Household tissue. Solid lines: cabinet empty. Dot lines: cabinet full-packed (100% amount of substance). Dash-dot lines: 80% amount of substance. Dash lines: 50% amount of substance.

cond most effective is the tissue. With the cloth, the Q descends at full stuffing to 1.07 and also with the tissue to 1.42 (assuming 2nd-order behavior). Instead, with the fibre, we are left at 2.36 and with the foam, at 2.31. Nevertheless, also the last-mentioned are yet usable, at least in conjunction with active equalization.

Both the resonant frequency and Q decrease monotonically as the filling is increased, even if the cabinet were crammed full, like in the dot line cases. It is thus seen that:

The performed tests with four different damping materials in a 10-litre enclosure do not support the notion according to which the amount of damping material would have an optimal value above which the resonant frequency would again begin to increase for the reason that the material eats enclosure volume. Not even a relatively high-density filling, like tightly packed cotton cloth, seems to induce such a turning.

The stiffness of the air spring though stops decreasing when the process has become isothermal, but any increase in the stiffness, as the filling is increased, cannot be observed.

11.3 Usage of Electrolytic Capacitors

Electrolytic capacitors, or electrolytics, have customarily been used in the signal path quite carelessly, regardless of their very high dissipation factor and other inaccuracies. For cost reasons and in view of the existing voltage drive practice, this is also understandable; but there are also good reasons for which, when sound quality is paramount, electrolytics should be avoided in signal processing and transferring:

- In 10-100 µF electrolytics, typically used in crossover filtering, the dissipation factor is at 1 kHz generally around 10% and at 10 kHz already several tens of percents. Crossover frequencies being, say, in the 2 kHz region, such extra resistance can already hamper the functionality of the filtering.

- A high dissipation factor and the large variations occurring in it make circuit optimization troublesome since, in order to achieve reasonable operation accuracy, the resistive component cannot be ignored and, on the other hand, its effect can be predicted only coarsely.

- The higher the dissipation factor is, the greater are also possibilities for the occurrence of hysteresis and other distortion effects; and the linearity of a resistance that is formed by a liquid may not be the best possible, even though the matter is not easily proved by measuring.

- The tolerance of electrolytics is generally ±20%, which should be regarded as insufficient for hifi use. From specialized manufacturers, there are though available 10% types, but this too is more than usually in other components.

- Two-sided, or bipolar, electrolytics that are intended for loudspeaker use are generally available for a voltage of 100 V, at most. When using a powerful amplifier, this can be too little, for the instantaneous voltage values in a capacitor can rise much higher than the voltage delivered by the amplifier. This happens especially when a capacitance reacts with inductance. With a series connection, the voltage strength can, however, be increased since AC voltage divides evenly between capacitors of equal size. (By contrast, DC voltage does not, in general, divide evenly but in proportions determined by the leakage currents.)

- Electrolytics are relatively unreliable components. Their lifetime is though prolonged as the operating voltage is decreased, but aging may manifest itself only as a creeping increase in losses, that easily goes unnoticed.

- A common phenomenon, notably in bipolar electrolytics, is that when charging an electrolytic that has been almost empty for a few minutes, the capacitance rises, during the charging phase, even to a double value with respect to normal. As a consequence, there occurs a momentary aberration of the signal every time the absolute value of the capacitor's voltage reaches a new maximum for a few minutes.

The last-mentioned property is explainable by considering the structure and equivalent circuit of an electrolytic, shown in essential parts in Fig. 11.4.

In actuality, every aluminum electrolytic capacitor consists of two series-connected capacitors, whose dielectric is made of an oxide layer grown on the surface of the aluminum electrodes. The paper-impregnated conductive liquid, that is left between the dielectric layers, forms the common node of these capacitors and also most part of the series

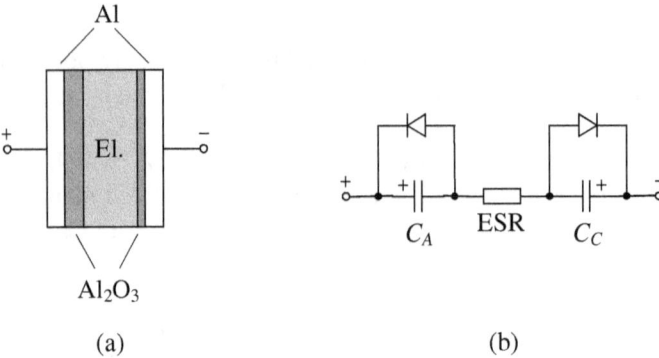

Figure 11.4. a) Functional layers of an aluminum electrolytic capacitor. Two aluminum oxide layers form the dielectrics, between which there is a conductive liquid, i.e. electrolyte. b) Fundamental equivalent circuit of the Fig. 'a' structure. The diodes model oxide leakage currents during wrong-polarity bias. The effective series resistance (ESR) is mainly due to the electrolyte's resistivity.

resistance (ESR).

Normally, the oxide layer on the anode side is much thicker and determines the capacitance of the entire series connection. The capacitance on the cathode side (C_C) is very high, relative to the former, and thus does not show much externally. Bipolar electrolytics differ from this only so that, in them, both capacitances have been made equal, resulting in symmetric structure.

Aluminum oxide somewhat leaks if the voltage across it assumes wrong polarity, that is, if the electrolyte becomes positive relative to the aluminum. This leakage has been represented in the equivalent circuit by diodes, that by no means are ideal but quite varying in conductivity.

In an ordinary polar electrolytic, the cathode oxide is so thin that it withstands voltage only up to 1.5 V or so. This also establishes the limit for the device's tolerance of negative voltages, for when the anode side becomes negative, its leakage diode starts to conduct.

When charging an initially chargeless bipolar electrolytic, at first the voltage attempts to divide evenly between both oxide capacitors, but because the capacitor next to the minus pole is charged to the wrong polarity, the associated leakage diode turns conductive and, in practice, shorts this capacitor. The plus side's capacitor, instead, gets charged normally, but because its capacitance is double, relative to the device's nominal value, the total capacitance is, in principle, also doubled during the charging.

When, after the charging phase, the electrolytic starts being discharged, the said short circuit is removed, and the capacitance normalizes because the potential of the electrolyte becomes negative with respect to both terminals, thus keeping the capacitor voltages in correct polarity.

When the same electrolytic is charged anew, the voltages of the capacitors remain correctly polarized up to the point where the external voltage exceeds the formerly reached maximum, after which the same diode again turns conductive and the capacitance again doubles. When the electrolytic is charged in the other direction, the leakage limit is met at the same voltage magnitude but is now due to the other diode.

However, with time, the negative potential of the electrolyte lessens by itself, correspondingly decreasing the threshold of the leakage. The recovery time may vary from minutes to hours.

Figure 11.5 displays some charging curve families, measured with a parameter analyzer, from ordinary bipolar electrolytics; revealing plainly the leaking above a certain threshold voltage. (For practical reasons, the charging has been made relatively slowly, but this should not have any conclusive import.)

When charging a capacitor by a constant current, the voltage should rise linearly with a slope I/C (I = charging current), but in the measured electrolytics, the curves bend even more than can be expected based on the simple model presented above. The results are not necessarily all typical but nor anything rare.

By close inspection, it can also be observed that even the graphs obtained without leakage are not completely straight but slightly arced.

In Fig. c, the charging almost entirely ceases as the leakage threshold is exceeded. The short-circuiting of one oxide capacitor thus also seems to somehow hamper the functioning of the other. Whether it be due to a general feature or some quality problem, this caprice too does not increase confidence in electrolytics in critical applications.

11.4 Project CS-12

In the following, there is presented, as a building project and design example, a 12-litre 2-way speaker (Current Speaker 12), that relies on actively operating response compensation and is therefore well suited for use with the amplifier presented in section 10.8. Even though it has been difficult to find tweeters that are well suited for current-drive, the result of the project is, nevertheless, a fully decent unity out of those wares that are commonly available.

In the model units, the bass-mid-frequency region has been served

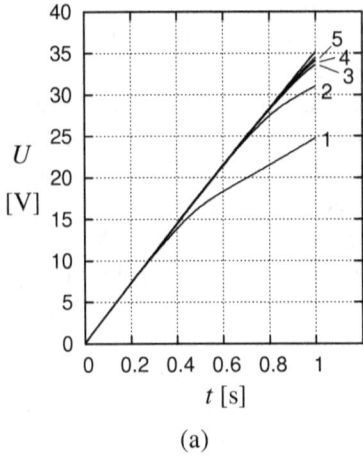

(a)

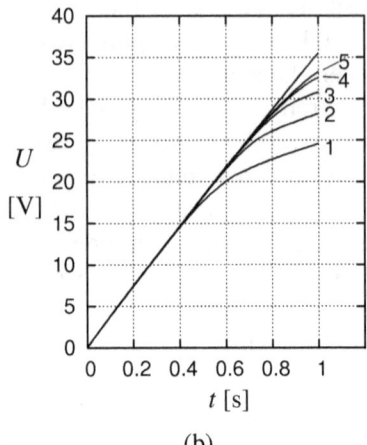

(b)

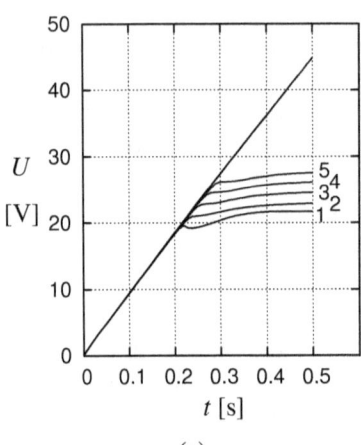

(c)

Figure 11.5. Voltage vs. time dependences measured from a well known manufacturer's new bipolar electrolytics when charging the capacitor repeatedly by a constant current. Between the numbered charging instances, the capacitor was emptied through a resistor. The uppermost graph has been obtained after loading the capacitor close to its rated voltage, before the emptying and measurement. a) An electrolytic of 47 µF and 63 V at 2 mA current for a time of 1 s. b) The same measurement from a different sample. c) An electrolytic of 10 µF and 100 V at 1 mA current for a time of 0.5 s.

by a Vifa driver P17WJ-00-04, whose diaphragm is of filled polypropylene. Afterwards, the manufacturing of this driver has been discontinued, and the availability has not been very good even initially. However, in the place one can use type PL18WO09-04 (also by Vifa), whose diaphragm is of coated paper and which corresponds, in both frequency response and impedance, well enough to the device on which the design is based.

In the choice, the following grounds have been used:

- Low enough resonant frequency and Q value; in PL18WO09-04, about 40 Hz and 2.6. In P17WJ-00-04, the resonant frequency was of the same range, but the Q value was about 1.5.

- The impedance has to be 4 Ω, so that less current would flow via the tweeter at its resonant frequency.

- The current-sensitivity approximately matches that of the tweeter in the vicinity of the crossover frequency (1500 Hz) which is important when passive attenuation circuits are not used.

- The increase in current-sensitivity with increasing frequency (horn effect) is lesser than usually which makes it possible to use 1st-order division.

Then, of course, it must be required, like usually, that the frequency response does not exhibit immoderate unevenness.

The treble is reproduced by Scan-Speak's Revelator-family tweeter D2905/990000, that has a 28 mm textile dome, a rear chamber, and a gradual-sloped aluminum flange. The driver is relatively pricey, but on the other hand, the purpose was not to save money but to offer to the willing a first time possibility to acquaint themselves with law-of-nature-compliant music transference on electro-dynamic loudspeakers.

The choice was affected by the following properties:

- Lowest resonant frequency (500 Hz) of tweeters on the market (at that time)

- Sensitivity suitable for the purpose due to the directional flange

- Does not contain ferrofluid, so detriments arising from the fluid's vibrations and viscosity variation are dispensed with. (Power handling is based on coppered pole pieces and narrow air gap.)

- The depth of the diaphragm relative to the mounting plane helps, for its part, in the leveling of the acoustic source points.

If the price is deemed excessive, there may be used, as a substitute, a somewhat less expensive type, D2905/970000, which is otherwise similar in structure, but the flange is conventional. The sensitivity is at 2-4 kHz 2-3 dB lower than in the Revelator, but on the other hand, the sensitivity of the PL18WO09-04 would also seem to be slightly lower than in the type used in the model units, so the replacement does not necessarily bring any disadvantage.

The schematic, that is shown in Fig. 11.6, involves nothing special except the tweeter's shunting network, whose resonant frequency is set close to that of the tweeter. The network is used to flatten the dip introduced in the loudspeaker's frequency response due to the fact that in this resonance region the phase responses of the drivers inevitably come at 180° from each other. Due to the rather high Q value of the tweeter used (ca 3.8), the depth of response dip would be, without the shunting network, around 7 dB.

Half of the mentioned 180° stems from the phase difference of the currents and the other half from the resonance-induced phase lead, that shows up in Fig. 2.3b (curve C).

The problem also occurs as well in voltage-operated loudspeakers [1], even though the matter is not generally recognized, but the upper-frequency driver is thought to behave flawlessly somewhere down to its resonant frequency.

Series resistors that improve the current-drive index have not been used here. If one wishes to make use of them, all element values have to be optimized anew.

The simulation circuit, that incorporates the measurement-based driver impedance models, is shown in Fig. 11.7.

r_2 represents the DC resistance measured from the filter coil L_1 (wire thickness 1 mm), r_1 representing other losses of the coil.

r_5, in turn, stands for the measured resistance of the ferrite core coil L_2, in the shunting network. L_2 can be made even of thin wire because the value of r_5 is not critical. R_4 just has to be chosen so that the sum $R_4 + r_5$ becomes 22 Ω.

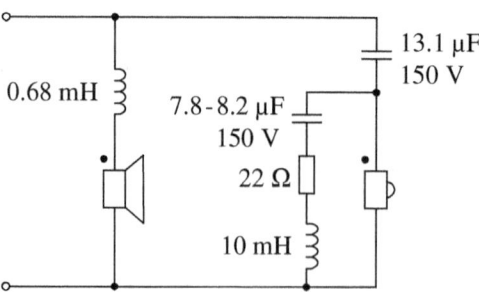

Figure 11.6. Schematic of the example speaker CS-12. To implement the capacitances, parallel connecting of capacitors is necessary.

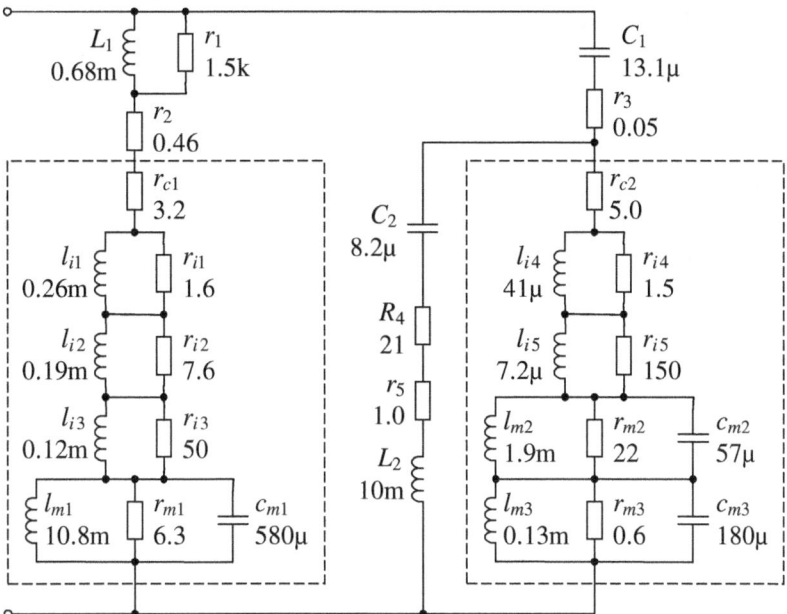

Figure 11.7. Simulation model used for the speaker. Physical circuit elements have been labeled by upper-case letters and those related to equivalent circuits by lower-case. Driver equivalent circuits have been distinguished by dashed lines.

The purpose of r_3 is to represent the losses of the polyester capacitor C_1 and the resistances of the wires.

The target value of C_2 depends a little on the tweeter's resonant frequency, that is best to be measured. If the frequency is somewhat below 500 Hz, like in this case, a suitable value is 8.2 µF; but if the frequency is 500 Hz or over, 7.8 µF is more appropriate. The resonant frequency of the shunting network should be preferably about 70 Hz higher than that of the tweeter.

In practice, at least C_1 has to be composed of several capacitors since the value does not fall close to commercial standard values.

For L_1 and C_1 is recommended 2% accuracy to keep the crossover filtering well controlled. This is achieved by using the measurement methods presented in chapter 13. Precision doesn't hurt in the shunting network parts either.

The bass-midrange driver's impedance has been modelled by three LR links and a resonance network corresponding to the final enclosing.

The values thus obtained yield a resonant frequency of 63.6 Hz and a Q value of 1.46 (formulas (2.19) and (2.20)). When using PL18WO09-04, however, the Q remains considerably higher; and the values needed for equalizer tuning should always be defined by measuring from the appropriate speaker.

The tweeter inductance has been modelled by two LR links, which provide, in this case, good agreement even up to one megahertz. (Sometimes it may be needful to know the impedance behavior also in the ultrasonic region.)

The fundamental resonance of the driver is represented by resonator l_{m2}-r_{m2}-c_{m2}. The circuit also includes another resonator, tuned to about 1 kHz frequency. The purpose of this extra network is to simulate the prominence observable in the impedance in this region.

The modelled driver impedance curves are shown in Fig. 11.8. At the crossover frequency used, the impedance magnitudes are close to each other which is of benefit in design.

The Scan-Speak's inductance doesn't yet become much visible at audio frequencies due to the copperings of the magnetic circuit. Hence, the driver's current frequency response doesn't rise either, with respect to the voltage frequency response, more than very gently.

The driver currents are shown in Fig. 11.9. The division has been set to 1.5 kHz in such a way that the currents equal unity, that is, are unattenuated, at the crossover frequency. Then, the phase difference between the currents should be, in principle, 120°; but the effect of the shunting

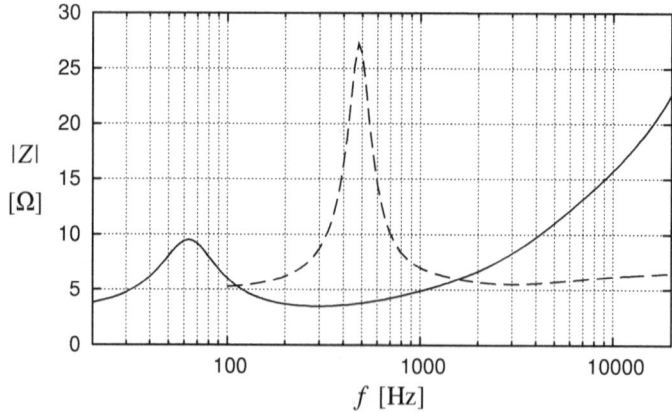

Figure 11.8. Driver impedance magnitudes simulated with the measurement-correspondent models shown in Fig. 11.7. Solid line: bass-midrange driver, enclosed. Dashed line: the tweeter.

network shows at the crossover frequency yet so that the phase difference comes out to be 124°.

The bass-midrange driver's current rises, at highest, a good decibel
above the applied current, and the tweeter's current, in turn, a scant decibel. With a tuning of more than 90°, overshoots are inevitable but stay
here yet moderate.

The alignment has been made this tight for a number of reasons: Increasing the impedance of L_1 and C_1 improves the current-drive indexes,
and the tweeter current stays in the stop-band sufficiently low, like also
the bass-midrange current at the highest frequencies. (It is nevertheless
safer not to use very high-powered amplifiers with this design.) Decreasing C_1 also helps to curb the above-described response dip.

In the tweeter current, one can detect at 70 Hz a small prominence,
that is due to the woofer's impedance peak. However, the use of the pole
shifting equalizer compensates this rise because the total current of the

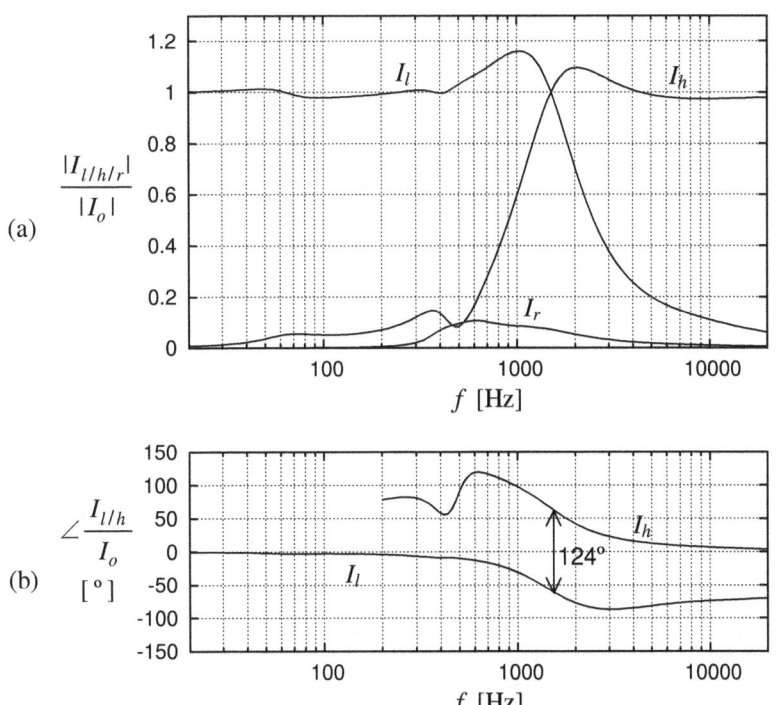

Figure 11.9. a) Current amplitudes vs. frequency: I_l = bass-midrange driver
current, I_h = tweeter current, I_r = shunt resonator current. b) Phase angles of the
driver currents with respect to the applied current (I_o).

speaker decreases in this region.

The current flowing through the shunting network remains, even at highest, below 11% of the total current, so the operation mode of the drivers does not, as a whole, degrade significantly due to this wastage current.

Figure 11.10 shows the simulated acoustic frequency responses, that are obtained by examining the currents through capacitors c_{m1} and c_{m2}, as explained in section 7.1. (The baffle step and non-idealities of the diaphragms naturally do not show up here.)

The bass region is dominated by the resonance peak, whose position

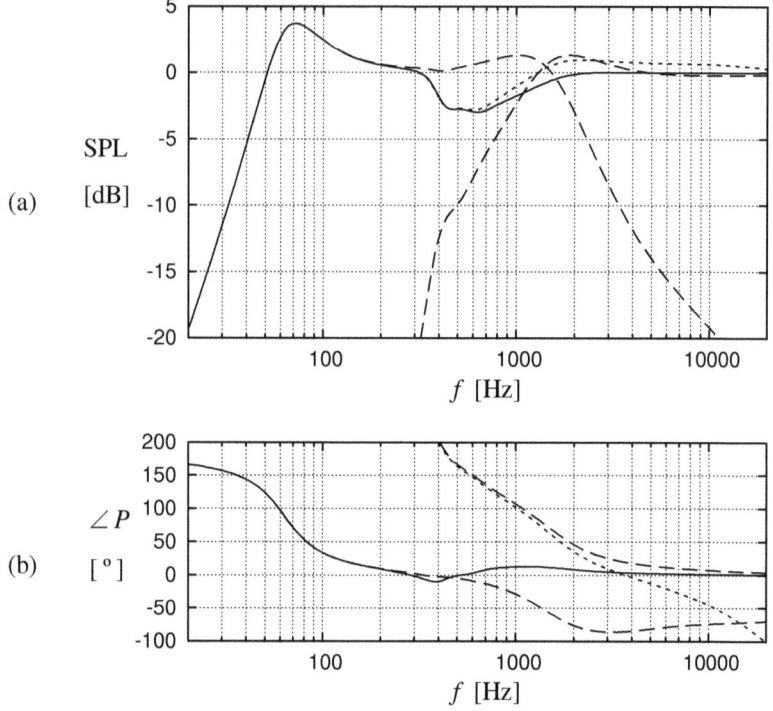

Figure 11.10. Simulated frequency responses, assuming equal current-sensitivities for the drivers. a) Amplitude responses of the drivers (dashed lines) and the sum response at two different positionings: Solid line: acoustic source points at the same distance from the listener. Dotted line: tweeter source point 5 mm farther back with respect to the woofer. b) Dashed lines: driver phase responses without delays. Solid line: total phase response without delays (acoustic source points in alignment). Dotted line: tweeter phase response when the source point is 5 mm back from the reference plane.

and height correspond to the parameters obtained before.

Despite the shunting network, an attenuation of 3 dB remains at the mid-frequencies, as the acoustic phases of the drivers become opposite, as shown in Fig. b. The attenuation does not, however, affect the power response, so the general balance should not suffer greatly.

The acoustic crossover frequency is slightly lower than the electrical because the nearness of the resonance raises the tweeter's response a bit yet at 1.5 kHz.

With the source points leveled, the acoustic phase difference of the drivers is, at 1.5 kHz, 128°, that is, a little greater than the difference in the currents. To keep the phase difference more moderate, it is expedient to situate the tweeter yet somewhat farther back than is required for the leveling of the source points.

By a shift of 5 mm, the amplitude response changes from that shown by the solid line to that shown by the dotted line, that is, rises about a decibel at most. The tweeter's phase lowers then to that shown by the dotted line in Fig. b, and the phase difference decreases at the crossover frequency to 120° which reduces a little the sensitivity to listening direction changes.

The total phase shown in Fig. b represents a minimum-phase system, like all realizations using 1st-order crossover filtering and having the acoustic source points leveled. The phase linearity of the system is also rather good above 200 Hz since deviations from a linearity-adhering phase (in this case, from zero) are of the order of 10 degrees.

As frequency decreases beyond the resonance region, the phase rises 180°, as is characteristic of a 2nd-order high-pass system. However, the use of a pole shifting equalizer shifts this phase step somewhat lower in frequency while also making it more gradual.

Consequently, as frequency decreases, the phase lead increases, so it can be said that low frequencies are reproduced somewhat in advance of higher ones* even though, due to the periodicity of the sine wave, it is impossible to define any unambiguous delay between different frequencies.

In addition to the woofer, there is generally acting, in the signal path, also other high-pass filters, which all contribute to the advancing of the phase of the lowest frequencies. The more steeply the infrasounds are cut at different stages of recording and reproduction, the more the phase

---

* According to prevailing thought, low frequencies are always left behind, in time, from higher ones. This is, however, an erroneous view, that is based on wrong interpretation of the group delay concept where the group delay is perceived as a kind of frequency-specific delay. The issue has been explained in section 14.5.

response rises also at audible frequencies although usually the matter is not much recognized.

The impedance curve of the speaker is shown in Fig. 11.11. The simulated impedance agrees well with the measurement values which demonstrates the functionality of the modelling.

At the crossover frequency is developed a peak, that extends to 11 Ω and stems from the used 120-degree tuning.

The effective nominal impedance could be regarded as being 6 Ω.

By feeding the equivalent circuit by a step-like current and observing, again, the currents through c_{m1} and c_{m2}, one can find the system's acoustic step response, that is shown in Fig. 11.12. The response is not totally flawless, due to the mid-frequency attenuation, but stays, however, very uniform compared with the results generally obtained from conventional 2-way speakers.

The pressure due to the tweeter reaches its apex first, after which the bass-midrange unit takes responsibility of the rest of the transient. However, from this, one must *not* draw such a conclusion that the operation of the bass-midrange driver would exhibit some delay with respect to the tweeter, since the asynchrony of the response peaks stems from the nature of the test signal itself and is not tied to so-called group delay.

Namely, the step function can be thought to be composed of a DC component of half the height and sine components that represent all frequencies. Those sine components having the highest frequencies naturally reach their apex first, after instant 0, and then the other ones, in the order of the period length. A phase-linear system only reproduces these components in the same order as they occur in the excitation signal; without needing any delays for explanation.

Loudspeakers are not intended to reproduce DC, so an ideal step res-

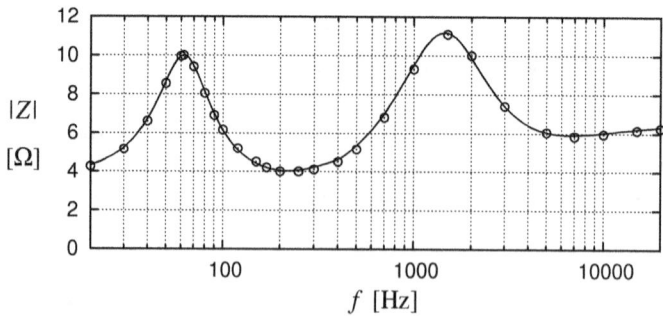

Figure 11.11. Magnitude of the total impedance as simulated (curve) and as measured (circlets).

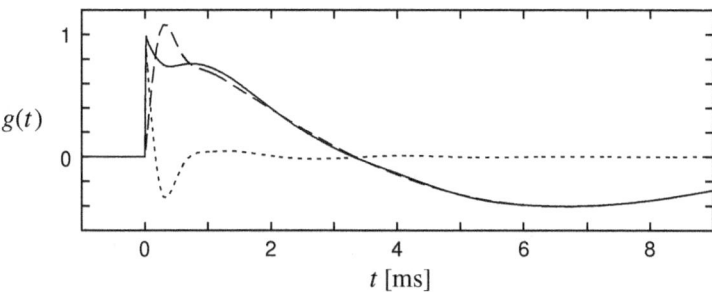

Figure 11.12. Simulated step response with the acoustic source points leveled. Solid line: total response. Dashed line: contribution of the bass-midrange unit. Dotted line: tweeter contribution.

ponse exhibits first a steep rise, then a slow and steady fall to the negative side and a return toward zero. An undamped bass resonance shows up as undulation in the tail.

The net area between the curve and time axis is always zero since the response represents the diaphragms' acceleration and the time integral of this, i.e. the velocity, must also eventually become zero.

Suggested cabinet dimensions, when using the D2905/990000, are given in Fig. 11.13. With 19 mm wall thickness, the internal volume will be 12 litres. At 2.5m distance, the most suitable listening height is,

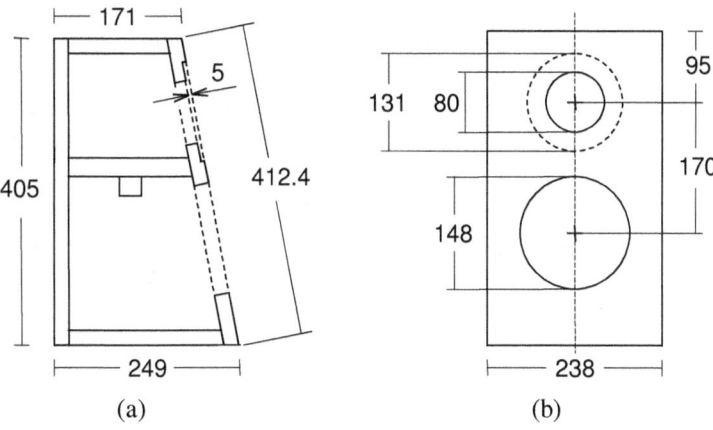

Figure 11.13. Cabinet construction. a) Side view. Walls have been supported with two crossing bracing bars, that are fastened by screws in addition to glue. b) Front panel as seen perpendicular to the surface.

with these dimensions, 15-20 cm above the cabinet's top surface.

The tweeter is embedded in the front panel's plane, but the woofer is not since thereby the inclination need can be reduced. With a rubber gasket, one can gain yet a bit more distance compensation, but, due to diffraction detriments, it is better not to raise the woofer very far out.

When using the D2905/970000, the needed cavity width is 105 mm and the depth, 4 mm. The diaphragm is then 10 mm farther out, so the listening height recommendation has to be correspondingly lowered if more front panel inclination is not desired. A shift of 1 cm in the depth of either driver corresponds, according to the geometry, a height difference of 15 cm at the 2.5 m distance; so the best listening height presumably shifts, in this case, close to the top surface's level.

The D2905/990000 has a fixed gasket. For the other drivers, one can use a rubber ring that is cut to size from a sheet of about 1 mm thickness. The screw holes to the gasket are best made by an ordinary paper hole punch.

The somewhat loose soldering tabs of the Scan-Speak may yet be supported with a glue drop, so that they surely do not resonate in the pressure of the bass tones.

The coils L_1 and L_2 should not be situated close to each other, so that the ferrite of L_2 would not increase the inductance of L_1. To minimize the transformer effect, i.e. the mutual inductance, it is also good practice to orient the coil axes perpendicular to each other, as is also generally recommended.

Further, if the speaker is going to be used on a steel stand, it is better not to situate the coils at the very bottom of the cabinet because a base made of magnetizing material can increase the inductance and losses of the coils.

As a stand, one can use even an ordinary step stool if the height is suitable. Actual speaker stands are generally too unstable with a load of this weight.

A common problem especially in tweeters is that in the air gap there has been caught small steel particles, that, when touching the voice coil, introduce extraneous distortion. Unfortunately, not even devices of the highest price class have been free from these residuals. With ferrofluids, the problem can be made invisible but not inaudible.

Due to the above, it is always best to check all tweeters before application.

The mobility of the voice coil can be examined by pressing, with a toothpick or alike nonmagnetic object, very carefully, the furrow around the dome, on different sides. The diaphragm should then swing at least a hair's breadth without any scrape. All units do not pass this test, and in many cheapie products, the voice coil may be on some side even totally stuck.

In liquidless drivers, it is possible to clear the air gap by oneself; but realigning the voice coil so that nothing is damaged and the coil moves freely may require, depending on the type, considerable caution and patience.

The cabinet is filled tightly with cotton cloth, that is best obtained from sheets. The cloth is rent into roughly 20 cm wide shreds, that are squeezed into tangles and crammed in with moderate force. One cabinet takes 4-5 single sheets, so it may be worthwhile to eye some offers.

Around the woofer is left, however, an inch of free space, minding also the hole in the center of the magnet. The cloth tangles hold in place pressing against each other, but it is nevertheless advisable to avoid turning the speaker front side down.

The front mask is most easily made out of plastic foam sold for the purpose; by attaching together, with elastic and rinsing-tolerant glue, two layers, of which the inner one forms only a frame. If desired, the mask can yet be thinned in the tweeter area by scissors.

Figure 11.14 shows a finished speaker with paint polish.

Figure 11.15 displays the measured current frequency responses at about the ideal listening height. The need for baffle step compensation becomes well apparent from the results, but the response dip in the 500 Hz region shows up less than anticipated.

The prominence around 2 kHz stems from a mild unevenness in the drivers' acoustic phase difference, most probably related to the chamber structure of the tweeter. In this region, the response can, however, be

Figure 11.14. The CS-12 completed. Stripes at the edges are mask fastening stickers.

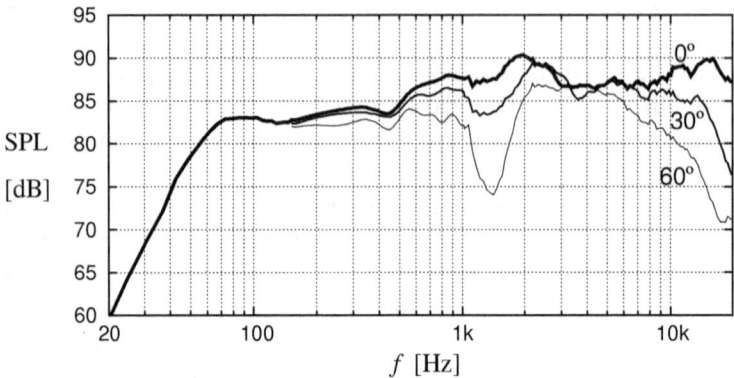

Figure 11.15. Frequency responses forward and 30° and 60° sideward, measured at 2.5 m, about 15 cm above the cabinet's top surface level (courtesy of Gradient oy). The decibel scale represents sound pressure at 0.354 A current, at 1 m distance. The bass region has been measured separately from the near field.

affected by adjusting the listening height (see Fig. 11.10a).

The attenuation in the side direction near the crossover frequency is mostly due to the increase in the bass-midrange driver's phase lag, in accordance with the minimum-phase nature, as the amplitude response falls off; and partially also due to the changing of the acoustic source point difference.

From the bass response, one can assess that the Q value is somewhat lower than the 1.46 given by the modelling. The difference stems from the fact that the response measurements have been taken some years after the design, and the Q has actually crept a little for some reason. UV radiation is not to suspect, however, for the driver has been kept protected by a black double-layer foam mask.

11.5 Project CS-8

As the second building project is presented a 1.5-way speaker that operates on the principle shown in Fig. 6.7, has a volume of 8 l and is correspondingly named CS-8. Besides the baffle step, the bass resonance has also been compensated with passive circuits, so the feeding amplifier can have flat frequency response.

The driver used is a 4½-inch, 4-ohm device Vifa PL11WH09-04*, whose operation range extends on current-drive up to 10 kHz, making it possible to omit the tweeter. The lower −6 dB cut-off frequency obtained is 56 Hz, which is, in this size class, a fairly competitive result.

The driver's rated resonant frequency is 67 Hz and the mechanical Q value 1.8, which numbers also hold true surprisingly well. Two drives accrue effective area 116 cm^2, which is yet fairly adequate for civilized listening.

The schematic is in Fig. 11.16. Driver A gets current mostly only at low frequencies, while driver B operates in the whole range. To adjust the frequency balance, A receives, however, a little current also at high frequencies via the 33 Ω resistor.

The resonance compensation has been realized on the principle familiar from section 8.1. The series resonance network has been applied only across driver B which is sufficient in this case because the resonance prominence of A remains quite minor even as such.

The overall equivalent circuit is shown in Fig. 11.17. The voice coil inductance has been modelled by the LR link combination shown already before in Fig. 7.5a. The air spring, in turn, has been modelled on the principle outlined in Fig. 7.8, except that, for practicality, inductances L_a have also been represented by mirroring current sources.

l_m and c_m represent the spring constant and moving mass of a free

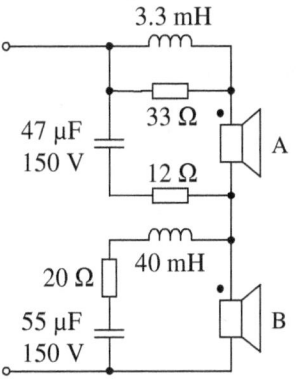

Figure 11.16. Schematic of the CS-8 example speaker. Driver B serves all frequencies; A mostly only the low ones.

---

* After writing this chapter, Tymphany has assigned this driver under "OEM" status, which means that the type is made unavailable for hobbyists and even small businesses. Unfortunately, it also seems at the moment that suitable substitutes are hard to find. Therefore, this section mostly serves as a design example unless things get better for DIY users. A possible replacement should be 4-ohmic and have comparable resonant frequency, mechanical Q, and frequency characteristics.

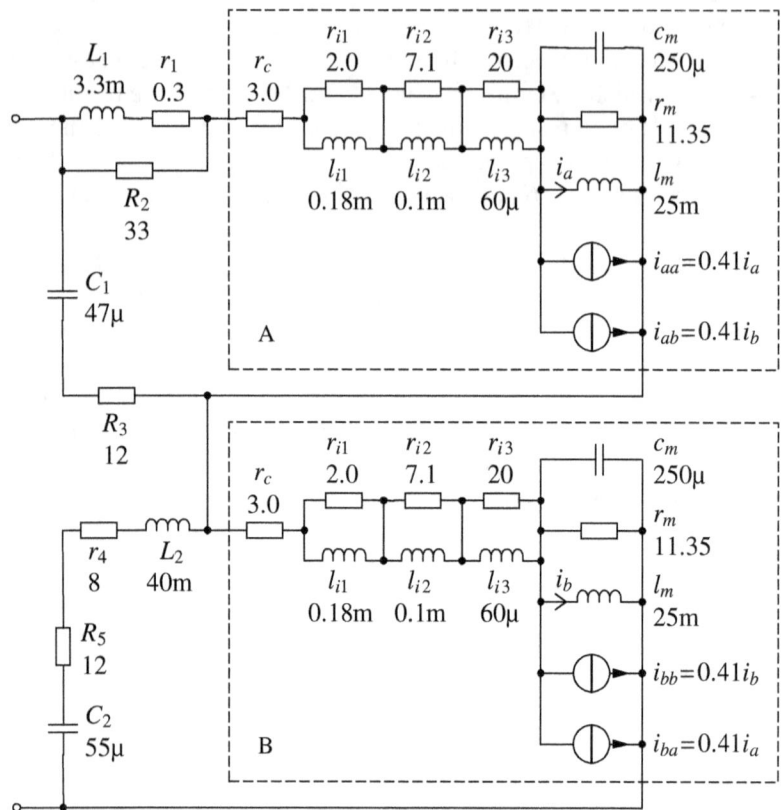

Figure 11.17. Detailed simulation model of the speaker. Again, physical elements have been labeled by upper-case and others by lower-case. The current-dependent current sources model the effect of the air spring.

driver, being almost the same as in Fig. 7.5a. Current-dependent current sources i_{aa} and i_{bb} represent the air spring force due to each driver's own displacement, while the current-dependent current sources i_{ab} and i_{ba}, in turn, represent the air spring force caused by one driver's displacement on the other.

The resonant frequency of the enclosed drivers was measured to be about 85 Hz. By simulation, one can reach the same value by setting the coefficient of the aforementioned current sources to 0.41. (The coefficient is equal for all the sources because the drivers are identical.) This means that the air spring stiffness is, in practice, 0.41-fold compared to the suspension stiffness of the driver when the other cone is immovable.

The Q value of the enclosed drivers was found to be 1.5, which enables passive equalization without significant cost to the current-drive index.

Both L_1 and L_2 are here with ferrite core. r_1 represents the winding resistance of L_1, and r_4 respectively that of L_2. These coils too should not be situated side by side and preferably not in the same orientation.

Inductance L_2 is so high that the coil has to be wound by oneself. However, it doesn't have to be large in size, and is best made out of 0.4-0.5 mm diameter wire. One only has to see to it that the sum of resistances R_5 and r_4 becomes 20 Ω.

To construct the coil, it is best to first dismantle some existing coil and count the number of turns needed for it. If the dismantled coil had inductance L_0 and turns number N_0, the desired inductance (L_2) is established on the same core with the turns number $N_0\sqrt{L_2/L_0}$. Difference in wire thickness is quite irrelevant, but at first it is better to provide some extraneous turns, so that the final measurement-based adjustment can be performed by lessening them.

For the resistors, 5 W power handling is enough if they are surrounded by free air. The tolerance of all elements should be 5%.

Figure 11.18 shows the impedance fitting result of an enclosed driver (solid line) when both drivers are active without filtering components. The position of the peak is set in place by the above-described coefficient and the height by resistance r_m.

The Figure also shows the total impedance, that stays in the audible range yet rather moderate.

The simulated frequency responses of the loudspeaker and both drivers are shown in Fig. 11.19. Again, the drivers are assumed to operate

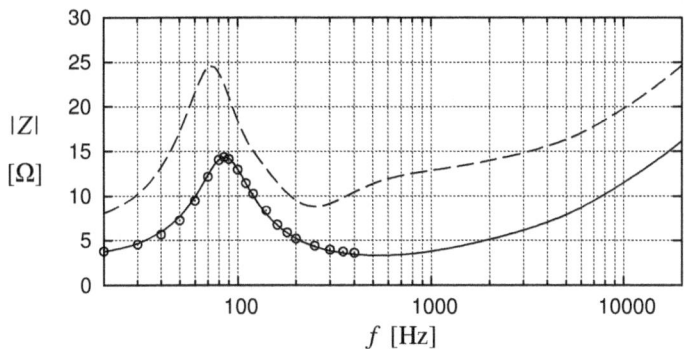

Figure 11.18. Solid line: modelled impedance of one enclosed driver (both ones receiving the same current). Circlets represent corresponding measurements. Dashed line: simulated impedance of the entire speaker.

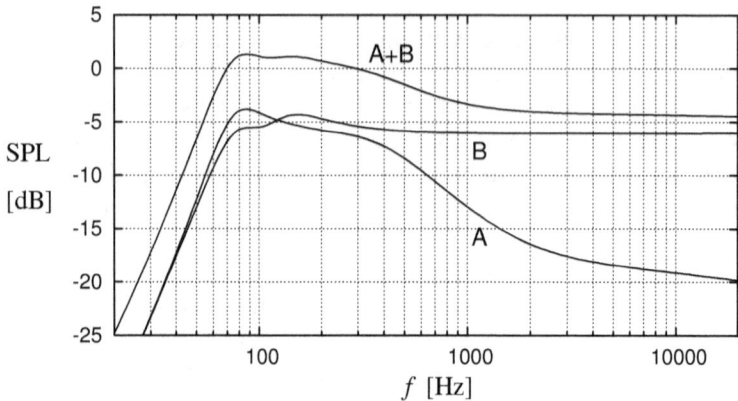

Figure 11.19. Simulated frequency responses for drivers A and B and their combination.

in an infinite front panel and with ideal diaphragms.

The response of driver B has been modified only in the resonance region, whereupon the responses are quite well balanced at low frequencies. The response of driver A, in turn, has been shaped so as to provide the needed baffle step correction in the total response. 'A' has, however, a little share also in treble reproduction which has been found necessary despite the fact that current-drive in itself already tends to reinforce this region.

The total response rises at low frequencies a little above the 0 dB level, which represents unaccentuated and unattenuated reproduction. This gives certain advantage especially to the operation mode of driver A, as the attenuation need is relieved, as explained in section 8.1.

Figure 11.20 shows the corresponding step response simulation. It can be seen also here that high frequencies are mostly the responsibility of driver B, whereas at low frequencies (far from instant 0), A and B operate pretty much congruently.

The virtual attenuation of high frequencies in Fig. 11.19 shows up in the step response as rounding at the onset. However, this roundness as well as the mentioned attenuation are removed when a real enclosure is used, instead of the infinitely wide one.

Enclosure dimensions are shown in Fig. 11.21. With the recommended 19 mm material thickness, the volume will be, excluding the space taken by the magnets, a good 8 litres.

Driver B is mounted in the upper hole, which is closer to the panel

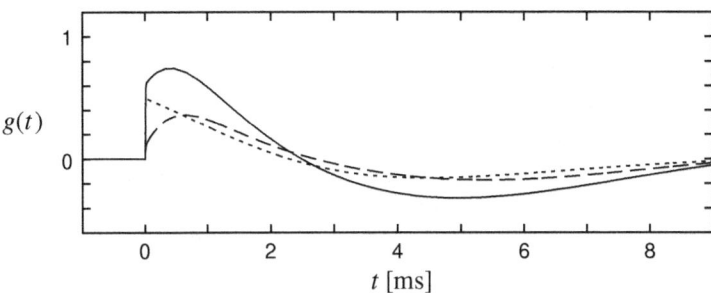

Figure 11.20. Simulated step responses for driver A (dashed line), driver B (dotted line), and both together (solid line).

center. Thereby, the source point of high frequencies is made, average-ly, as far as possible from the cabinet edges.

The air holes in the driver chassis are so narrow that they may be blocked too much when using proper board thickness. Hence, it is best to bevel or round the front panel holes from inside to necessary extent. The screw spots can be left unbeveled, however, to ensure sufficient thickness in them.

Embedding of the drivers can also be established with a lifting plate that is glued over the front panel and has holes of the size of the flange. This also gives a little more stiffness.

The cabinet is crammed full of cotton cloth rent into shreds, so as to

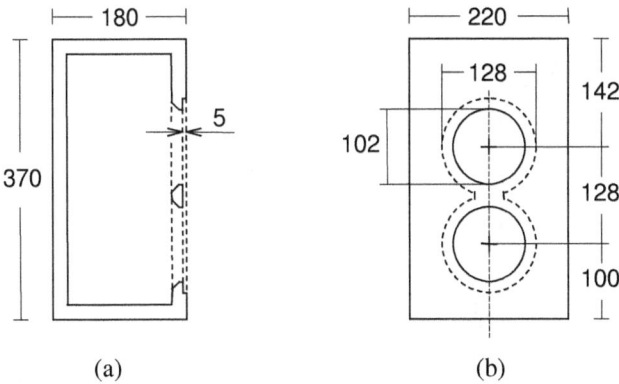

Figure 11.21. Cabinet construction. a) Side view. To improve air mobility, the inner edges of the holes have been beveled. b) Front view. The upper hole is intended for driver B.

make the Q value low enough. Around the driver is left empty space so much that the device can be mounted freely. It is important to also fill the space between the drivers tightly, for this has a great significance to the resonance parameters.

The speaker is recommended to be directed straight to the listener. As an aid, one can use e.g. supports placed underneath. The speaker can also be kept upside down if doing so brings some advantage. The position may have significance at least for floor reflections.

For the front cover is recommended foam mask, similarly as in the preceding project.

Figure 11.22 shows a finished pair of speakers. The filtering circuits have been housed in a separate box, so that one doesn't have to be careful of them when stuffing the damping material. The solution also facilitates possible servicing. One should, however, see to it that such a box is not allowed to develop resonations as it vibrates with the cabinet.

The measured current frequency responses are shown in Fig. 11.23. The on-axis response is, as a general nature, slightly rising which partially compensates the fall in the power response as the directivity of the drivers increase.

The measured bass response agrees well with the simulation result, and the attenuation slope seems to be 14 dB per octave in both cases.

At the high end, reproduction extends yet to frequencies above 10 kHz, even though with only about half of the amplitude. The peak seen at 9 kHz does not have a counterpart in the driver's characteristics, so the issue is probably about unit-to-unit divergence.

Figure 11.22. The CS-8 completed. The rear box containing the filtering components could be smaller. The terminal posts are discerned under the box.

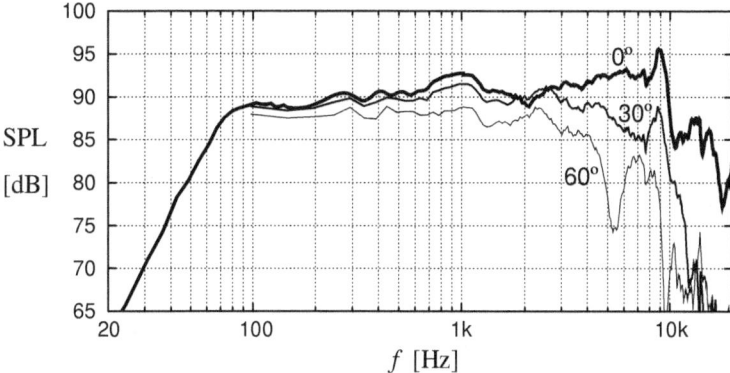

Figure 11.23. Frequency responses forward and 30° and 60° sideward, measured at 1 m distance, at the height of the cabinet center. (The bass region has been measured separately from the near field.) The scale represents sound pressure at 0.354 A current, at 1 m distance.

11.6 A Minimum-Phase Active Filter

In the circuit of the CS-12 speaker, we already used a shunting network by which the current flowing through a driver was constrained in its stop-band. Because the network only acts in a rather narrow region, it does not, in effect, increase the order of the crossover filtering, and so the minimum-phase property, a virtue of 1st-order filtering, is retained.

A corresponding scheme can also be utilized in an active crossover filter, thus enabling one to gain more attenuation in the stop-band without losing the minimum-phase nature and hence the accuracy of phase reproduction.

Figure 11.24 gives a block diagram of such an active filter, tuned to 1 kHz crossover frequency. As a starting point, there is used normal 1st-order filtering, which is aided by one band attenuator in the low channel and two in the high channel.

First, there is a common band-boost stage, by which the signal is augmented some two decibels in the crossover region. This emphasis is used to approximately compensate the dip introduced as the band attenuators inevitably increase a little the phase difference between the signals at the crossover frequency.

The band attenuator stages and the compensation stage are 2nd-order systems, that can be defined by a characteristic frequency (center fre-

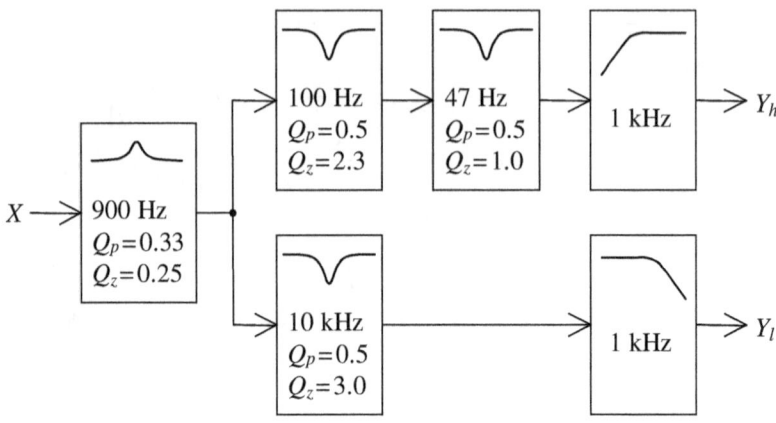

Figure 11.24. Example of a minimum-phase crossover filter. The frequencies may be scaled as needed.

quency) and two Q values, as in Table 1 in section 2.2. Band-cut is the case when the Q value of the zeroes is greater than that of the poles, that is, when $Q_z > Q_p$; and band-boost, in turn, when $Q_z < Q_p$. The height of the prominence or dip is determined by the ratio of the Q values and the width by their magnitude.

Tuning of the filter is finding a compromise between the attenuation obtained for the stop-bands and the phase difference accumulating at the crossover frequency. Figure 11.25 shows the result obtained with the given parameters.

A decade away from the crossover frequency, the attenuation achieved is 35 dB on both sides, and the attenuation of the high-pass channel stays also at the lowest frequencies above 35 dB. In the low-pass channel, that has only one band attenuator, the response yet undergoes a rise at the highest frequencies, but in this region the matter is not of much practical significance any more.

The total phase shift stays very close to zero because the system is minimum-phase and the amplitude response is virtually constant.

The crossover frequency phase difference comes out to be 111°, so even regarding the lead caused by the resonance of the upper-frequency driver, the acoustic phase difference remains, at the crossover frequency, still below 120°. With a division of 1 kHz, suitable transducers for the upper band are mostly small 2-3-inch cone drivers, whose resonant frequency can easily be made lower than 200 Hz.

A circuit topology suitable for band-cut is shown in Fig. 11.26a and

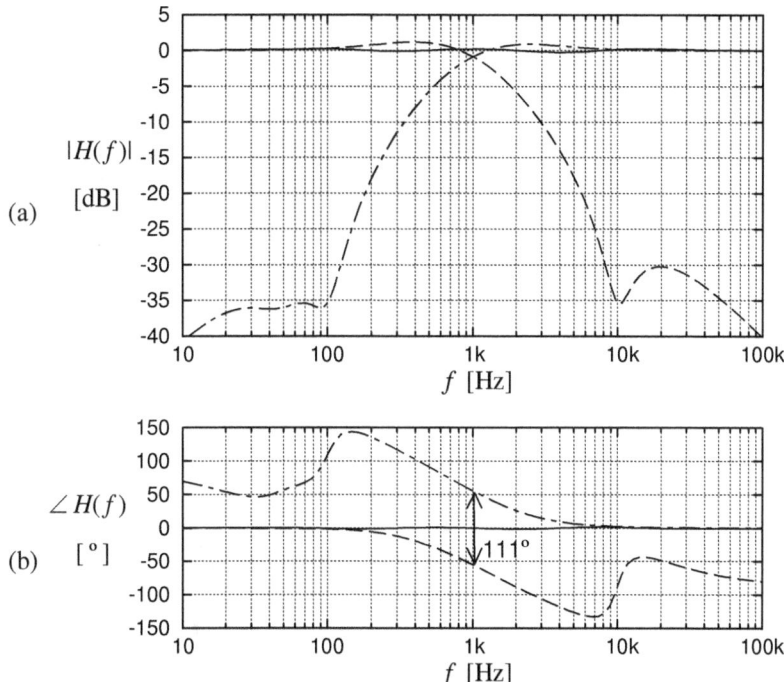

Figure 11.25. a) Amplitude responses of the Fig. 11.24 filter. Low-frequency channel (dash line), high-frequency channel (dash-dot line), and the sum (solid line). b) Corresponding phase responses.

one for band-boost in Fig. 11.26b. The circuits are non-inverting, and far from the center frequency, their gain is unity. The first-mentioned can also well be used to damp the bass resonance, instead of the filter in Fig. 8.3.

The transfer function of the Fig. 'a' band-cut circuit is

$$\frac{U_o}{U_i} = \frac{s^2 + \left(\dfrac{1}{R_3 C_1} + \dfrac{1}{R_3 C_2}\right)s + \dfrac{R_1 + R_2}{R_1 R_2 R_3 C_1 C_2}}{s^2 + \left(\dfrac{1}{R_1 C_1} + \dfrac{1}{R_3 C_1} + \dfrac{1}{R_3 C_2}\right)s + \dfrac{R_1 + R_2}{R_1 R_2 R_3 C_1 C_2}} \quad (11.3)$$

The transfer function of the Fig. b band-boost circuit is

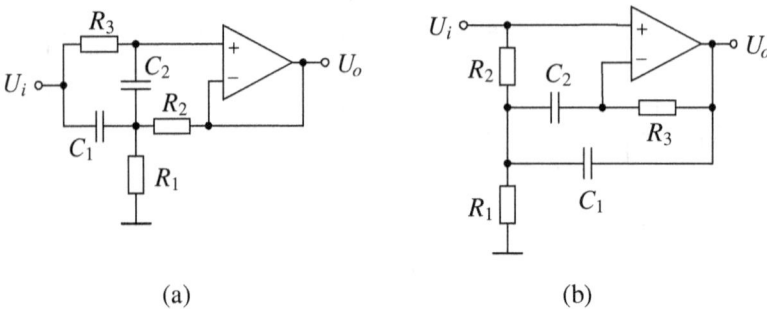

Figure 11.26. a) A band-cut circuit. b) A band-boost circuit that functions inversely with respect to the former.

$$\frac{U_o}{U_i} = \frac{s^2 + \left(\dfrac{1}{R_1 C_1} + \dfrac{1}{R_3 C_1} + \dfrac{1}{R_3 C_2}\right) s + \dfrac{R_1 + R_2}{R_1 R_2 R_3 C_1 C_2}}{s^2 + \left(\dfrac{1}{R_3 C_1} + \dfrac{1}{R_3 C_2}\right) s + \dfrac{R_1 + R_2}{R_1 R_2 R_3 C_1 C_2}} \qquad (11.4)$$

The expressions are reciprocals of each other, so, when using the same values in both circuits, the responses exactly cancel each other.

A stage with the desired operation parameters can be designed by first assigning values for C_1 and C_2 (It is useful to try different ratios.); then solving R_3 from the 2-term coefficient of s, using equation (2.17); next solving R_1 from the 3-term coefficient of s; and lastly solving R_2 from the constant term.

Due to their simplicity and ease of use, the circuits are also suited for many kinds of frequency response correction.

A complete circuit diagram corresponding to Fig. 11.24 is shown in Fig. 11.27. Output buffers have not been drawn since their necessity depends on the loading presented by the next stage.

Using the filter is naturally not tied to current-drive, but if errors in phase reproduction are altogether audible, they are most certain to show up in that mode.

The responses can be scaled on the frequency axis just as the responses of any RC circuit. Multiplying all resistances by a number divides all frequencies by the same number. The same rule also applies to capacitances.

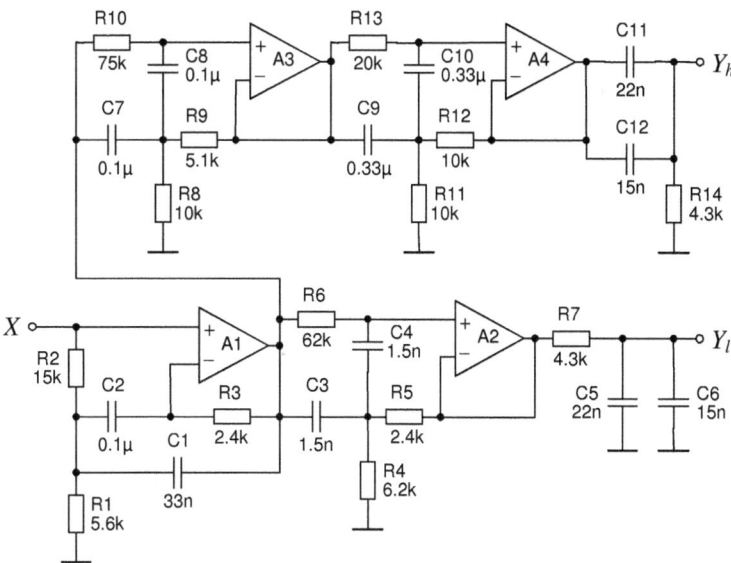

Figure 11.27. Schematic of a minimum-phase example crossover. A suitable amplifier circuit is TL074.

[1]   W. M. Leach, "Loudspeaker Driver Phase Response: The Neglected Factor in Crossover Network Design", *Journal of the Audio Engineering Society*, vol. 28, June 1980, p. 410-421.

12
PROTECTIONS

Using current-drive facilitates in many ways the design of equipment overload protections. By and large, the protection need of amplifiers is lessened because possibilities for the occurrence of overlarge currents are reduced. For loudspeaker protection, again, current-drive introduces new possibilities since limiting devices can also be connected in parallel with the driver. It is also a benefit that protection devices used in series mode do not degrade sound quality as they do under voltage drive.

12.1 Amplifier Protection

In the overload protection of transconductance amplifiers, as well as ordinary amplifiers, one must take care e.g. that the output voltage of the power stage does not rise or fall in special circumstances beyond the supply voltage range and that the output current does not in any conditions exceed the output transistors' capacity, above which the internal bonding wires of the device usually melt.

The first-mentioned requirement can usually be satisfied with a pair of diodes, as in Fig. 10.9. To keep the voltage drop of the diodes low, it may be necessary to use Schottky diodes.

To limit current into a safe range, there are usually used extra transistors that open above a specified limit and take away drive current from the output transistors, thus preventing excessive increase of the output current. In a current-output amplifier, however, any danger of overcurrent due to external fault conditions does not necessarily exist.

We will first consider the short-circuit possibilities in the basic circuit of Fig. 10.1b, in which three terminals are generally accessible for the user: both ends of the speaker output and the ground.

The most usual case is that the output nodes become shorted. The circuit then turns into a voltage follower stage that is loaded by a low resistance, R. So, such a short-circuit does not pose any immediate danger when the input signal level remains unchanged, but the amplifier nevertheless warms more than usually because the voltage across the conducting output transistor increases. The lowness of voltage gain may also make the amplifier oscillate which yet increases power losses; so when prolonged, the condition may lead to the activation of the possible thermal shutdown feature.

The shorting of resistor R, in turn, causes the feedback to disappear, in which case even a small signal makes the output voltage swing from edge to edge. This is, however, harmless for the amplifier, even though not necessarily for the tweeter or the hearing of those involved.

Instead, joining the output of the power stage to ground enables the generation of high currents and is as harmful or harmless as in a voltage amplifier. In practice, this case should be, however, less probable than the shorting of the speaker lines.

Similar conclusions also apply to the single-supply topology shown in Fig. 10.6; so here again, danger is introduced mostly only by joining the output point to ground.

In the grounded-load topologies (Figures 10.5 and 10.8), the user is only able to short the speaker output but not the power stage. Also here, a speaker short results only in increased warming and possible instability, so such an amplifier may do even without overcurrent protection, thus also dispensing with the possible adverse effects of these circuits.

Current-drive also enables the using of relays for signal breaking for different reasons; without any detriment to circuit operation stemming from contact resistance build-up in the tips.

12.2 Speaker Protection

The power protection of a current-driven speaker driver can be implemented both in series mode, by breaking the circuit, and in parallel mode, by shorting the driver. Connecting a dummy load in place of the driver is not imperative since although in the former case the amplifier's load impedance increases and in the latter case decreases, neither should harm an appropriately designed amplifier. Parallel mode power limiting can also be implemented without muting the driver totally, enabling the program to continue without interruption.

The simplest means of protection is without doubt an ordinary fuse. Because the thermal time constants of voice coils are of the order of 1-10 s, it is mostly only the fastest types (FF) that are useful. Fuses have the advantage that they react to current, that represents the voice coil's heating power frequency-independently. There is, however, the problem that, to ensure fast enough reacting, the fuse has to be underrated which makes it prone to blow at unnecessarily low power.

Abrupt signal breaking can also be implemented with a relay, in the way shown in Fig. 12.1. The voltage across the driver is rectified by a diode bridge that feeds the relay's coil via a leveling filter. As the average level of the voltage rises enough, the relay operates and replaces the driver by a dummy load R_2, that roughly matches the driver's impedance.

R_1 should be at least 50-fold and preferably 100-fold relative to the driver's impedance in order that the harmonic distortion introduced by the circuit and the effect on the current-drive index would remain minute. However, because the winding resistances in existing 5-6 V relays are typically only 100-150 Ω, the capacitor assumes only a small portion of the peak voltage, and hence the circuit is not suited, with normal relays, for drivers having very low power handling.

With 10 000 µF capacitance, the charging time constant is about 1 s. The discharging time constant, after the signal disappears, is determined only by the capacitance and coil resistance and is slightly longer.

The voltage across the dummy load keeps the relay active until the signal has returned to a permitted level. However, a problem is that the power handling required of the load resistor should be much greater than that of the driver. Arranging a resistor pack of hundreds of watts for the purpose is not often practical, but fortunately there are also other possibilities.

By leaving R_2 out, the load impedance of the amplifier disappears, at least in the driver's frequency band, as the relay operates. The amplifier then exhibits hard voltage clipping but with small power losses. If the

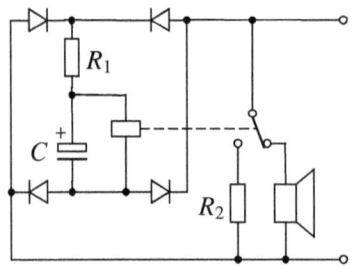

Figure 12.1. Driver protection using a relay. The dummy load R_2 can also be omitted or replaced by a short. The driver symbol can be interpreted to also include possible response equalization circuits.

harmonic overtones produced by the clipping present a danger to some other driver, it too can be disconnected by the same relay.

The rise in voltage, as the relay operates, also results in that the relay remains activated and releases only after the signal (and possible direct voltage) has been removed.

The protection can also be realized by short-circuiting the feeding nodes, so that the voltage drops to zero, and the relay remains activated only for a designated time, which is determined, in addition to the time constant, by the difference between the relay's operation and release voltages. Consequently, if the overloading continues, the relay changes its state repeatedly, the release period being shortened when the over-voltage is increased.

Protectors that react to voltage as such are not able to estimate voice coil warming very accurately since the voltage also contains the electro-motive forces, that however, do not directly warm the voice coil. Design is also hampered by the fact that typical values for relay threshold volta-ges are not specified, not to mention the tolerances.

Current-drive makes it possible to also use transistors as the limiting element since current passed past the driver is away from the driver it-self. Figure 12.2 shows a MOSFET-based protection circuit, by which one can avoid some of the problems related to relays.

R_2 and C make up a leveling filter that produces from the rectified voltage an approximate short-term average. R_1, which has much lower value than R_2, is needed in order that C would be charged toward the average voltage, instead of the peak value. Averaging operation is better since warming is determined by the RMS value, and in practical wave-forms, the ratio of peak values to the RMS value varies more than the ratio of the RMS and average values. (The RMS is always greater than the average.)

R_3 is higher than R_2, so that the capacitor is not vainly loaded. Then,

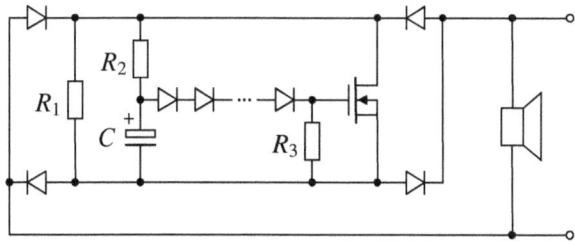

Figure 12.2. Simple MOSFET power limiter. The number of diodes connected in series depends on the desired operation level.

the reaction speed of the protector is determined by time constant R_2C.

The voltage drop developed in the diode chain is set to correspond to the difference between the capacitor voltage required for the activation of the protector and the MOSFET's threshold voltage.

With the diodes, the switching can be made sharper, and at the same time, the dependence on threshold voltage deviation is reduced. However, with the lowest power levels, the diodes cannot be used because the mentioned voltage difference becomes diminutive.

As the transistor, it is best to choose a type of at least 10 amperes even if the expected currents were smaller, since the higher the transconductance of the transistor, the more definitely the circuit works. The threshold voltage should be low and the deviation moderate. However, as MOSFETs are quite inexpensive, some sorting of the threshold voltages is affordable.

In the experiments was used a 12 A, 60 V type, MTP3055VL, which proved to be up to the task.

The heat sinking need for the MOSFET can be estimated by the maximum current of the amplifier and the maximum voltage of the driver. However, after the protector turns active, most part of the power is dissipated in the amplifier, whose thermal design should anyway take into account the possibility of a short.

If the impedance of the object to be protected is heavily frequency-dependent or the impedance otherwise represents poorly the voltage dropped in the voice coil resistance, the observed voltage can be filtered before the rectification and averaging. An example of such a circuit is in Fig. 12.3.

R_1, R_2, R_3, and C_1 constitute a symmetrized filter by which it is pos-

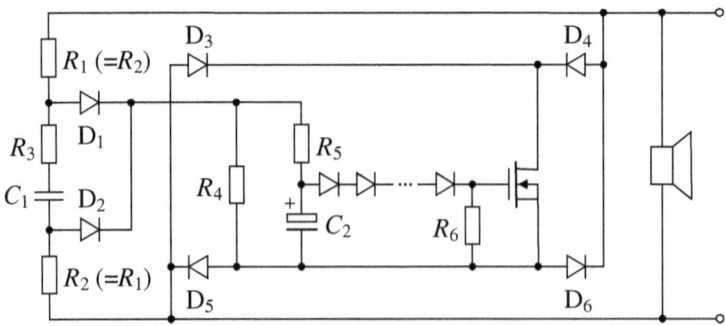

Figure 12.3. MOSFET power limiter with equalization. The protector can also be extended across the driver's peripheral circuits.

sible to attenuate the high frequencies 0-6 dB; e.g. to compensate voice coil inductance. During one half-cycle, the driving circuit gets its voltage from this filter through D_1 and D_5, and during the other half-cycle, through D_2 and D_6. D_1 and D_2 can be ordinary small-signal diodes, as also the ones leading to the gate.

The driving circuit must not load the filter too much, and this, again, must not establish too low an impedance in parallel with the driver. If R_1 and R_2 are, say, some hundreds of ohms, R_4 should thus be at least some kilo-ohms and R_5, respectively, still higher.

The protector's operation may yet be sharpened and the threshold voltage dependence lessened by using a drive transistor in front of the power transistor.

An N-type MOSFET can be driven with a PNP transistor, as in Fig. 12.4a. The leveling filter is now referenced to the positive node, and the protector turns on when the PNP transistor's base-emitter voltage, i.e. the voltage across R_3, reaches a certain limit (about 0.6 V), so that R_4 begins to gain current and opens the MOSFET.

Deviation in the base-emitter voltage of bipolar transistors is minute, so the operation accuracy is good even at low powers. The current gain factor is not of much significance here because the base is mostly driven by voltage.

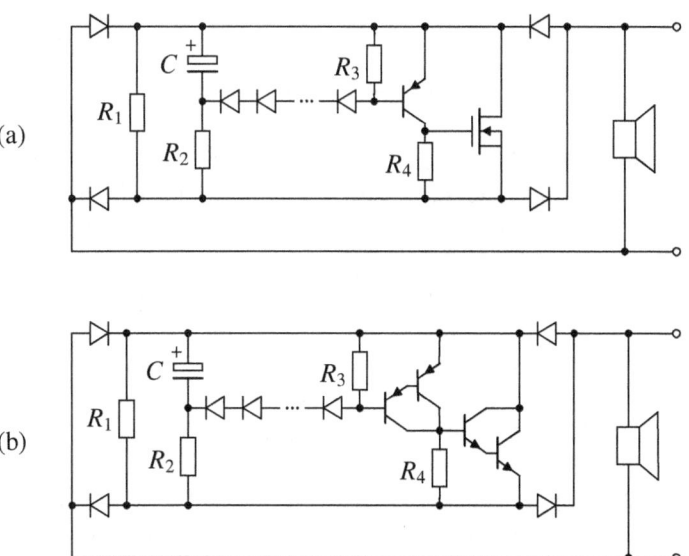

Figure 12.4. a) MOSFET power limiter with PNP pre-stage. b) Limiter realized with two Darlington transistor pairs.

The base current required for activation should be, however, at least an order of magnitude smaller than the current through R_3, which, in turn, should be much smaller than the capacitor's charging current, that again, has to be small relative to the current in R_1. Therefore, the base current can be, in practice, only some microamperes; so, to avoid interferences, high current wires may not be good to be situated very close to the drive transistor though we are not dealing with any precision circuit.

The waveform of the limited voltage is rather different between the circuits of Figures 12.4a and 12.2 because the drive transistor's collector current and the MOSFET's gate voltage proportional to it depend a little on the instantaneous signal value. The first-mentioned circuit works by softly cutting the signal peaks exceeding a certain limit, whereas the latter tends to cut the bottom part of waves, leaving the peak shape quite intact.

Using a pre-stage also enables the use of a bipolar Darlington pair as the power transistor, as in Fig. 12.4b. To get enough base current for this pair, a Darlington transistor can also be employed in the pre-stage. The base voltage required for activation is then 2-fold compared with Fig. 'a'.

The equalizing filter described above can also be appended in the circuits of Fig. 12.4 if only the loading capability of the different stages is minded.

12.3. Voice Coil Temperature Indication

The temperature of a current-driven voice coil can also be monitored directly by resistance measurement. By passing a small direct current (e.g. 20 mA) through the driver, the DC voltage component appearing at the terminals indicates the actual temperature of the voice coil, once the resistance at room temperature is known. With passive speakers, however, the method is only suitable for woofers unless a separate DC wire is provided for the tweeter.

Figure 12.5 shows a possible realization of the method. The scheme is also applicable for ground-connected loads.

The current source can well be a mere resistor from the feedback node to either supply rail. R_1 does not degrade the amplifier's performance in any way, especially if the supply voltage has been regulated.

The direct current component due to the amplifier itself should be nulled out or possibly adjusted so that any separate current source is not

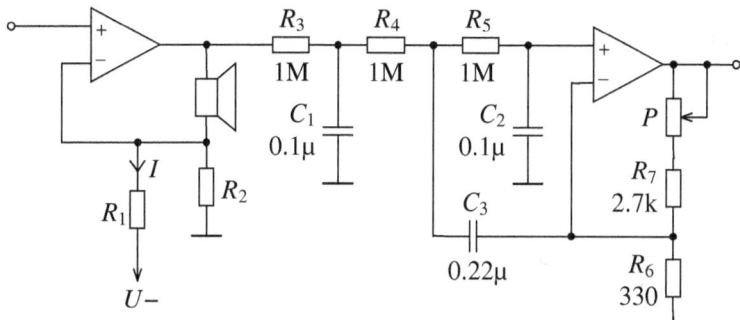

Figure 12.5. Monitoring voice coil temperature by measuring resistance change. The voltage drop caused by the direct current I, flowing through the speaker, is extracted with a low-pass filter, whose gain is adjustable by potentiometer P.

even need. However, in some cases, the temperature dependence of the amplifier-based current may pose a problem, so that rather high current has to be used which charges the driver's excursion capability.

The AC content is removed from the speaker voltage by a 3rd-order low-pass filter, that also amplifies the remaining DC component by the amount needed. With the values given, 20 Hz frequency is already attenuated by 72 dB, relative to zero frequency; the settling time into 5% limits being 0.8 s.

The voice coil resistance and corresponding voltage increase linearly with temperature (0.4%/°C), so the obtained signal may be used e.g. to drive a temperature indicator or to trigger an alarm in case a safe limit is exceeded.

12.4 Overexcursion Indication

As was found in section 3.9, at the lowest frequencies the driver's excursion limit is encountered much earlier than the power limit. Thus, it is also useful to monitor the displacement if the speaker's capacity is to be exploited maximally without yet entering the region of heavy distortion.

According to equation (3.4), displacement is obtained from current by a second-order low-pass function. Consequently, to detect overexcursions, one only needs a suitable low-pass filter accompanied with a peak value indicator. Naturally, the signal is taken from a point after the volume control and possible equalization stages.

Such a circuit, mostly suitable for active speakers, is shown in Fig. 12.6. Filtering is performed by a well-known Sallen-Key circuit, a kind of which can also be discerned in Fig. 12.5. The succeeding inverter enables peaks of both polarity to be observed.

When the characteristic frequency and Q value of the filter match the parameters of the driver, the voltage obtained (U_o) follows the displacement of the diaphragm, provided that distortion is not yet immoderate. The peak value stored in C_3 can thus be used, for instance, to turn on a warning light when the linear range of the driver is exceeded.

However, the overshooting cannot necessarily be prevented even if the program signal were cut off at the same moment as the limit is reached, for the phase difference between the displacement and current can be so large that removing the signal may even increase the diplacement momentarily.

The transfer function of the Sallen-Key circuit is

$$\frac{U_o}{U_i} = \frac{\dfrac{1}{R_1 R_2 C_1 C_2}}{s^2 + \left(\dfrac{1}{R_1 C_1} + \dfrac{1}{R_2 C_1}\right)s + \dfrac{1}{R_1 R_2 C_1 C_2}} \qquad (12.1)$$

Tuning can be carried out by first choosing some practical values for R_1 and R_2 and then solving C_1 and C_2 as follows:

$$C_1 = \frac{(R_1 + R_2)Q}{R_1 R_2 \omega_0} \qquad (12.2)$$

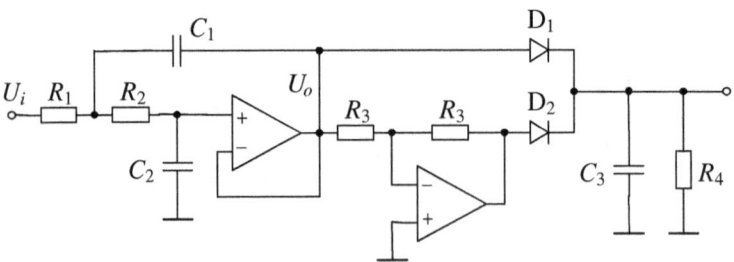

Figure 12.6. Circuit for displacement monitoring. The low-pass filter is tuned according to the driver used. Holding time of the peak value is determined by time constant $R_4 C_3$.

$$C_2 = \frac{1}{R_1 R_2 C_1 \omega_0^2} \qquad (12.3)$$

where ω_0 $(=2\pi f_0)$ and Q are the desired resonance parameters.

By trying different resistance values, one can eventually find such capacitances that fall close to commercial standard values.

13
MEASUREMENT TECHNIQUES

13.1 Resonance Parameter Extraction

The resonant frequency and mechanical Q value of a drive unit can be determined from the resonance region impedance curve, assuming that voice coil inductance can be regarded as negligible at the frequencies in question.

To extract the resonant frequency f_0, it is enough, in principle, to only find the point where the impedance magnitude reaches its maximum. However, especially when the Q value is low, the maximum point is not always easy to be distinguished. Then, one can use a two-point method, in which one finds, on both sides of the resonance point, the frequencies where $|Z|$ is e.g. 80% of its maximum value. Due to the logarithmic symmetry, f_0 is the geometric mean, i.e. the square root of the product, of these frequencies.

Q_m can be extracted using relation (2.15) because the driver's motional impedance forms a 2nd-order band-pass function. Thus, one only has to find the motional impedance's 3-dB limit frequencies f_1 and f_2.

Figure 13.1 shows the impedance diagram, familiar from Fig. 3.7, at the frequencies in question, where the direction angle of Z_m is $\pm 45°$.

By applying Pythagoras' theorem, we can now write:

$$|Z(f_1)|^2 = |Z(f_2)|^2 = [R_c + \tfrac{1}{2}(Z(f_0) - R_c)]^2 + [\tfrac{1}{2}(Z(f_0) - R_c)]^2$$

which yields:

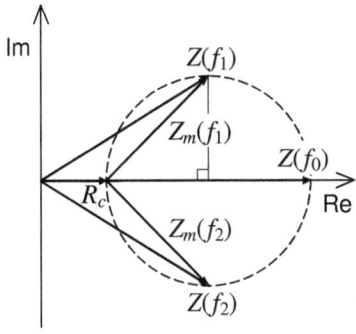

Figure 13.1. Vector diagram illustrating the determination of motional impedance bandwidth. Z_m =motional impedance, Z =total impedance.

$$|Z(f_1)| = |Z(f_2)| = \sqrt{\frac{Z(f_0)^2 + R_c^2}{2}} \tag{13.1}$$

By finding the frequencies that satisfy this condition, Q_m is obtained from the formula

$$Q_m = \frac{f_0}{f_2 - f_1} \tag{13.2}$$

There also exists other methods for solving Q_m but no simpler than this.

With current-driven speakers, it is relevant that impedance is measured using constant current, instead of constant voltage. Figure 13.2 depicts suitable measurement connections when using a normal voltage-output signal generator.

When measurement currents less than 10 mA are sufficient, one can use the simple series resistor method (Fig. 'a'), in which the generator's

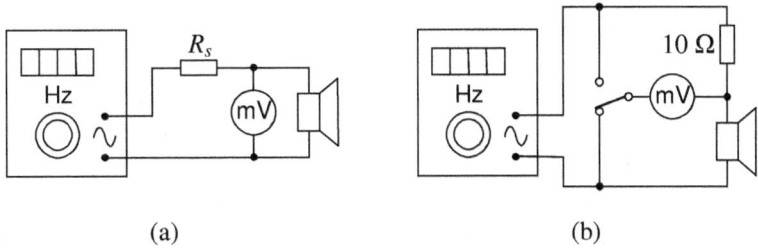

(a)   (b)

Figure 13.2. Impedance magnitude measurement circuit for low currents (a) and higher currents (b). At low frequencies, an ordinary multimeter can be used as the voltmeter.

output level is held constant.

Resistance R_s should preferably be at least 100-fold relative to the driver's nominal impedance. The generator's output resistance, usually 50 Ω, also becomes summed to the circuit's total impedance.

The circuit is calibrated by substituting in place of the test object a known resistance of the same range and by adjusting the voltage across the resistor to correspond to the desired current. After the calibration, the impedance magnitude is obtained conveniently as the ratio of the voltage and current.

By using only small series resistance and adjusting the current separately for each frequency, somewhat higher currents can be obtained. During calibration, the voltmeter can be connected across the resistor and during impedance measurement, across the speaker, as in Fig. b.

13.2 EMF Extractor

The electromotive force extraction device, presented in the following, is useful when resonance measurements have to be performed often and at moderately high currents. In addition to a transconductor stage, the unit incorporates circuitry by which the impedance's electromotive component can be separated from the resistance. Notably, the Q value determination is thereby made easier.

The unit operates on mains voltage, that is yet led up to the circuit board, so the project is suitable only for experienced builders who master electrical safety. One must avoid handling the board always when the mains cable is connected.

The schematic is shown in Fig. 13.3 and the corresponding parts list in Table 5. The input is connected to a sine wave generator to drive the transconductance amplifier that feeds the examined driver. The relative magnitude of the electromotive force is seen from a DC ammeter, that can be an ordinary multimeter* if it is equipped with 2 mA and 200 µA ranges.

First, the input signal is attenuated and stripped of any DC component and high frequencies. The sensitivity has been set so that 5 V input voltage gives rise to 50 mA test current. If higher sensitivity is desired, R_2 can be increased and C_2 correspondingly decreased.

The power amplifier used (A_1) is OPA547, which, with light cooling,

---

* Many portable meters require the user to change the range e.g. in 10-minute intervals to keep them from turning themselves off. For continuous-nature use, it is meaningful to choose, where possible, a meter that does not have this childish feature.

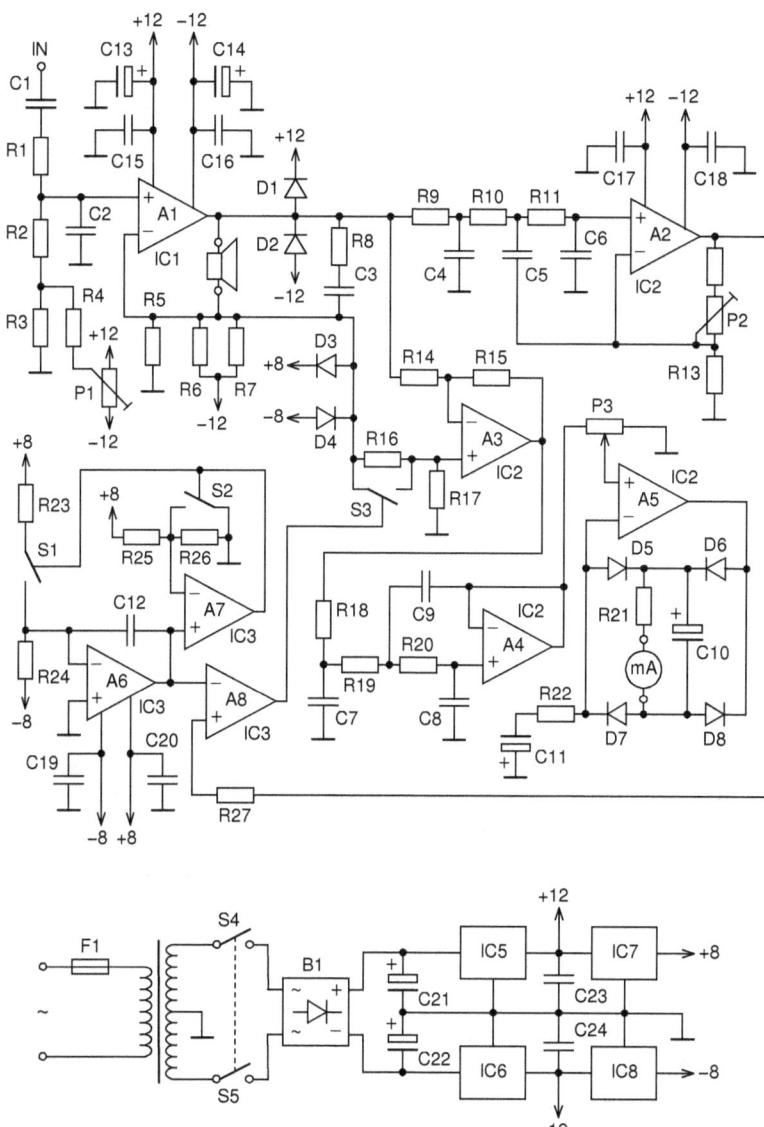

Figure 13.3. Circuit diagram for the electromotive force extractor. Integrated circuits IC_1 and IC_2 operate on ±12 V, while circuit IC_3 and the circuit containing switches S_1, S_2, and S_3 (IC_4) operate on ±8 V. The operation switch (S_4-S_5) has been situated in the transformer's secondary circuit.

Table 5. Parts list for the Fig. 13.3 circuit diagram. Resistor values are from the E24 series and have 1% tolerance.

R1 .... 24kΩ		C1 .... 33µF; 50V; bipolar	
R2 .... 2.4kΩ		C2 .... 33nF	
R3 .... 220Ω		C3 .... 0.15µF	
R4 ....220kΩ		C4 .... 0.47µF	
R5 .... 10Ω; 0.25W		C5 .... 1µF	
R6 .... 1.2kΩ		C6 .... 0.47µF	
R7 .... 1.2kΩ		C7 .... 33nF; 5%	
R8 .... 10Ω; 0.25W		C8 .... 4.7nF; 5%	
R9 .... 1MΩ		C9 .... 68nF; 5%	
R10 .... 1MΩ		C10 .... 100µF; 50V	
R11 .... 1MΩ		C11 .... 2200µF; 50V	
R12 .... 2.4kΩ		C12 .... 10nF; 5%	
R13 .... 150Ω		C13, C14 .... 10µF; 50V	
R14 .... 10kΩ		C15, C16 .... 0.1µF	
R15 .... 10kΩ		C17, C18, C19, C20 0.1µF	
R16 .... 100kΩ		C21, C22 .... 470µF; 50V	
R17 .... 100kΩ		C23, C24 .... 1µF	
R18 .... 8.2kΩ		D1, D2 .... 1N5818	
R19 .... 10kΩ		D3, D4 .... 1N5818	
R20 .... 10kΩ		D5, D6, D7, D8 ...... 1N4148	
R21 .... 1kΩ		IC1 .... OPA547T	
R22 .... 180Ω		IC2, IC3 .... TL074	
R23 .... 10kΩ		IC4 .... CD4066	
R24 .... 20kΩ		IC5 7812	
R25 .... 10kΩ		IC6 7912	
R26 .... 10kΩ		IC7 .... 78L08	
R27 .... 100kΩ		IC8 .... 79L08	
P1 .... 20kΩ; lin		F1 .... 50mA	
P2 .... 1kΩ; lin		B1 .... rect.bridge; 1A	
P3 .... 10kΩ; lin; multi-turn		Transformer .. 2x12V; ≈5VA	

can deliver current up to hundreds of milliamperes.

Resistors R_6 and R_7 introduce a direct current of 20 mA through the driver. The voltage component caused by this current is detected and amplified by a low-pass filter, which is of the same type as that in Fig. 12.5. Thus, the DC voltage seen at the output of A_2 is directly proportional to the voice coil's resistance.

A_6, A_7, C_{12}, R_{23}-R_{26}, and CMOS switches S_1 and S_2 constitute a triangular wave generator, whose operation is based on alternate charging and discharging of C_{12} by a constant current. The output of A_6 produces isosceles triangle wave reaching approximately from 0 to +4 V at 4-5 kHz frequency.

A_8, that functions as a comparator, compares the triangle wave's voltage with the aforementioned DC voltage and generates thereby a pulse signal with a pulse width directly proportional to the voice coil resistance.

To find out the driver's EMF, the output voltage of A_1 has to be subtracted by the voltage across R_5 and the voltage dropped in the voice coil resistance. The last-mentioned can be taken into account by multiplying the voltage across R_5 by the factor $1+R_c/R_5$, where R_c is already known. The subtraction is performed by a difference amplifier formed by A_3 with R_{14}-R_{17}.

Gain from the output of A_1 to the output of A_3 is always -1. Gain from the common point of R_5 and the speaker to the output of A_3 is, in turn, $+1$ when switch S_3 is open, and $+2$ when the switch is closed.

Depending on the duty cycle (ratio of on-time to total time) of the 4-5 kHz pulse signal controlling the switch, the voltage of R_5 is thus amplified by a factor whose time average is within the range 1-2. By setting the duty cycle to match the ratio R_c/R_5, the output of A_3 yields a signal that represents the electromotive force, though containing yet pulse frequencies due to the switching action and a DC component due to the resistance measurement current.

R_c can be at most as high as R_5, that is, 10 Ω in this case.

The mentioned pulse frequencies are filtered out by a 3rd-order low-pass filter (A_4 with its peripheral elements), that is Butterworth-tuned and does not affect the amplitude of the frequencies examined.

After the level control, P_3, the EMF signal is converted into a measurable direct current by a rectifying transconductor circuit. Thereby, the instrument can be used yet at frequencies below 10 Hz where multimeters do not function properly anymore. All DC voltage accrued along the way is stored in electrolytic C_{11}, so that the meter's reading is directly proportional to the electromotive force.

The reading is intended to be kept below 2 mA. At higher currents, A_5 may become saturated, whereupon the result is no longer valid.

Diodes D_3 and D_4 have been included to ensure that the potential of switch S_3 does not exceed the chip's supply voltage when voltages are turned on.

The input capacitor C_1 has to be bipolar because the generator's output may exhibit an unknown DC voltage. In corresponding places, one often sees used two ordinary electrolytics connected face-to-face in series. However, such a connection does not prevent one of the electrolytics from charging to wrong-polarity voltage.

Figure 13.4 shows the top layout view for the circuits of Fig. 13.3, using the board that is shown in appendix H. The board also includes

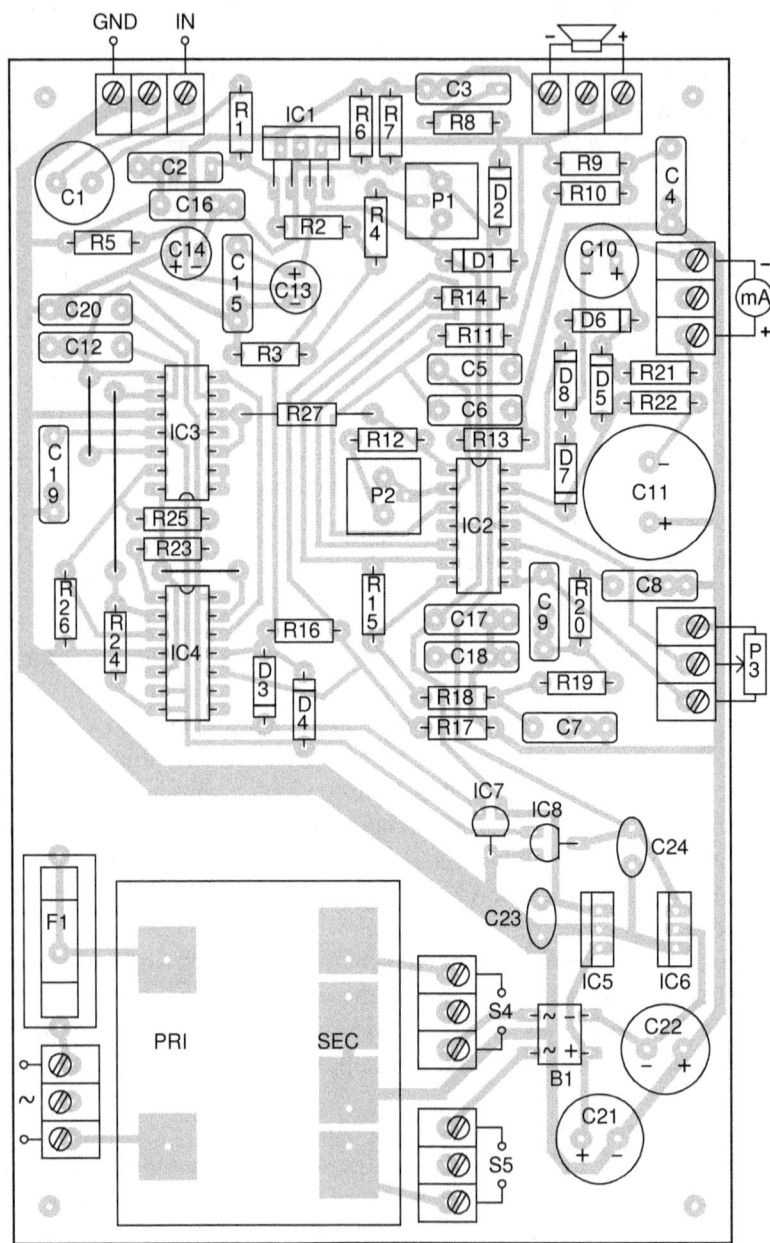

Figure 13.4. Layout design. IC_1 needs a small heat sink.

space for the transformer. Some capacitors have several pitch options. For external connections, one can use 3-way screw terminals.

Tuning is performed by setting known, accurate resistors of at most 10 Ω in place of the driver and adjusting the duty cycle, observed at the output of A_8, to correspond to the resistance connected. The case is best seen with an oscilloscope, but a multimeter can also be used. At low resistances, one should adjust trimmer P_1, and at high resistances, trimmer P_2. Achieving accuracy requires a few iteration rounds.

When tuning with a meter, the input is fed by a signal having suitable level and frequency in the tens of hertzes range (but not exactly the mains frequency), and the direct current obtained is monitored while trying different less than 10 Ω resistances in place of the driver. Because pure resistance does not develop any electromotive force, the meter should always read nearly zero. The reading is adjusted to its minimum by P_1 at low resistances and by P_2 at high ones. It is useful to concentrate the best accuracy in the 3-7 Ω range.

The elimination of the voice coil resistance from the impedance representative signal facilitates a little the finding of the resonance point, but the real benefit is gained in the Q value determination. There is no need to measure the resistance separately, and equation (13.1) is simplified to the form:

$$|Z(f_1)| = |Z(f_2)| = Z(f_0)/\sqrt{2} \tag{13.3}$$

while formula (13.2) remains unchanged. Consequently, by setting the reading at the resonant frequency to e.g. unity, f_1 and f_2 are found at a value of 0.707.

The EMF extraction enables an even more direct way to find out the Q. The method is based on the asymptotes of the 2nd-order band-pass function (dashed lines in Fig. 2.3a) and hence involves a small approximation but gives very consistent results with the above-described bandwidth method.

First, f_0 is found in the normal fashion. After this, the signal frequency is set exactly to value $f_0/5$ and the meter reading is adjusted exactly to a value of 20 or 200 μA. Finally, the frequency is restored to f_0, whereupon the digits representing the Q value can be directly read from the display.

Despite their protection structures, CMOS circuits are susceptible to being damaged due to electrostatic discharges (ESD), so it is well not to touch the pins of IC_4 without protective measures. Also, in the acquiring of such circuits, one should use shops where ESD issues are taken seriously. Otherwise, there is a great risk of getting malfunctioning stuff.

Figure 13.5 shows a finished unit opened. A plastic box is sufficient since very weak signals are not treated.

13.3 Capacitance Measurement

Capacitance measurement is nowadays also provided in many basic-level multimeters. However, the accuracy is often 5% added by a couple of counts, and even this generally applies only to new meters. In many critical places, such measurement uncertainty is no more acceptable, so we need a better means to discover the actual values of the capacitors to be employed.

The method presented in the following is based on the cancellation of the voltmeter's errors in a two-phase measurement. The accuracy achieved is determined mostly only by the accuracy of the used reference resistor and the resolution of the voltage measurement and can easily be better than 0.5%.

The measurement circuit and the phasor diagrams related to it are shown in Fig. 13.6. In addition to a frequency-accurate sine generator and a multimeter, one only needs a series resistor (R_a) and a reference resistor (R_r), whose value approximately matches the impedance of the

Figure 13.5. The EMF extractor completed. Ventilation holes are provided easily by using raising rings between the top and bottom pieces, so that a gap is left between them.

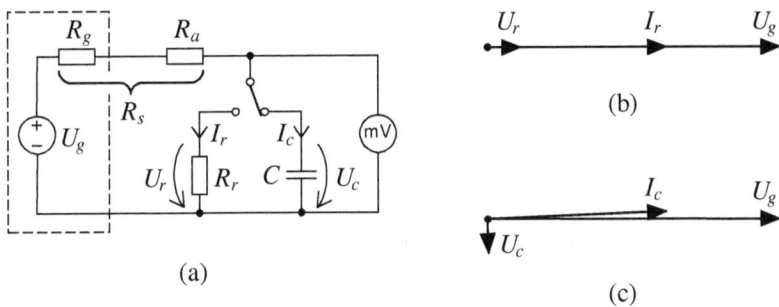

(a)

(b)

(c)

Figure 13.6. a) Circuit for accurate capacitance measurement. U_g and R_g are the generator's idling voltage and output resistance. The meter should be tapped directly on the terminals of R_r or C to prevent wire resistances from affecting the result. (The toggle switch is only a graphical one.) b) Phasor diagram when the reference resistor R_r is connected. c) Phasor diagram when the capacitor is connected.

examined capacitor (C) at the measurement frequency used.

To achieve the best accuracy, resistor R_a is chosen so that $|U_r|$ and $|U_c|$ are some $1/50$ of $|U_g|$. If U_g is 5-10 V, the meter can thus be kept in the 200 mV range. Naturally, the frequency must lie in a range where the meter is capable of operating at least satisfactorily.

When R_r is connected, we obtain from Fig. 'a': $U_g = U_r + R_s I_r$ where R_s is the total series resiatance. Because $I_r = U_r/R_r$, we further obtain:

$$U_g = U_r(1+R_s/R_r) \tag{13.4}$$

When the capacitor is connected, we can write: $I_c = (U_g - U_c)/R_s$. In addition, for a capacitor it always holds: $j\omega C = I_c/U_c$, so

$$C = \frac{I_c}{j\omega U_c} = \frac{U_g - U_c}{j\omega U_c R_s} \tag{13.5}$$

However, $|U_c|$ is very small compared to $|U_g|$, and the respective phasors are virtually perpendicular to each other (Fig. c), so $U_g - U_c$ can be replaced by mere U_g without noticeable magnitude error. With result (13.4), we now get:

$$C \approx \frac{U_r\left(1+\dfrac{R_s}{R_r}\right)}{j\omega U_c R_s} \tag{13.6}$$

By taking absolute values,

$$C \approx \frac{|U_r|\left(1+\dfrac{R_s}{R_r}\right)}{2\pi f |U_c| R_s} \tag{13.7}$$

In practice, the absolute values can be replaced by the measured RMS values.

As U_r and U_c are nearly equal, their relative measurement errors are also virtually equal and hence cancel each other out. Likewise, the error introduced in the determination of R_s is almost entirely canceled because $R_s/R_r \gg 1$. Consequently, by using a precision resistor of say 0.1% as the reference, there is not left any significant error sources in the expression.

The generator's distortion is quite irrelevant here since the effect of the harmonic multiples on the result remains, in practice, negligible. If a function generator with a frequency counter is not available, accurate frequencies can also be obtained from certain test discs or a computer's sound card. For lack of better, it is also possible to use a tuning fork and a loudspeaker in setting the frequency.

When needed, the series resistance R_s can be easily determined. By measuring the generator idling voltage U_g and U_r, R_s is obtained from the formula

$$R_s = R_r\left(\frac{U_g}{U_r}-1\right) \tag{13.8}$$

that directly follows from equation (13.4).

13.4 Inductance Measurement

The above-described measurement concept can also be applied for inductances, even though quite similar accuracy is not necessarily achieved in practice, due to the considerable resistance of inductors.

Here, R_a is chosen so that the measured voltages are roughly $1/100$ of the generator voltage. Thereby, the approximation error is kept small while the meter resolution remains yet moderate.

The inductor's DC resistance (designated R_l) should preferably be known at least with two-digit accuracy. This is easily achieved by a voltage division measurement performed with a battery and a series resistor of about one kilo-ohm.

The measurement frequency has to be so high that the coil's reactance, ωL, is many times greater than the resistance.

Equation (13.4) applies as such also here.

When the inductor is connected, in place of the capacitor, we can write: $I_l = (U_g - U_l)/R_s$ where I_l and U_l are the current and voltage of the coil. The impedance of the coil becomes then

$$Z = \frac{U_l}{I_l} = \frac{U_l R_s}{U_g - U_l} \qquad (13.9)$$

From the phasor diagram in Fig. 13.7a, it can be reasoned, however, on similar grounds as with equation (13.5), that $(U_g - U_l)$ can be reduced to mere U_g. Hence:

$$|Z| \approx \left| \frac{U_l R_s}{U_g} \right| = \frac{|U_l| R_s}{|U_r| \left(1 + \dfrac{R_s}{R_r} \right)} \qquad (13.10)$$

where the absolute values of the voltages may again be replaced by the RMS values.

Also here, the relative voltage errors as well as the error in R_s are canceled, so the accuracy obtained for $|Z|$ is good.

The inductance L is solved by applying Pythagoras' theorem to the impedance diagram depicted in Fig. 13.7b. We have:

$$R_l^2 + (\omega L)^2 = |Z|^2$$

from which:

$$L = \frac{\sqrt{|Z|^2 - R_l^2}}{2\pi f} \qquad (13.11)$$

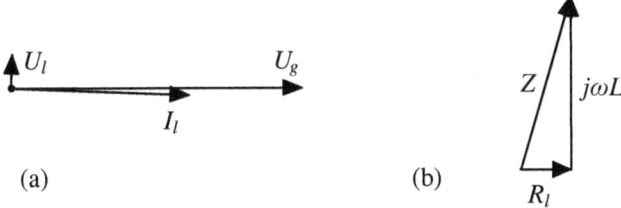

(a)   (b)

Figure 13.7. a) The phasor diagram when an inductor is used in place of the capacitor in the circuit of Fig. 13.6a. b) The decomposition of coil impedance into resistance R_l and reactance ωL.

The effect of R_l inaccuracy is also left diminutive if $|Z| \gg R_l$.

The differentiator formed by the inductance and resistance R_s accentuates the harmonic frequencies, so distortion can become significant here. If the generator distortion approaches one percent and involves relatively high frequencies, it is advisable to filter the measured signal a little to attenuate the distortion components. This can be done by a simple low-pass circuit (Fig. 13.8), whose corner frequency is set a little higher than the measurement frequency. It is important, however, that the same filter is used for both U_r and U_l.

13.5 Lossless Current Meter

Sometimes it may be necessary to get measured currents flowing in the loudspeaker, e.g. to ensure that the circuit operates in the designed fashion. However, all multimeters, regardless of the price level, have so high load resistance in current measurement that they are ill-suited for the purpose. For example, in the 200 mA range, the resistance is often over 1 Ω which already skews the result significantly, as the circuit impedances are of the order of 10 Ω. Moreover, only in few meters the frequency band covers the whole audio range.

Current can, however, be also measured without loading the circuit, by making use of a familiar concept relating to operational amplifiers, the principle of virtual ground, in which the negative feedback causes the amplifier's both input terminals to stay at the same potential. In fact, it is indeed strange why this means is not used, but all meters are still based on the mere series resistor method. The display may have a legion of digits, of which, however, practically none is significant unless the Thévenin impedance of the measured object is very high compared to the meter's resistance. (Curiously, issues related to current seem to be in the world somehow haphazard also more generally, not just the driving of loudspeakers.)

Figure 13.9 shows a practical circuit by which one can measure AC currents up to 12 mA at all audio frequencies without loading the test

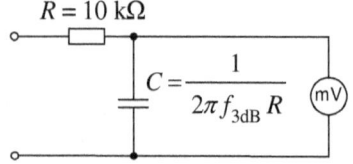

Figure 13.8. Filtering of distortion components from the measured voltage. C is determined according to the desired 3-dB corner frequency.

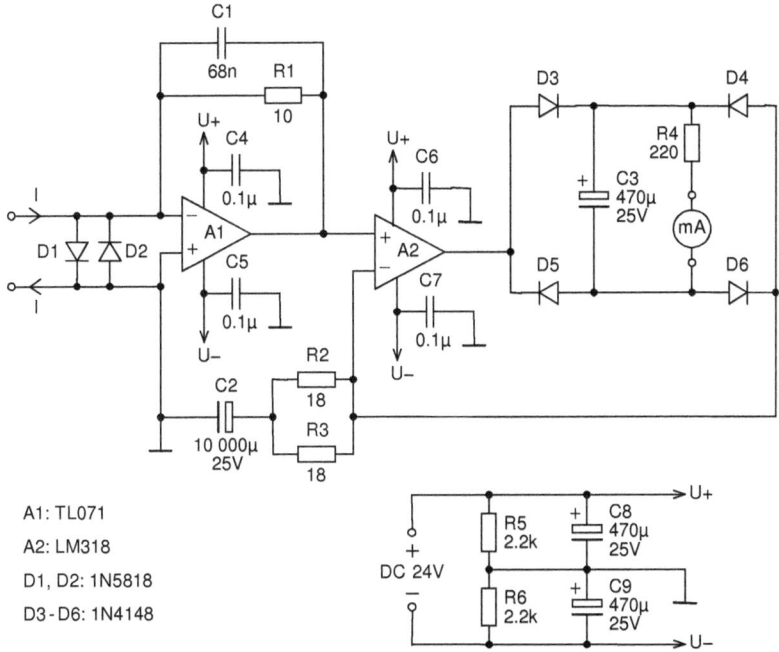

Figure 13.9. Non-loading AC current meter and a suitable splitting network by which a one-sided voltage source can be used to provide the needed two-sided supply voltage. Capacitors C_4-C_7 should be situated close to the respective amplifiers. The polarity of C_2 is insignificant.

object.

A_1 forms a current/voltage converter, also seeing to it that the measurement terminals stay practically at the same potential. The feedback resistance R_1 has to be relatively low in order that the input terminal difference voltage, that is proportional to the amplifier's output voltage, would remain small enough also for the highest operation frequencies. The offset voltage of A_1, seen at the input terminals, is irrelevant when measuring AC current. The purpose of C_1 is to ensure the stability of the stage when the Thévenin impedance of the measured object is capacitive.

A_2 with its associated circuits makes up an AC/DC converter, a kind of which was already used in the Fig. 13.3 EMF extractor. The display device can be a multimeter set at 20 mA DC range.

Schottky diodes D_1 and D_2 protect against overcurrents (up to some limit) and assume all current when the instrument is off. The diodes also enhance stability by preventing the occurrence of high voltage amplitudes.

The meter with its power supply forms a closed system, that does not accumulate electric charge. Thus, the current coming from the test object and the one returning to it (I) are automatically equal at every moment, without any need to care about the matter. So, the grounds of the meter and measured circuit can freely be separate.

The supply voltages are most easily obtained from a regulated off-the-shelf mains adaptor. However, there is no such ones available for two-sided voltages, so the needed middle potential has to be arranged by oneself. The current consumption being relatively small, the ground level can be established by a passive voltage division, as in the Figure. The mains adaptor should be rated at 24 V and 100 mA.

Calibration is not necessarily needed since accuracy is determined by the accuracy of resistors R_1, R_2, and R_3. The display meter conducts a direct current whose magnitude is 1.11 times the rectified average of the measured current. So, the reading obtained indicates the RMS value of a sinusoidal current.

The frequency range extends to 20 kHz when the current is at least 1 mA. At lower currents, the limit decreases, being about 2 kHz at 0.1 mA. The reason is that, when the drive voltage becomes very low, the slew rate of A_2 is no more sufficient in the crossing of the diode voltage drops. If this is to be improved, an amplification stage can be added in front of A_2. Then, R_2 and R_3 must be multiplied and C_2 divided by the gain of the stage.

In battery-operated measurement equipment, the use of the virtual ground concept is limited by current consumption, that is always a little greater than the measured current. However, currents of a few milliamperes are not yet a problem, and the method is most useful just in the low ranges, that otherwise call for high series resistances.

13.6 Coil Loss Measurement

Sometimes it can be helpful to know a coil's dissipation factor or loss angle also for rather high frequencies, where losses are often dominated by other factors than the winding resistance. Figure 13.10 depicts a suitable measurement scheme, that employs an oscilloscope operating in the XY mode.

First, the distortion of the sine wave generator is reduced by a low-pass filter formed by C_f and the generator's own resistance R_g. The cor-

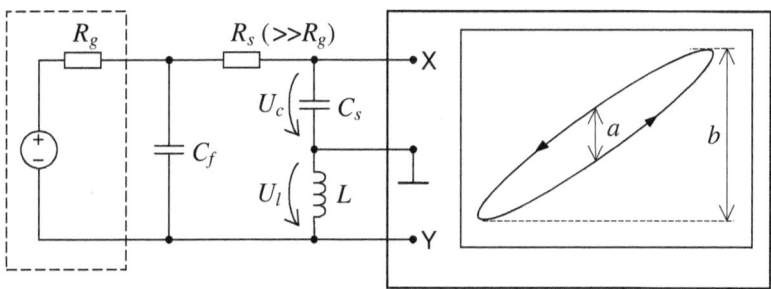

Figure 13.10. Coil dissipation factor measurement with an oscilloscope. When C_s is ideal, ratio a/b indicates the sine of the loss angle.

ner frequency $(1/(2\pi R_g C_f))$ is preferably set a little lower than the measurement frequency.

Capacitor C_s and the examined inductor L are exposed to the same current, and their respective voltages are connected to the oscilloscope's horizontal and vertical deflections. When the coil and capacitor function ideally, U_c and U_l are exactly in opposite phase, and the screen shows only a straight, inclined line. When the coil is lossy, U_l lags with respect to U_c, and the screen shows an ellipse, like the one drawn, whose width is proportional to the dissipation factor.

The mentioned phase lag, i.e. loss angle, results in that when the X signal crosses zero, the Y signal is not yet zero but equals the sine of the loss angle, multiplied by the amplitude. Consequently, the dissipation factor, defined as the tangent of the loss angle δ, is the tangent of the angle whose sine is a/b, that is,

$$\tan \delta = \tan(\arcsin \frac{a}{b}) \qquad (13.12)$$

When δ is small, we can further approximate:

$$\tan \delta \approx \sin \delta = \frac{a}{b} \qquad (13.13)$$

The dissipation factor of C_s should be very low because it too affects the result. Thus, only polypropylene capacitors are fit for the purpose.

When connecting the equipment to the mains power, one must see to it that the ground pins of the generator and oscilloscope will not be in contact with each other.

14

MYTHS AND ATTITUDES

14.1 The Whole Picture of Electrical Damping

In common hifi parlance, there is a lot of talk about electrical damping and damping factor, which are thought to be as if representative of the amplifier's ability to control voice coil movement. In pseudo-scientific renditions, electrical damping is often considered even indispensable in controlling the driver's time behavior, without comprehending that the transient behavior and frequency behavior of a linear system are only one and the same thing looked from different perspectives.

Moreover, when referring to electrical damping, it is often not defined that the issue applies only to the resonance region, and so there is easily created a perception that control is totally lost unless the amplifier's output resistance is small. To dispel such myths, it is necessary to first consider a little the practical significance of electrical damping at different frequencies.

We will take as an example a smallish woofer in closed enclosure, with $Bl = 6$ Tm, $m = 0.008$ kg, $k = 2000$ N/m, $b = 1.5$ Ns/m, and $R_c = 6$ Ω. The amount of electrical damping can be simulated with the model of Fig. 7.2b, applied in Fig. 14.1a.

Current source I_x represents an unwanted force that for any reason acts upon the diaphragm; the force that should be canceled by means of electrical damping. The voice coil movement caused by this force gives rise to a physical current, I_0, which in turn introduces a damping force, I_d, that tends to resist the motion in the resonance frequency region. With the switch open, the operation corresponds to current-drive mode,

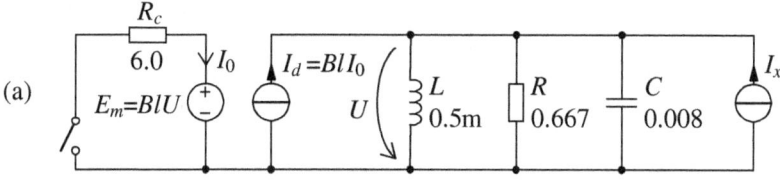

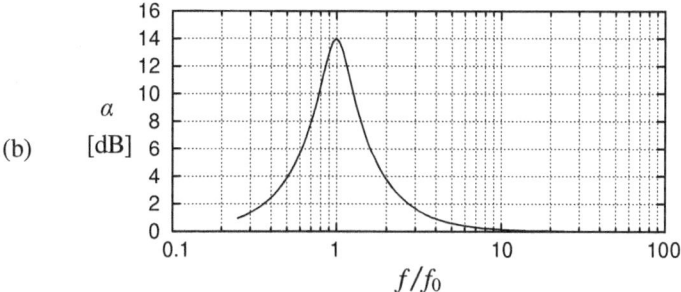

Figure 14.1. a) Model for the simulation of electrical damping in the example driver. I_x represents an interfering force, whose mechanical effect is partially attenuated when the switch is closed, i.e. when the driver is short-circuited. b) Obtainable electrical damping (α) vs. normalized frequency (f_0 =resonant frequency).

and with the switch closed, voltage drive mode.

Without restricting the generality of the analysis in any way, I_x can be interpreted to be sinusoidal since all other waveforms can always be composed out of harmonic sine components. By comparing e.g. the velocity-representative voltage U in both modes, the attenuation obtained in the voltage drive mode is as shown in Fig. b.

At the resonant frequency, electrical damping is generally more than ten decibels, but already one octave away from the resonant frequency, the attenuation has dropped in the example below 4 dB. Two octaves away, the attenuation is already less than a decibel, that is, practically negligible. The range reproduced by a driver can, however, extend as far as about two decades from the resonant frequency.

In the Q value adjustment of the fundamental resonance, electrical damping is of course applicable, but there are no other essential effects related to it. For instance, in the cabinet noise penetration through the cone, this property is of no help.

It is said that the effect of electrical damping can be demonstrated by flicking the diaphragm, whereupon its movement should be lesser when the amplifier is turned on or the terminals are short-circuited. A small difference in behavior can indeed be observed because the frequencies generated are low. If, however, the excitation of the diaphragm could be done at a frequency of, say, over 200 Hz, connecting the driver would no more bring about any difference.

By the damping factor of an amplifier is meant the ratio of the nominal load impedance to the output resistance, to which the cabling resistance also summates. So, the figure is the reciprocal of the current-drive index defined before, being, however, somewhat misleading since, in reality, nothing is damped in proportion to this factor, especially at typical values, that in transistor amplifiers are several tens.

Considering damping factor, one cannot help constantly coming across a mystical story about a speaker diaphragm that, when the signal somehow abruptly terminates, fails to get itself stopped but continues its travel capriciously, beginning then to act with the voice coil as a generator and producing a back voltage and back current and who knows what energies, which then somehow get the amplifier out of kilter, giving rise to distortion if there is not enough damping factor. And what is most lamentable, on grounds of this tale and its different variations, the damping factor is only striven to be increased when the direction should be totally the opposite.

It is altogether pointless to seek to investigate the effects of the generated electromotive forces in the time domain since, when the system is linear, there does not exist any possibility that one of the system's components could behave, as a function of time, somehow uncontrolled or introduce distortion in any situation. Motional EMF is only one linear component among other components and can be *completely* modelled with the parallel RCL network seen in Fig. 7.1 as well for both time and frequency domain analyses. If, again, due to *Bl* variation or some other reason, the motional EMF is no more behaving linearly, distortion is, especially on voltage drive, inevitable, and high damping factor is not of help then either. Further, when it is kept in mind that sound arises from diaphragm acceleration and not from displacement, one can leave all "the diaphragm continues its travel" type contemplations at their due value and focus on, instead of imaginary controlling, establishing a pure acceleration force.

At times, distortion is said to be developed in that the current component caused by the driver's back voltage introduces a voltage drop in the amplifier's output resistance, and this change in the output voltage is then considered to be as though an aberration in the signal. Even if we ignored that, in actuality, a driver responds to current instead of voltage, such functioning of the output resistance cannot introduce any distortion since (the EMF being linear) there is not any factual distortion mechanism involved. By connecting the equivalent circuit of Fig. 7.1 to an ideal voltage source through some output resistance, varying this resistance results only in different Q values and corresponding frequency and impulse responses; but with distortion this issue has as little to do as any change in a speaker's frequency response.

Much effort has been devoted to guarding the output voltage from both linear and nonlinear EMF effects, yet always closing the eyes from the most important: at the moment when the current component caused by a nonlinear EMF flows in the voice coil and the speaker pours forth a corresponding cry, *the damage has already taken place*, and it is absolutely futile to yet try to fix the matter by steadying the output voltage, for thereby one only maximizes the resulting detriment.

In discussion forums, one sees the need of high damping factor being also justified by various energies and power dissipations. There is talk about the cone's kinetic energy or inertia and how the driver feeds power or energy to the amplifier's output circuit and where these energies are spent and how they deform the transients.

Power issues are definitely pivotal in the thermal design of amplifiers and in the determination of the safe operating area of the output transistors; but to the correctness of the actuation of the driver, the momentarily flows of energies and powers are irrelevant.

14.2 Slew Rate Distortion

One salient subject of debate, especially in the high-end community, has been the so-called dynamic distortion of amplifiers, i.e. distortion related to steep signal changes, and its bearing on the perceived sound. Concerning time domain phenomena, there are also a lot of belief-based perceptions, and because the potential discernability of amplifier distortions at least does not lessen with current-drive, there is reason to investigate a little to what extent the concerns are justifiable.

The distortion stemming from the finiteness of the amplifier's slew rate (maximum rate of change of the output voltage), usually called TIM distortion, arises in the amplifier's input stage, i.e. the differential stage, when the voltage across the input terminals, the difference voltage (U_d in Fig. 10.1), becomes so large that the input stage is no more able to handle it but becomes saturated. When this occurs, the output voltage changes at a constant speed; and the amplifier is unable to respond to excitation until the difference voltage has returned into its linear range, that is generally something less than a volt.

What, then, effects the increase in the difference voltage? From Fig. 7.11b, it can be seen that at high frequencies the differential amplifier functions like an integrator, that is, the output voltage is the short-term time integral of the difference voltage. Again, because the output voltage follows the input voltage (holds roughly also on current-drive), the difference voltage is, at high frequencies, proportional to the derivative, i.e. the rate of change, of the applied signal and therefore increases with frequency.

Slew rate distortion can be minimized by input stage design and by keeping the difference voltage small which requires high differential gain. It is, however, vain to seek to increase the speed more than need be.

Figure 14.2 shows a simple circuit by which one can measure change rates from a line level signal. The indicator device can be a multimeter that has 10 MΩ internal resistance. The circuit responds only to positive changes, but negative ones can also be examined by reversing the signal polarity.

R_1 and C_1 make up a differentiator whose output voltage in volts corresponds to the rate of change in V/μs. The differentiator is followed by two successive peak value detectors because by one the hold time cannot be made long enough for observations. C_2 is charged rapidly to the peak value of the derivative but is discharged into the inverting input of A_1 in milliseconds. In C_3, the voltage stays so long that it can be read.

This is not any sort of precision instrument, due to the short duration of the measured peaks and the offset voltages of the amplifiers (which though can be nulled), but nevertheless works appropriately in the range 0.02-0.5 V/μs.

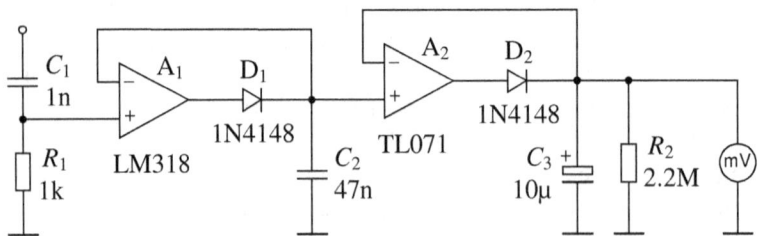

Figure 14.2.   Circuit for peak rate of change measurement. A suitable supply voltage is ±5-±6 V.

The highest rate of change the author discovered by examining a few CDs was about 0.17 V/µs. Often the peaks seem to be associated with cymbal strokes. 0.02-0.03 V/µs readings are found even in rather quiet passages.

The output voltage of the CD player used has been rated at 2 V, so the amplitude peaks can extend to $\pm 2\sqrt{2}$ V. Thus, the maximum speed, normalized into ±1 V range, comes out to be 0.06 (V/µs)/V.

The corresponding value in a 20 kHz sine wave is 0.126 (V/µs)/V, and this is also roughly what the CD format is capable of conveying. (Thus, there doesn't exist any sudden steps at least in recorded signals.)

Consequently, it is, in practice, extremely rare that the rate of change extends to even half of that what is reached with a 20 kHz sine wave at the maximum amplitude; so if the harmonic distortion and intermodulation (IM) distortion of an amplifier are totally negligible up to 20 kHz, then there cannot occur, in practice, TIM distortion either; except, of course, by heavily overdriving the amplifier, whereupon the matter is, however, hardly of interest any more.

It should be noted that the existence of the difference voltage in itself does not denote any misfunctioning or distortion. The difference voltage can be even of the same order of magnitude than the input voltage, and the system may still function completely linearly if only all its parts are linear. To keep harmonic distortion down, the difference voltage should, however, stay relatively low which is realized when the bandwidth is large.

It is also vain to suspect that the feedback signal somehow wouldn't keep up in rapid alterations or that the feedback loop would develop some exotic delays, that would then blur transient reproduction. Amplifiers operate in a minimum-phase manner and are generally not able to produce any inappropriate phase errors, let alone delays. Should such anomalies occur, they would be compelled to show up in the amplifier's phase response.

14.3 The Plague of Compression

Along with voltage drive, and most evidently also by its contribution, there has rooted, in the audio field, also another practice that severely spoils sound quality: *compression*, which means rapid altering of the gain factor in order to reduce level changes in the signal. In addition to compression, it is also common to employ *limiting*, that functions otherwise similarly but still faster and only during instances of peaks.

With few exceptions, all produced recordings are today more or less compressed, and the same also applies to radio channels, who flatten the sound still more to sound syrupy and as loud as possible. In this "loudness war" of record producers and also radio channels, nowadays escalated even to crazinesses, sound quality has become a minor point; and what counts is only the loudness of the sound with respect to competitors, no matter what kind of buzz clutter may result. Although the mastering engineers would at times doubt the reasonableness of the whole race, the producers demand bigger sound; and, what is most tragicomical, it is not always minded even if the signal were heavily clipped big part of the time.

By compression is effected so-called reduction of dynamics, that is, the reinforcement of quiet points relative to louder ones. However, there seems to be prevalent an astonishingly regular belief that the changes in level relations would be the only effect that compression has; as though nothing else noticeable would happen in such a treatment. It hasn't been at all regarded that the alteration of the gain factor is by nature a nonlinear action, and a fundamental feature of nonlinear systems is always the production of distortion.

In compression distortion, we are above all dealing with amplitude modulation. Suppose, for example, that a piece of music contains some beat at a frequency of say 5 Hz and this beat makes the compressor's gain fluctuate within the limits of, say, only 1 dB and at the same time some musical instrument produces a tone of 100 Hz. Assuming yet the gain fluctuation to be sinusoidal, we may now apply the equation describing amplitude modulation, (4.3), using a modulation index of 0.06, that corresponds to ±0.5 dB variation.

Thus, in accordance with Fig. 4.11, there is established around the 100 Hz signal two distortion components, whose frequencies are 95 and 105 Hz. The magnitude of the components is 3%, so the total distortion will be a good 4%.

The fact that the gain variation is in practice other than sinusoidal can only make things worse. If the shape of the variation is, say, ramplike, as possible with a rhythmic signal, the line spectrum of the ramp is just reflected on both sides of the aforementioned 100 Hz frequency. The total distortion does not essentially change from the above, but the distortion products spread then farther away, at frequencies 95, 90, 85... Hz and at 105, 110, 115... Hz. Likewise, a sine wave of e.g. 500 Hz is scattered in this case at frequencies $500 \pm n \cdot 5$ Hz.

Thus, compression is a powerful producer of nonharmonic distortion, dispersing the energy of all sine wave components

into a frequency band that has twice the bandwidth of the gain fluctuation. A gain fluctuation of just one decibel effects a distortion of the order of 4%. Typically, the change can be, however, even ten decibels in milliseconds, so the distortion figures rise at times to really tremendous heights.

Consequently, such claims that compression, when "correctly" or moderately used, would not produce distortion, are mostly based on wishful thinking.

How, then, can it be possible that such a maltreatment of signal, that has already made natural-sounding recordings an endangered species, has been accepted as a common practice? There are several reasons:

- Indifference: The general public has not learned to demand naturality from sound in the same way as there is expectations for picture. The sound is just either heard or not heard, and it seems to suffice that the words can be recognized. If a picture is grainy or the hues incorrect, it is common to seek to figure out the fault or to complain somewhere; but when the sound is commensurately awry, similar actions are not easily undertaken.

- The dominance of plastic jar equipment: Recordings and broadcasts are preferably processed to be listened through cheap-grade stereo packages or button earphones, that are manifoldly deficient in reproduction, whereupon the result obtained with higher-quality gear is not cared about.

- Distortion components close to the signal frequency do not sound as offensive as those farther away. However, those over 15 Hz away are already sensed as gruffness; and when the attack time of a compressor is, say, of the order of 5 ms, the modulation products easily extend somewhere 100 Hz away from the signal.

- Of the different kinds of distortions, familiar to laymen is only the overdrive-induced clipping distortion, that consists of relatively high frequencies. Instead, modulation products mainly appear in the same region as the signal itself, and so they are not so easily interpreted as distortion as are the crackles produced by clipping. Low-frequency distortion may even please some and sound as a trendy effect.

- The masking effect of voltage drive distortions: Voltage drive introduces, along with other distortion effects, also amplitude modulation; and these, together with cabinet noise penetration, reduce the relative import of the additional distortion brought by compression.

Compression distortion most plainly shows up in speech voice and on current-drive and makes e.g. a radio sound broken and burdensome since to us there has developed a certain perception about how a human voice ought to sound. The distortion products are most severely heard at lower mid-frequencies as roughness, growling, and throatiness. The accentuation of the quiet points, again, brings to the fore all breath intakes, swallowings, and rasps, that don't at all belong to normal communication and give the impression of talking directly to the ear. For the selfsame reason also the s-letters tend to fizz; and accordingly there has then arisen the necessity to employ particular s-attenuation processors (de-essers), that wobble the gain factor still more when certain treble frequencies appear in the signal.

Squeezing taken to the extreme is found, for instance, in many commercials whose sound is mere harsh fizz.

In radio practice, there is used, in addition to actual compression and limiting, also numerous other peak leveling techniques and processings, so the use of FM broadcasts for any quality-demanding purpose is not a good idea.

Wouldn't it be in order also to have some discussion about to what extent e.g. radio stations may modify a transmitted musical piece without it being classified as a modified version? If the signal contains, on average, say a quarter of inappropriate frequencies and the work already sounds considerably different than was the original intent, does this do justice for the performers? A lot of music is bought on the grounds of what is heard in the radio, and the present practice is surely not in favor of an artist or record that strives for naturality.

The compression of acoustic music is comparable, for instance, to a practice where a TV camera would be equipped with a lens that magnifies some part of the picture. No matter that shapes and relations get warped when one only gets visible something as big as possible. As far as is known, this has not yet been undertaken although it would be all logical, after crumpling the sound.

Compression is sometimes rationalized by stating that in a vehicle or other noisy environment quiet passages would not otherwise be discerned adequately. However, the only reasonable practice for these cases would be to leave such operations as the duty of the listener, whereupon everyone could adjust his dynamics and distortion exactly as he pleases, just as is done with the frequency response. Electronics are cheap, and new features are easily provided if there is demand.

Compressing and its opposite, expanding, are also used in one form or another in all noise reduction systems. By expansion during playback, it is quite possible to restore the dynamics that was reduced in the recording phase; but cancellation of the distortion components would require that the expander should be able to produce exactly equal but opposite sidebands with respect to those produced by the compressor. Due to phase errors and tracking inaccuracies, this is, however, very difficult and hardly occurs in practice. Noise reduction has been extensively used in recordings made with analogue recorders; so such, generally old material is neither best for the assessment of current-drive systems.

14.4 What RMS Power?

Concerning both amplifiers and loudspeakers, it is common to talk about RMS power, and many manufacturers also wish to attach to their power specifications the marking RMS. However, many perhaps don't come to think what this abbreviation actually means. Although the issue is not directly related to the current-drive topic, there is something parallel since here too we see how common inexact practices and perceptions are, in the field of audio technology.

RMS stands for *Root Mean Square* and denotes the square root of the mean of the signal's square, corresponding to a DC voltage or current that feeds to the load the same power as the voltage or current in question. So, the RMS value of quantity x is defined as

$$X_{RMS} = \sqrt{\frac{1}{T} \int_0^T x^2 \, dt} \qquad (14.1)$$

where the integration over time T and division by it implies averaging.

Thus, for AC voltages and currents, the RMS value, also called the quadratic mean, is a very useful concept, but the RMS value of power does not indicate anything meaningful. Take the simplest of examples:

We will feed a resistor of 1 Ω by a sinusoidal voltage of 1 V RMS, so that the current has an RMS value of 1 A. Readily, we then know that the average power dissipated in the resistor is 1 W. But what about RMS power?

By marking $u = i = \sqrt{2} \sin(\alpha)$, one obtains as the instantaneous power $p = u \cdot i = 2\sin^2(\alpha)$. By integrating over a half cycle, we may write as the mean square of power $(1/\pi)\int_0^\pi p^2 d\alpha = (1/\pi)\int_0^\pi (2\sin^2(\alpha))^2 d\alpha$. By using trigonometric identities, the integrand reduces to simple cosine functions, and the solution obtained is 3/2. Consequently, the power in accord with the RMS formula is in our example $\sqrt{3/2}$ W ≈ 1.22 W!

However, this is not quite likely to be what those speaking about RMS power really mean. It is more probable that a sine wave power of some duration is meant; although factually, the designation RMS by no means implies what waveform or conditions are being considered.

The international standard IEC 60268-5, that addresses loudspeaker measurements, also does not recognize RMS power. Somewhere, it has just been invented to tack on this stylish-sounding acronym, and so the ball has begun to roll. It would be easier, however, that it would somehow be told what kind of signal is considered.

14.5 The Real Meaning of Group Delay

Group delay is also a concept beset by illusions. Group delay is defined as the opposite of the phase curve's derivative, as given by equation (B30) in appendix section B8; and in describing a system's time behavior, the quantity has only certain limited import. In hifi culture, however, it is customary to interpret group delay pretty liberally.

Group delay is often conceived to be as though a frequency-specific general delay, which, however, it is not. In fact, it is impossible to even define unambiguously a frequency-specific delay, due to the periodicity of the sine wave. There doesn't exist any means to distinguish which sine wave top at the output represents which top at the input, and hence it is not necessarily even very meaningful to speak about delays experienced by single frequency components.

A constant group delay has generally been considered to mean that phase relations between different frequency components would remain intact. The notion is, however, wrong, as can be detected on the grounds of Fig. 14.3.

In the Figure, the derivative of the phase curve (and hence the group delay) is the same at frequencies f and $2f$ and also between them. However, the points 1 and 2 are not located on the same origin-crossing line, so a signal containing frequencies f and $2f$ does not preserve its waveform.

Moreover, even if the frequencies f and $2f$ were reproduced flawlessly in terms of mutual phase, the group delay can be different at these frequencies.

Thus, group delay does not tell anything how individual frequency components are delayed when travelling through the system. Instead, the phase shift and changes in mutual phase relations experienced by different frequencies manifest in the phase response curve itself.

Hence, also the group-delay-derived notion that bass frequencies in a loudspeaker would always lag behind other frequencies, on the time axis, doesn't hold true as such.

Further, group delay may also assume negative values at some frequencies and hence also wouldn't be quite reasonable as a frequency-specific delay.

What sense does group delay then have? The meaning mostly relates to the transfer of information carried by the signal, for instance, in the envelope curve (curve along the wave tops) of wave packets or bursts containing several cycles. Such envelopes may be delayed by a different amount than the individual waves [1].

Group delay is, as the name implies, a *group* delay, that indicates the delay of the envelope curve of a group of waves appearing in some frequency band, *in the special case* that the amplitude response and group delay are constant in that frequency band. If, instead, one or both of the conditions are not satisfied, that is, if the amplitude response or group delay is not constant in the band in question, the shape of the envelope curve changes, and the group delay then no more has any clear physical interpretation.

As flat as possible group delay may still be a property that is worth pursuing, for the sake of overall timing accuracy, but is not any proof of phase correctness.

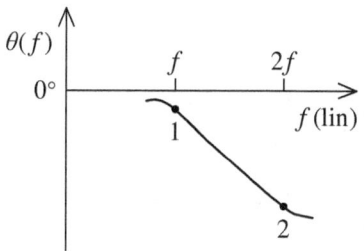

Figure 14.3. Example phase response that produces constant group delay from f to $2f$, but nonetheless distorts the mutual phase of frequencies f and $2f$.

When discussing delays, there is also the problem of proportioning them. A change of 1 ms in group delay in the treble region is certainly more significant than an equal change at bass frequencies.

Instead or in addition to group delay, there could be used a quantity that might be called say *phase projection* and is defined as:

$$\theta_p(\omega) = \theta(\omega) - \frac{d\theta}{d\omega}\omega \qquad (14.2)$$

Graphically, θ_p represents the phase axis intercept of the phase curve's tangent, as depicted in Fig. 14.4.

By presenting θ_p as a function of frequency, one can obtain an illustrative perception of how well the phase relations between frequency components are really preserved. The lower $|\theta_p|$ is in a given frequency range, the better the phases match within that range; and the alignment between two frequencies can be assessed by the net area left between them. When $|\theta_p|$ approaches 360°, the alignment becomes good again.

14.6 Music Counts

Given the in many ways detrimental effect on sound quality by the present speaker driving practice and also by compression, it is also justified to ask what import all this may have had to the musical elections of consumers and the evolution of musical culture universally? Surely, becoming distorted doesn't do good for any type of music; but it is justifiable to believe that the audibility and harmfulness of these distortions are largely dependent on the basic nature of the music, and therefore some types may suffer conclusively more than others, whose attraction is fairly unaffected by the quality of the reproduction chain.

The greatest loser is liable to be acoustic instrument produced, tradi-

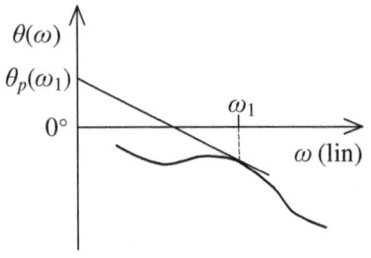

Figure 14.4. Phase projection θ_p indicates how far from the origin the tangent of the phase curve intersects the phase axis. The closer θ_p is to zero or $\pm n \cdot 360°$, the better.

tional music containing natural timbres and elements which, when corrupted, charge the pleasantness and plausibility of reproduction. Instead, e.g. from the sound of an electric guitar, it is more difficult to conclude which distortions are deliberately produced by the player himself and which, again, are excess introduced afterwards. The imprint that voltage drive brings to sound is just a kind of electricness, fizziness and artificiality, so it is quite obvious that music that has originally been produced electrically and synthetically or even by machine tolerates the additional electricness introduced by the reproduction chain better than say hall-recorded concert music.

The clotting of sound and lack of resolution in the loudspeaker again does not favor polymorphic, melodiousness- and nuance-based music but tends to steer customs in plainer, more striking, and same pattern repeating direction. When none of artistically meritorious musical elements work quite properly, what are we left with? Mostly only rhythm. No wonder the dominant substance in the mass production music of our times has come to be beat and drumming, in the background of which there is offered various electrical and synthetic sounds. Generally, this rhythm has also been computerized and syncopated to be thumping or whipping. Nowadays, even children's songs are customarily provided with is pounding.

Of course, the absurd loudspeaker operating practice of our times does not alone explain any culture changes; but it is, however, very justifiable to suspect that this major technical aberration has at least been strongly influencing in the background on what kind of perception there has formed to people about different types of orchestras and forms of music, in the era of electric communication.

As remarkable and far-reaching as the ramifications of voltage drive might be, there pertains to our culture, however, a certain, in societary effects perhaps yet more significant mistake: it is the notion or attitude according to which it does not matter what kind of music one listens, as long as one just likes or supposes to like what one hears. This viewpoint has been adopted without any due justifications only in the last century and is at total variance with what in all past ages in fairly all advanced cultures has been known and understood about the power of music as the moulder of man's character and value system and hence as the upholder of the very empire.

According to the said conviction, which contradicts as well ageless wisdom, normal faculties of observation, as also examined information, music has only a *subjective* and brief, hearer's taste dependent, inconsequential emotional effect and nothing else. Yet in numerous investigations, music has been found to also have direct, will-independent effects and even plants being sensitive to it. A certain kind of sound or music in suitable duration has enhanced growth and made stems turn toward the loudspeaker; whereas a certain kind of sound or music has made plants suffer, turn away from the speaker and eventually die.

The reactions observed in plants and animate life are perhaps the most direct evidence of that music also has upon living organisms an *objective*, i.e. stance-independent, influence that is by nature positive or negative, life-enhancing or life-opposing and -damaging.

Music exerts, and has always exerted, and objective influence also upon humans; constructive or destructive, harmony- or disharmony-advancing, spiritually upward-leading or spiritually downward-leading, or then some kind of mixture of both; and it is not specifically difficult to discriminate where on this axis each style about falls. The determining factor is the using of musical elements; not so much the lyrics although they too have their own suggestion effect.

During recent decades, downward-leading music has been made and used much more than upward-leading music. Current-drive technique might, however, have potential to change this ratio to be a bit more positive.

In accord with the prevalent negligibility axiom of music, it has also been accustomed to think that each cultural or lifestyle trend as though develops its own music. Yet already in the ancient civilizations of the antique, like China, India, Egypt, and Greece, it has been known that the case is just vice versa, that is, music determines the course of the entire evolution; and in securing the welfare and stability of the empire, the purity of music has been regarded even as the most crucial factor.

For instance, the culture of Greece lived its pinnacle period in the 400s BC; but at the end of the century, their music gradually began to change from the traditional and classic into rule-breaking and radical; just as once again has happened in our times. So the cultural and military might of Greece began to decline, when coming to the 300s BC, accompanied by numerous civil wars, and Macedon had it easy to conquer the territory in 338 BC.

Observing recent history also clearly indicates the case to be so that a new musical innovation always brings along a new behavioral and value system innovation; and not so that first there would arise some new lifestyle, whose advocates would then by degrees develop a fitting musical genre for themselves.

The power of music in the governing of people is well reflected in a statement by Scottish parliament member Andrew Fletcher from 1704:

"I knew a very wise man ... that he believed if a man were permitted to make all the ballads, he need not care who should make the laws of a nation." A suitable saying would also be: As in music, so in life. Regrettably, however, often we are not even allowed to determine what we listen and when, or to what kind of culture's or ideology's disseminators our moneys go through various storage medium levies and other collective charging automatics.

The general 'who cares' attitude toward sound has also made possible the forced music played in public premises, regular in stores and supermarkets and even in hospitals, sports events, and airplanes. The feeding of forced music in effect implies a decision that you don't have a right to think your own thoughts in that place, but rather you have to submit to those thought models that the producers of the music represent and that they strive to plant into your consciousness. Likewise, it is implied that you don't have a right to maintain your own state of mind in that place, but you are required to accommodate your own state of mind to that state of mind in which the makers of that music have made their produce and which they desire to manifest by it. Besides that the reproduction systems of public spaces are generally unfitting for any music use altogether, the played material is also more often than not from that negative end.

Here and there businesses direct their forced music also into publicly-owned pedestrian paths. By what right?

Forced music should be viewed at least with equal seriousness than we have already learned to view passive smoking, that has gained much attention also from the legislator. The adverse health effects of both are known, but only the latter receives attention, whereas the former is not discussed nor acted upon because else there would arise an unthinkable situation: the negligibility axiom of music, assumed in the foundations of our culture, would partly have to be called into question!

All of us probably have some experience about what kind of sound can be produced by rubbing polystyrene foam or what kind of shivers the grating of e.g. train wheels may sometimes induce. What would we become like if we were subjected to listen to the squeaking of polystyrene foam many hours daily? According to modern conviction, sound is only acoustic vibration and wave motion that cannot have any straight effect on humans unless the decibels reach up to the limits of hearing damage. And yet we can experience spine-chillingly that such influence *does* exist.

Paradoxically, never before has music been so easily available in its various forms, and never before has it been pumped into us so extensively through all channels; and yet never has common knowledge and contemplation about the real nature and practical effects of music been at a lower ebb than these days.

By music, a civilization is possible to be elevated to the blossom of a golden age. By music, a civilization is also possible to be bombarded into anarchism and self-destruction. In which direction the development shall lead – to that each one is able to affect by his choices, knowing one's responsibility.

---

[1]   A. Bruce Carlson, "Communication Systems", McGraw-Hill, 1981, p. 177-178.

APPENDIX A
Introduction to Complex Numbers

Using complex numbers is necessary and very handy in making AC-technology and signal processing understandable, and without complex numbers it is difficult to gain overall conception even of the operation of primitive filters. In the following, the fundamentals of complex arithmetic are presented for those not familiar with the art or who may have forgotten it.

Complex numbers are two-dimensional numbers that have, in addition to the usual real dimension, also an other, imaginary dimension that is perpendicular to the former. The basic unit on the imaginary axis is the *imaginary unit j*, that corresponds to unity on the real axis and has the basic property:

$$j^2 = -1 \tag{a1}$$

The real axis and the imaginary axis define the *complex plane*, and each complex number corresponds to a point on this plane. In rectangular form, a complex number z is

$$z = a + jb \tag{a2}$$

where a is called the *real part* and b the *imaginary part* of z. Complex numbers are often viewed as vectors in the complex plane, like in Fig. A1.

When b is zero, z is real, that is, the set of complex numbers includes the set of real numbers. When a is zero, z is said to be purely imaginary.

Complex numbers are added like vectors by separately summing the real parts and the imaginary parts:

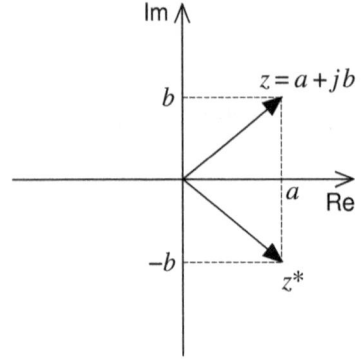

Figure A1. Graphic representation of a complex number z and its conjugate, z^*.

$$z_1 + z_2 = (a_1 + jb_1) + (a_2 + jb_2) = a_1 + a_2 + j(b_1 + b_2) \qquad (a3)$$

Subtraction is performed accordingly.

For the product of complex numbers, we get

$$z_1 z_2 = (a_1 + jb_1)(a_2 + jb_2) = a_1 a_2 - b_1 b_2 + j(a_1 b_2 + a_2 b_1) \qquad (a4)$$

where property (a1) has been used.

Normal laws of associativity, commutativity, and distributivity, familiar from real arithmetic, also apply as such for complex numbers.

The complex number that has the same real part but opposite imaginary part than z is called the *complex conjugate* of z and designated z^*. So, if $z = a + jb$,

$$z^* = a - jb \qquad (a5)$$

Thus, z^* and z are always mirror images of each other about the real axis, as shown in Fig. A1.

The following relations are easily proved:

$$z + z^* = 2a \qquad (a6)$$

$$zz^* = a^2 + b^2 \qquad (a7)$$

Therefore, the sum and product of complex conjugates are always real. The following rules are also valid:

$$(z_1 + z_2)^* = z_1^* + z_2^* \quad ; \quad (z_1 - z_2)^* = z_1^* - z_2^* \qquad (a8)$$

$$(z_1 z_2)^* = z_1^* z_2^* \quad ; \quad (z_1/z_2)^* = z_1^*/z_2^* \tag{a9}$$

Property (a7) is useful in the division of complex numbers because the denominator can be made real by multiplying both the numerator and denominator by the complex conjugate of the latter:

$$\frac{z_1}{z_2} = \frac{z_1 z_2^*}{z_2 z_2^*} = \frac{z_1 z_2^*}{a_2^2 + b_2^2} \tag{a10}$$

A complex number can be expressed by polar coordinates as follows (Fig. A2):

$$z = r[\cos(\alpha) + j\sin(\alpha)] \quad , \quad z \neq 0 \tag{a11}$$

where r is the *absolute value* of z (length of the vector) and α is the angle of z relative to the positive real axis and usually limited to the interval $(-\pi, \pi]$. Using Pythagoras' theorem and rule (a7), the absolute value becomes

$$r = |z| = \sqrt{a^2 + b^2} = \sqrt{zz^*} \tag{a12}$$

Angle α, which is also called the argument of z, satisfies relations

$$\tan(\alpha) = \frac{b}{a} \quad ; \quad \sin(\alpha) = \frac{b}{r} \quad ; \quad \cos(\alpha) = \frac{a}{r} \tag{a13}$$

When determining the angle, one must, however, be careful about the correct quadrant.

Multiplication and division become particularly simple in the polar form. If $z_1 = r_1[\cos(\alpha_1) + j\sin(\alpha_1)]$ and $z_2 = r_2[\cos(\alpha_2) + j\sin(\alpha_2)]$, it can

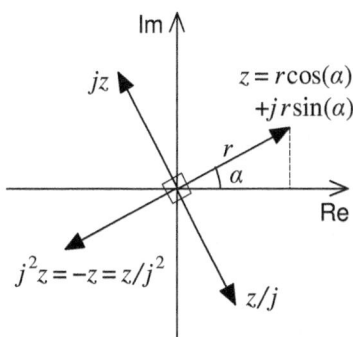

Figure A2. Complex number z in polar coordinates and the effect of multiplication and division by j.

be shown, using general formulas of trigonometry, that

$$z_1 z_2 = r_1 r_2 [\cos(\alpha_1 + \alpha_2) + j \sin(\alpha_1 + \alpha_2)] \tag{a14}$$

Thus it is seen that

$$|z_1 z_2| = |z_1| \, |z_2| \tag{a15}$$

and

$$\angle(z_1 z_2) = \angle z_1 + \angle z_2 \tag{a16}$$

where symbol $\angle$ denotes the argument. Therefore, in the multiplication of complex numbers, one only needs to multiply the absolute values and add together the angles.

For division, it holds accordingly:

$$|z_1 / z_2| = |z_1| / |z_2| \tag{a17}$$

and

$$\angle(z_1 / z_2) = \angle z_1 - \angle z_2 \tag{a18}$$

In division, one thus only needs to divide the absolute values and subtract the denominator angle from the numerator angle.

For the imaginary unit, it always holds: $|j| = 1$ and $\angle j = \pi/2$ (90°), and therefore, due to above rules, multiplying a number by j keeps the absolute value unchanged but increases the angle by 90°, whereas dividing by j deducts the angle by 90°, as shown in Fig. A2.

The exponential function of complex numbers is defined as follows:

$$e^z = e^{a+jb} = e^a [\cos(b) + j \sin(b)] \tag{a19}$$

It can be shown that with this definition the exponential function fulfils the same laws of calculus as the corresponding real function. Thus, for example

$$\frac{de^z}{dz} = e^z \tag{a20}$$

and

$$e^{z_1 + z_2} = e^{z_1} e^{z_2} \tag{a21}$$

By substituting $z_1 = a$ and $z_2 = jb$, we obtain from equations (a21) and (a19) so-called *Euler's formula*:

$$e^{jb} = \cos(b) + j \sin(b) \tag{a22}$$

which, together with equation (a11), yields an important result:

$$z = r[\cos(\alpha) + j\sin(\alpha)] = re^{j\alpha} \qquad \text{(a23)}$$

A complex number can thus also be expressed as an exponential function.

We can also see that

$$|e^{j\alpha}| = 1 \qquad \text{(a24)}$$

$$\angle e^{j\alpha} = \alpha \qquad \text{(a25)}$$

and

$$|e^{z}| = e^{a} \qquad \text{(a26)}$$

APPENDIX B
Properties of Linear Systems

In order to understand the behavior of loudspeakers and various filters quantitatively, in both frequency and time domains, and to be able to establish desired frequency characteristics for equipment, we need some insight into the regularities governing signal processing and a uniform language to describe analogue filtering operations mathematically. In ordinary books dealing with electronics and speaker building, this background is usually not taught, and hence too often the theoretical knowledge of hobbyists remains fragmented or limited to inserting numbers to a given formula. The fundamentals of using complex number algebra and transfer functions are, however, so essential and indispensable tools for signal handling applications that it is not worthwhile for anyone to bypass the little effort that acquiring these practices may require.

The aim is not to treat the subject extensively and theoretically but only to the extent that is necessary to form a basis for understanding sound reproduction systems. For readers not conversant with phasor calculation, appendix A provides a tutorial on complex numbers and their use.

Digital filters also belong to so-called linear systems as well as analogue ones. With digital technology, it is possible to realize certain features, like delay and phase linearity, that are difficult to be implemented with analogue circuits. Concerning the issue of current-drive, however, digitalization has nothing special to give, so this side is not treated.

B1   *Definitions*

An apparatus that provides a linear relationship between its input signal $x(t)$ and output signal $y(t)$, is called a *linear system*. Linearity is achieved when the following two properties are in effect:

- Multiplying (amplifying) the input signal by a constant k causes the multiplication of the output signal by the same constant.

- If a sum of two signals is applied to the input, the output gives the sum of the output signals caused by each input signal individually.

In symbolic form, we can thus write: if

$$x(t) \rightarrow y(t)$$

then

$$kx(t) \rightarrow ky(t) \tag{b1}$$

where arrows indicate cause-effect-relationship. Further, if

$$\begin{cases} x_1(t) \rightarrow y_1(t) \\ x_2(t) \rightarrow y_2(t) \end{cases}$$

then

$$x_1(t) + x_2(t) \rightarrow y_1(t) + y_2(t) \tag{b2}$$

The latter property is also known as the *superposition principle*, and it has a pivotal significance in the analysis of linear systems because the effect of each input component can be examined separately and independently.

Figure B1 depicts a linear system using block diagram notation. *H* denotes the law by which the output depends on the input. It is customary to mark quantities that represent functions of time with lowercase letters and those that represent functions of frequency with uppercase. Symbol for the time variable is usually t (sometimes also τ), but convention for the frequency variable varies in literature, and Fig. B1 itemizes different customs. When dealing with spectrums or frequency distributions, either the cyclic frequency f or the angular frequency ω may be used, which are always tied by the relation:

$$\omega = 2\pi f \tag{b3}$$

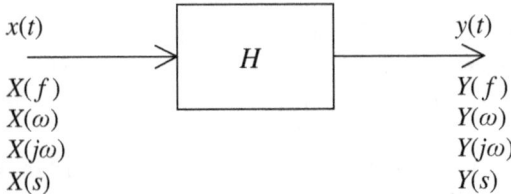

$x(t)$ $y(t)$

$X(f)$ $Y(f)$
$X(\omega)$ $Y(\omega)$
$X(j\omega)$ $Y(j\omega)$
$X(s)$ $Y(s)$

Figure B1.   Block diagram representation of a linear system, and different notation practices of signals.

Sometimes the imaginary unit j ($j = \sqrt{-1}$) is also appended to the variable because the factor $j\omega$ often appears in expressions. When discussing transfer functions in particular, the common complex frequency s is used, which, in the case of sinusoidal signals, is the same as $j\omega$.

In principle, whatever quantities may occur as input and output of a linear system. Often the issue is about voltages or currents or mechanical motion and in our case also acoustic pressure. Many kinds of regulation systems found in industry (e.g. control of temperature, pressure, or surface level) are also linear by nature. It does not even have to be a technical system. For instance, in economy, there might exist, between a product's demand and supply, a dependency that resembles a linear system.

The input signal is also called *excitation*, or *stimulus*, and the output signal is also called *response*. *Frequency response* refers to the complex-valued function that describes the frequency dependency of the ratio of output to input.

Besides obeying conditions (b1) and (b2), a linear system must in practice also be time-invariant which means that none of its characteristics change with time. If any of the system's operation parameters is not constant over the whole time of observation, none of the laws derived for linear systems are valid. Especially, in cases where system parameters depend on the input signal, even the linearity conditions do not hold, and this kind of equipment should be avoided in all parts of the sound recording and reproduction chain; but unfortunately, they are nevertheless used and even very commonly.

If the said linearity conditions are applied strictly, there should not appear any input-independent signals at the output, and hence the system should be noiseless. (This applies also to offset shift, which is noise at zero frequency.) Moreover, it should be able to handle signals however large. Of course, no physical system is that ideal, and in practice, it is fully sufficient that linearity is fulfilled accurately enough at the signal levels and frequencies the system is intended to handle.

In practice, linearity is usually not tested according to (b1) and (b2) because the matter can be examined with a bare sine wave, by making use of the following crucial property:

> **If a linear system is excited by a sine wave, the output also yields a sine wave with the same frequency.**

Any deviation from a sinusoidal response means, thus, that the system departs from linearity or is not time-invariant and therefore produces distortion. It is well to know, however, that the rule doesn't hold in the reverse direction. In other words:

> **Even if the response to a sine wave were quite purely sinusoidal, this doesn't guarantee that the system is both linear and time-invariant and therefore suitable for quality sound signal processing.**

The manifold noise reduction methods serve as an example of this. However, if the system can be considered time-invariant (like usually), the sine wave test, performed at several different frequencies and amplitudes, is a good indicator for linearity.

How well are linearity and time-invariance then possible to be accomplished in practice? At least in electronic circuits and filters, one can get very close to ideality, and with feedback it is possible to linearize amplifiers that may also contain quite nonlinear stages. If only linearly working components are used in the signal path, the result is also inevitably linear. If e.g. a circuit contains only linear resistors and capacitors and normally connected operational amplifiers, it is even impossible to accomplish markedly nonlinear behavior (assuming the circuit is stable and operation area limits are not exceeded).

Instead, in loudspeaker drivers and systems, the goals of low distortion and frequency balance are much more difficult to be achieved, by which reason loudspeakers' and their usage's effect on sound quality is generally more prominent than of other factors.

The primary performance of speaker drivers can be described, in both frequency and time domains, even by rather simple linear systems which is also a basic requisite for response simulation and the design of correction circuits. However, to consider minor response details and resonances in the modelling of the system is generally not worthwhile.

As presented in chapter 4, the possibilities of speaker drivers for linear and time-invariant operation are crucially better when they are operated directly, instead of the present, indirect way.

B2 Differentiators and Integrators

The simplest linear systems that still modify the signal somehow are differentiators and integrators. In accord with its name, a differentiator always yields the time derivative, or rate of change (multiplied by some real number), of the input; and correspondingly, an integrator always yields the time integral of the input signal. Differentiators and integrators are widely used as building blocks in electronic filters, but similar relationship is also found between certain quantities related to audio physics. For instance, the acceleration of an object is, by definition, the derivative of its velocity; and velocity, for one, is the derivative of the distance travelled. Differentiation an integration are always the reverse operations of each other, so distance is found by integrating velocity, and velocity is found by integrating acceleration. These relations, that also pertain to the motion of speaker diaphragms, are illustrated in Fig. B2.

If a differentiator is fed by a sine wave having frequency ω and amplitude 1, that is, if

$$x(t) = \sin(\omega t)$$

the output will be

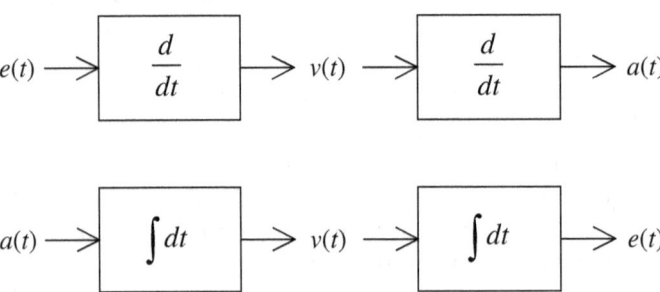

Figure B2. Relations between acceleration a, velocity v, and distance travelled e, expressed with differentiation and integration blocks. In integration, one must, of course, consider the whole time during which the integrated quantity differs from zero.

$$y_D(t) = \frac{dx}{dt} = \omega \cos(\omega t) = \omega \sin(\omega t + \frac{\pi}{2}) \tag{b4}$$

which is a sine wave whose amplitude is directly proportional to frequency and whose phase is $\pi/2$ radians, or 90°, ahead of the input. Likewise, if the same excitation $x(t)$ is given to an integrator, we get:

$$y_I(t) = \int x dt = -\frac{1}{\omega} \cos(\omega t) = \frac{1}{\omega} \sin(\omega t - \frac{\pi}{2}) \tag{b5}$$

which is a sine wave whose amplitude is inversely proportional to frequency and whose phase is 90° behind the input (the integrator is assumed to start from zero).

With this information, we are able to depict the frequency response of both systems by so-called *Bode-plots*. Figure B3a shows the gain of the differentiator, i.e. the absolute value of the frequency response function $H(\omega)$, and the generated phase shift θ as functions of ω. Figure B3b shows the corresponding graphs for an integrator.

The scale of the ω-axis is logarithmic, that is, a constant-length shift along the axis represents the multiplication of frequency by a constant

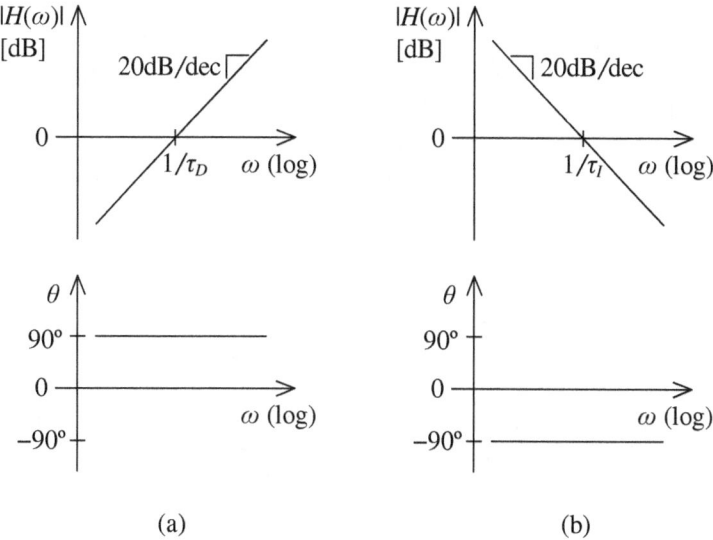

(a)   (b)

Figure B3. Frequency response diagrams of an ideal differentiator (a) and an ideal integrator (b).

factor. In Bode plots, the gain, or magnitude, is expressed in decibels (dB), i.e. a base 10 logarithm is taken and the result is multiplied by 20. Thus, 0 dB always corresponds to a gain of unity, and every 20 dB increase coincides with multiplication by 10.

Since both axes are therefore logarithmic by nature (although the division of the dB-scale is linear), the magnitude graphs (for the differentiator, $|H(\omega)| = \omega$; for the integrator, $|H(\omega)| = 1/\omega$) are here straight lines having a slope of 20 dB per decade of frequency.

The phase shift of the differentiator is +90° for all frequencies, that is, the system advances phase by a quarter cycle. Correspondingly, an integrator causes a quarter cycle lag at all frequencies ($\theta = -90°$).

Due to the above facts, it is evident that one parameter is enough to specify an ideal differentiator or integrator, a parameter that tells where the line intersects the 0 dB level. This crossing frequency can also be expressed, instead of ω, by its inverse τ, which is called *time constant*. In this case, these time constants (τ_D and τ_I) equal unity.

The input function used in equations (b4) and (b5) was a real sine wave, by which we could, in this simple special case, easily solve the output directly from the differential equation describing the system. Generally however, linear systems are analyzed using complex numbers and phasor calculation, so that both amplitude and phase information can be treated at the same time, in a compact and easily visualized form.

An especially useful concept for this is the complex sinusoid, which can be viewed as a rotating phasor in the complex plane. According to Euler's formula, familiar from complex arithmetic, any complex number can be expressed in the polar form $r \exp(j\alpha)$, where r is the absolute value (length of the vector), j is the imaginary unit, and α is the angle relative to positive real axis. If we now define the input function as

$$x(t) = e^{j\omega t} \tag{b6}$$

we obtain the phasor shown in Fig. B4. The length of the phasor is always 1, and it rotates counterclockwise as time t increases, with angular frequency ω.

The complex exponential can be differentiated and integrated like a real one, so we obtain now as the output of the differentiator

$$y_D(t) = \frac{dx}{dt} = j\omega\, e^{j\omega t} = j\omega\, x(t) \tag{b7}$$

which is just the input signal multiplied by the imaginary frequency $j\omega$. For the integrator, we correspondingly have:

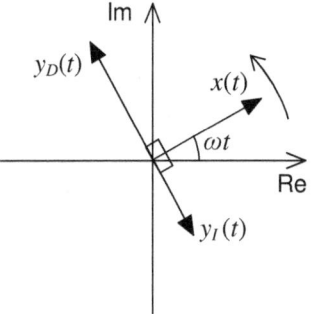

Figure B4. Differentiator and integrator phasor diagrams in the complex plane. Differentiator output $y_D(t)$ leads the input $x(t)$ by 90º, and integrator output $y_I(t)$ lags the input by 90º.

$$y_I(t) = \int x\,dt = \frac{1}{j\omega} e^{j\omega t} = \frac{1}{j\omega} x(t) \tag{b8}$$

which is the input signal divided by $j\omega$.

The output phasor is thus obtained by multiplying the input phasor by a frequency-dependent complex number. This coefficient is the system's frequency response $H(\omega)$. For the differentiator, it therefore holds: $H(\omega) = j\omega$, and for the integrator: $H(\omega) = 1/j\omega$. Because multiplying by j means only increasing the phase angle by $\pi/2$ and dividing by j means only decreasing the angle by $\pi/2$, the abovementioned output phasors are perpendicular to $x(t)$, as is consistent with results (b4) and (b5) and shown in Fig. B4.

Actual physical signals are, of course, not complex numbers but always real-valued. Rotating phasors are nevertheless easily interpreted as practical sine waves by observing only the real part or imaginary part of the phasor.

The operation of all analogue filtering circuits is based on the differentiator-integrator-relationship that exist between voltage and current in capacitors and inductors. The current (i) flowing into a capacitor is always the derivative of the voltage (u) multiplied by the capacitance C, that is:

$$i = C\frac{du}{dt} \tag{b9}$$

By feeding the capacitor by a complex sinusoid and designating the current and voltage phasors by I and U, we obtain, using equation (b7):

$$I = j\omega CU = \frac{U}{Z} \tag{b10}$$

where the frequency-dependent complex number $Z = 1/j\omega C$ is the capacitor's impedance. Thus, equation (b10) is the equivalent of Ohm's law for capacitors.

In inductors, the voltage is always the derivative of the current multiplied by the inductance L, that is:

$$u = L\frac{di}{dt} \tag{b11}$$

The relationship between corresponding phasor quantities is consequently:

$$U = j\omega LI = ZI \tag{b12}$$

where $Z = j\omega L$ is the impedance of the inductor. Thus, equation (b12) is the equivalent of Ohm's law for inductors.

It must be noted that the frequency response of an integrator is non-existent at zero frequency which implies that an ideal integrator is not stable at constant input. In real-world integrators, the gain must also be limited, so that it levels off or turns down as frequency drops sufficiently. Hence, technical integrators are always lossy, i.e. they don't remember events of distant past but always tend to return toward zero.

Correspondingly, practical differentiators are only able to work up to some boundary frequency, which is often determined by the requirement of stability.

B3 Time Domain Representation

A linear system can be described mathematically in several different ways. In the time domain, analysis can be performed using linear differential equations or the impulse response. The same information can also be portrayed by the step response.

We adopt as an example the passive circuit of Fig. B5, with current i as the input and voltage u as the output. Current i divides into three separate arms, and by stating the current of each arm in terms of u, we obtain, based on laws (b9) and (b11):

$$C\frac{du}{dt} + \frac{1}{R}u + \frac{1}{L}\int u\,dt = i$$

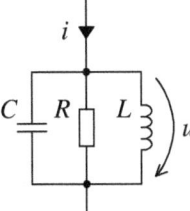

Figure B5. An electrical linear system used as an example. Current i is defined as the excitation and voltage u as the response.

Differentiating both sides of the equation with respect to time yields:

$$C\frac{d^2u}{dt^2} + \frac{1}{R}\frac{du}{dt} + \frac{1}{L}u = \frac{di}{dt} \qquad (b13)$$

This is a constant-coefficient 2nd-order differential equation that contains all information about the system of Fig. B5. Terms containing the output quantity are usually collected to the left-hand side, and terms containing the input quantity, to the right-hand side.

The order of the system (= degree of the highest derivative) is always the same as the number of reactive elements acting in the system. An impedance is reactive (includes reactance) if it is inductive or capacitive, that is, its imaginary part is non-zero. In realizable systems, the order of the right-hand side of the equation cannot exceed that of the left-hand side.

From the differential equation of the system, it is, in principle, possible to solve the output for any input function; but in practice, this is not quite straightforward nor easy; and therefore, other approaches are usually employed. Finding the impulse response from the differential equation is, nevertheless, clear-cut.

By impulse response (designated $h(t)$) is meant the output to an excitation that is illustrated in Fig. B6a and called the *unit impulse*, designated by $\delta(t)$. The unit impulse is a theoretical test signal that is zero everywhere else except at $t = 0$ where it is not explicitly defined, but the area it encloses is 1, as in Fig. B6a when ε approaches zero.

$h(t)$ is found out from the differential equation (DE) by the following procedure [1]:

1. The right-hand side of the DE is set to zero, and for this, so-called homogenous DE, the solution functions are obtained in a normal fashion. (The solving of homogenous constant-coefficient DE:s is explained in appendix D.)

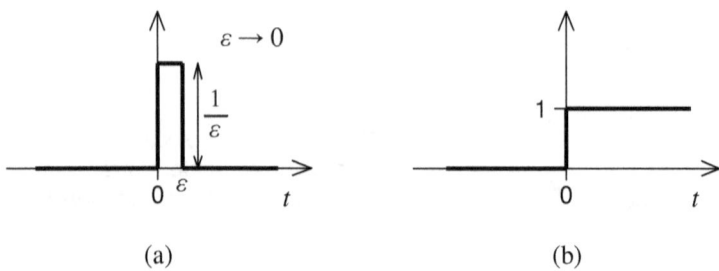

Figure B6. The unit impulse function (a) and the unit step function (b).

2. Coefficients for the solution functions are found by using generic initial conditions. Let the order of the DE be N, let the coefficient of the highest derivative on the left-hand side be a, and let the solution of the homogenous equation be $h_0(t)$. Then, the $(N-1)$th derivative of $h_0(t)$ at $t = 0$ will be $1/a$, and all the lower derivatives will respectively be zero. In our example case (b13), $N = 2$ and $a = C$, thus we obtain as the initial conditions: $h_0'(0) = 1/C$ and $h_0(0) = 0$.

3. The impulse response is now found from $h_0(t)$ by applying to it the same operations that on the right-hand side of the original DE are applied to the input quantity. In our example, then, a mere differentiation is performed, and therefore $h(t) = h_0'(t)$.

If, in the example of Fig. B5, we have $R > \sqrt{L/4C}$ (complex system poles), we obtain, through the above steps, the following expression for the impulse response:

$$h(t) = \frac{e^{-t/2RC}}{C}\left[\cos(\omega_d t) - \frac{1}{2RC\omega_d}\sin(\omega_d t)\right], t > 0 \qquad (b14)$$

where it has been marked:

$$\omega_d = \sqrt{\frac{1}{LC} - \frac{1}{4R^2C^2}} \qquad (b15)$$

Thus, $h(t)$ consists of sinusoidal components that decay exponentially with time and have angular frequency ω_d.

Figure B7 shows a graph of $h(t)$, using values $R = 10$, $C = 0.0005$ and $L = 0.01$. The Figure also shows the impulse response obtained for the

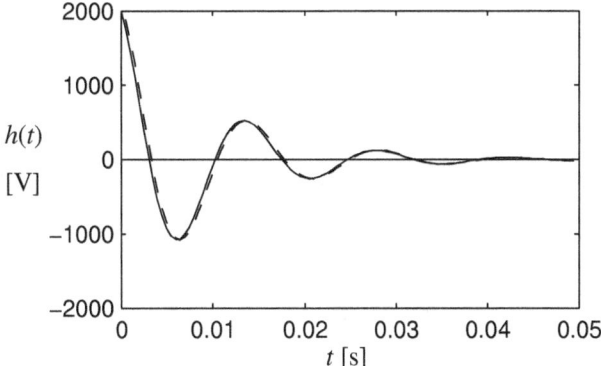

Figure B7. Impulse response of the example system with $R=10$, $C=0.0005$ and $L=0.01$; according to expression (b14) (solid line) and circuit simulation (dashed line).

circuit in question by a circuit simulation program, thus confirming the validity of expressions (b14) and (b15). The small deviation between the simulated and calculated responses stems from the fact that in simulation one cannot use an ideal unit impulse as the excitation, but instead an approximation, like that in Fig. B6a, has to be used. In this case, the height of the pulse was set to 2000 A and its duration to 0.5 ms.

In Fig. B7, the oscillations fade with time toward zero, and hence the system is stable (like passive circuits always). The alternative is that the amplitude of oscillation undertakes exponential growth which signifies an unstable system. If, in such a case, we limit the amplitude (by some nonlinear means) to a constant value, we have an oscillator.

If $R < \sqrt{L/4C}$ (poles real), the expression of $h(t)$ contains only exponential functions, and there will be no oscillations in the impulse response. Regardless of the system's order, the impulse response generally consists of only exponential, sine, and cosine functions because the solution of the homogenous equation is composed of them.

By integrating the unit impulse (whose area was 1), we obtain the rectangular signal shown in Fig. B6b, called the *unit step*. In practice, it is often easier to use a step, rather than an impulse, as the test signal because the energy of a limited-height and short-duration pulse is inevitably low.

By integrating both sides of the system DE (like e.g. (b13)), it can be seen that if the input quantity is replaced by its integral function, the output quantity must also be replaced by its integral function. The step response (designated $g(t)$) of the system is therefore found by integrating the impulse response, that is:

$$g(t) = \int_0^t h(\tau)d\tau \qquad (b16)$$

Terms of the form $\exp(\alpha t)\sin(\beta t)$ or $\exp(\alpha t)\cos(\beta t)$ (α and β constants), that generally appear in $h(t)$, are integrable by applying the method of integration by parts twice.

In our example case, we will however, get by more easily because the integral function of $h(t)$ is already known due to step 3 in the above procedure. By cancelling the differentiation performed in step 3, we obtain, when $R > \sqrt{L/4C}$:

$$g(t) = h_0(t) = \frac{e^{-t/2RC}}{C\omega_d}\sin(\omega_d t) \ , \ t \geq 0 \qquad (b17)$$

ω_d being the same as before. Figure B8 shows the graph of $g(t)$, using the same element values as before. A similar oscillation can be seen in the step response as was in the impulse response. Because the system is of band-pass type, the step response also converges to zero with increasing time.

When determining the impulse response by the described procedure, we make use of all information that is available from the DE. Because the DE contains all information about all properties of the system, the same also applies to the impulse response and the step response.

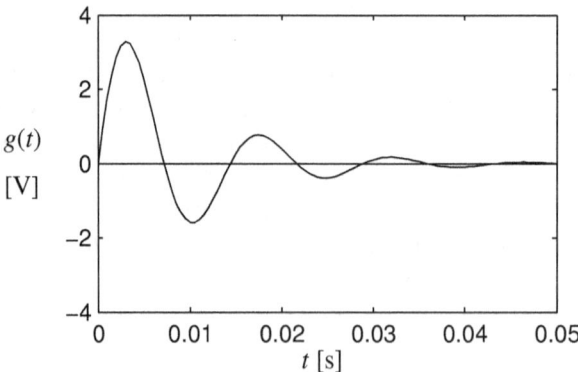

Figure B8. Step response of the example system with $R=10$, $C=0.0005$ and $L=0.01$; according to expression (b17).

**The impulse response is not only a detail in the time perfor-
mance of a linear system but a concept that unequivocally
defines the whole system. If, therefore, the impulse response
is known, the whole behavior of the system in both time and
frequency domains is, in principle, known.**

This being the case, it should be possible to solve the output $y(t)$ for
any input signal $x(t)$, using only the impulse response $h(t)$. This can be
accomplished by an integral operation called *convolution*, which is described by the formula:

$$y(t) = \int_{-\infty}^{+\infty} x(\tau)h(t-\tau)d\tau \tag{b18}$$

The output value at instant t is, therefore, obtained by integrating
over all time the product of the input and the time-reversed and by t
shifted impulse response. This operation is illustrated graphically in Fig.
B9. Figure 'a' depicts the original impulse response and Fig. b its mirror
image $h(-\tau)$. Figure c shows $h(t-\tau)$, that is obtained from $h(-\tau)$ by shifting it in time by amount t. (t is considered constant.) The point repre-

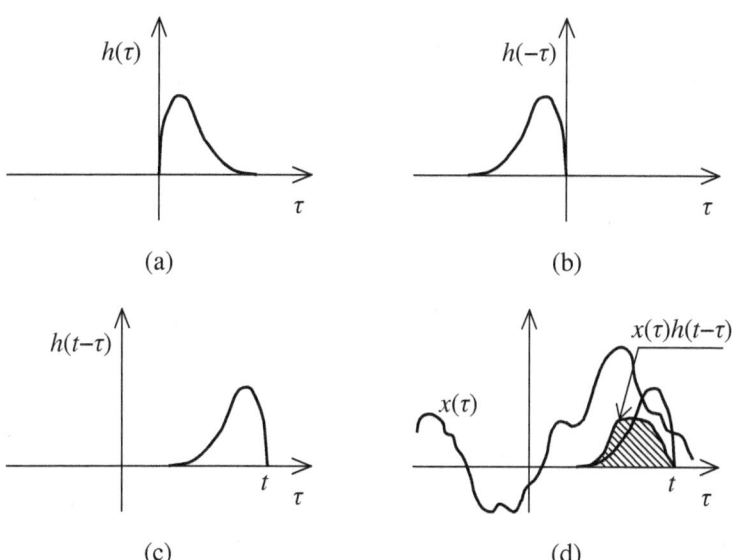

Figure B9.   Forming of convolution between two signals, $x(\tau)$ and $h(\tau)$. The
shadowed area in Fig. d gives the value of the convolution at instant t.

senting instant 0 in the original impulse response has now moved to instant t because $t-\tau$ becomes zero at instant $\tau = t$. In Fig. d, both $h(t-\tau)$ and the input signal are shown together with a curve that represents the product of the two. The area left under this curve is, according to equation (b18), $y(t)$. In practice, the upper limit of integration can be set to t because $h(t-\tau)$ is always zero when $\tau > t$.

With the convolution integral, one can thus compute the output value at a specified time instant. By stepping t, the shape of $y(t)$ can be found out, point by point. Solving the integral analytically is generally possible only in simple cases; but numerically used, equation (b18) allows one to perform transient analysis to any system whose impulse response is known, either by calculation or measurement.

B4 Frequency Response

Earlier, in the context of differentiators and integrators, we used the complex sinusoid as the input signal (equation (b6)) and thereby found out the frequency response function, $H(\omega)$. We will now apply the same to the example system of Fig. B5, described by DE (b13).

The derivative of the input signal i is now, based on equation (b7), the same as the signal itself multiplied by $j\omega$. Because the output signal u is also a complex sinusoid ($u(t) = H(\omega)i(t) = H(\omega)\exp(j\omega t)$), it too may be differentiated similarly. Equation (b13) can thus be written in this case as:

$$C(j\omega)^2 u(t) + \frac{1}{R} j\omega u(t) + \frac{1}{L} u(t) = j\omega i(t)$$

from which we further obtain (noting that $i(t) = \exp(j\omega t)$):

$$u(t) = \frac{j\omega}{C(j\omega)^2 + \frac{1}{R} j\omega + \frac{1}{L}} i(t) = H(\omega)e^{j\omega t} \qquad (b19)$$

The result can be generalized:

The factor determining the relation between the output and input phasors, i.e. the frequency response, $H(\omega)$, is obtained directly from the system's differential equation by replacing all differentiation operations by the coefficient $j\omega$.

Like the DE and the impulse response, $H(\omega)$ is also enough to define a linear system completely.

The absolute value of $H(\omega)$ determines the system amplification, and the angle (designated $\theta(\omega)$) determines the phase shift produced by the system. By expressing $H(\omega)$ in the polar form, we can write for the phasors, generally:

$$
\begin{aligned}
y(t) &= H(\omega)e^{j\omega t} \\
&= |H(\omega)|e^{j\theta(\omega)}e^{j\omega t} \\
&= |H(\omega)|e^{j(\omega t + \theta(\omega))}
\end{aligned}
\tag{b20}
$$

The output phasor is thus obtained from the input phasor by multiplying its length by $|H(\omega)|$ and shifting its phase angle by $\theta(\omega)$. These relations are illustrated in Fig. B10, $x(t)$ and $y(t)$ rotating in the complex plane counterclockwise, with angular frequency ω and mutual angle difference θ, which in the Figure is negative. $H(\omega)$ is a solid vector that depends only on frequency.

The effect of the system on a real sine wave is seen by considering the real part of both the input and output phasors. Having as the input $x(t) = \text{Re}[\exp(j\omega t)] = \cos(\omega t)$, we get as the output

$$
\begin{aligned}
y(t) &= \text{Re}[H(\omega)e^{j\omega t}] \\
&= \text{Re}[|H(\omega)|e^{j(\omega t + \theta(\omega))}] \\
&= |H(\omega)|\cos(\omega t + \theta(\omega))
\end{aligned}
\tag{b21}
$$

The result only confirms that the system alters the amplitude and phase of a concrete sine wave (here $\cos(\omega t)$) in accordance with the frequency response.

The expression of $H(\omega)$ is always convertible to a form from which

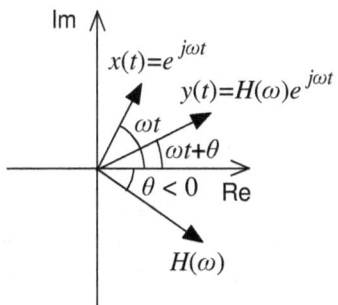

Figure B10. Phasor diagram depicting the operation of a linear system at one frequency. Output $y(t)$ is determined by multiplying input $x(t)$ by the frequency response vector $H(\omega)$. $H(\omega)$'s direction angle θ may, in principle, assume any values but is usually regarded to be in the range $\pm 180^{\circ}$.

the gain and phase shift can be determined. The terms in both the numerator and denominator are always either real or purely imaginary ($j^2 = -1$), and for the example system, we can now write:

$$H(\omega) = \frac{j\omega}{C(j\omega)^2 + \dfrac{1}{R}j\omega + \dfrac{1}{L}} = \frac{j\omega}{\dfrac{1}{L} - C\omega^2 + j\dfrac{\omega}{R}} \tag{b22}$$

Using Pythagoras' theorem, we now have (because $|j| = 1$):

$$|H(\omega)| = \frac{\omega}{\sqrt{\left(\dfrac{1}{L} - C\omega^2\right)^2 + \left(\dfrac{\omega}{R}\right)^2}} \tag{b23}$$

$\theta(\omega)$ is obtained by subtracting the denominator angle from the numerator angle in the expression of $H(\omega)$, yielding:

$$\theta(\omega) = 90° - \arctan\left(\frac{\omega}{R}\middle/\left(\frac{1}{L} - C\omega^2\right)\right) \tag{b24}$$

By using for the circuit elements the same example values as before ($R = 10$, $C = 0.0005$, $L = 0.01$), we obtain for the system, from expressions (b23) and (b24), the Bode plot shown in Fig. B11.

The behavior of the frequency response at low and high frequencies can also be directly concluded from the expression. When ω is small enough, the denominator is dominated by a real constant term, in this case, $1/L$. The numerator, in turn, is here directly proportional to frequency, and thus at low frequencies the circuit acts as a differentiator. When ω is high enough, the denominator is dominated by the highest-order term, in this case, $C(j\omega)^2$. At high frequencies, expression (b22) is thus reduced to the form $1/(Cj\omega)$, so in this region the circuit acts as an integrator.

The $H(\omega)$ obtained, (b22), is in fact the same as the impedance of the parallel connection of capacitor C, resistor R, and inductor L, as stems from the markings of Fig. B5. At low frequencies, the dominating element is the inductor; at the point of maximum gain, the resistor; and at high frequencies, respectively, the capacitor.

In Fig. B11, one can also see the general relation that exists between gain and phase curves. A change in the slope of the gain graph shows up in the phase graph as a level change of corresponding measure. As the degree of the gain slope changes by one unit (i.e. 20 dB per decade), a 90° step occurs in the phase. In Fig. B11, the slope of $|H(\omega)|$ changes

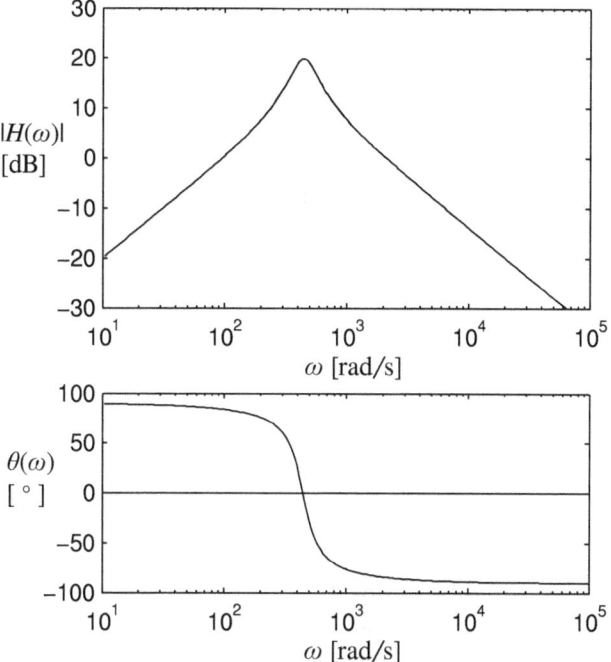

Figure B11. Gain and phase shift of the example system (Bode plot), using values $R=10$, $C=0.0005$, and $L=0.01$.

from a 1st-degree rise to a 1st-degree fall, so the shift in $\theta(\omega)$ is, correspondingly, $-180°$.

Besides the Bode plot, another way to depict frequency response is the *Nyquist plot*. Here, $H(\omega)$ is drawn in the complex plane, with ω as a parameter, so that both magnitude and phase are visible in the same diagram, although at the expense of frequency information.

The Nyquist plot of the example system is shown in Fig. B12. As frequency travels from zero to infinity, $H(\omega)$ draws a full circle, starting from the origin, circling in the direction of the arrows, and finally approaching the origin again. The diameter of the circle equals the resistance R.

That the figure indeed is a circle, can be proved starting from the expression for $|H(\omega)-R/2|$, which represents the distance of $H(\omega)$ from the real-axis point $R/2$. This distance (= radius) can be detected to be constant $(R/2)$ for all frequencies.

In this case, the values of L and C have no effect on the diagram itself. They only affect the frequency at which each point of the circle is

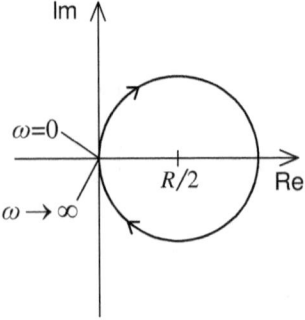

Figure B12. Nyquist plot of the system under consideration. Arrows on the locus indicate the direction of $H(\omega)$, as frequency increases. $H(\omega)$ draws a circle, with $R/2$ as the center.

reached.

B5 Transfer Functions

Above, we have discussed the frequency response, assuming the input signal is a steady-state sinusoid. The behavior of a system can, however, be analyzed even more generally with so-called *transfer function*. In practice, the transfer function is directly obtained from the frequency response function by replacing $j\omega$ by the complex frequency s, which may also include a real part. The efficient use of transfer functions is based on the Laplace transform and its inverse transform, that are not considered here otherwise than by stating that the transfer function is the Laplace transform of the impulse response. The frequency response function $H(\omega)$ can be regarded as a special case of the transfer function (designated $H(s)$), covering frequency analysis.

It is customary to write transfer functions in a form where the coefficient of the highest power of s in the denominator is 1. The transfer function of the example system is, consequently

$$H(s) = \frac{\dfrac{1}{C} s}{s^2 + \dfrac{1}{RC} s + \dfrac{1}{LC}} \qquad (b25)$$

Analogue transfer functions are, also generally, rational expressions of s, with real coefficients. An exception to this are so-called phase-linear systems, like, for instance, one that causes a mere delay to the signal. This kind of systems cannot be described accurately by any rational transfer function. In practice, this doesn't establish a big problem because implementing analogue delays is also difficult.

In the analysis of electrical systems, it is usually not necessary to write out the differential equation since the transfer function may be directly worked out from the circuit diagram, using phasor calculus; in which case we seek to solve the ratio of output to input by utilizing all information available from the circuit. Although the voltage and current phasors are, in principle, functions of time, their mutual relationships depend only on frequency. For phasor quantities, one may apply the generalized Ohm's law and the Kirchhoff's laws, just the same way as with DC-circuits. The impedances of capacitors and inductors may be written as $1/(sC)$ and sL, respectively. In some cases, it is easier to use the reciprocals of impedances, i.e. admittances, instead of the impedances.

Using admittances, we can write for the circuit of Fig. B5:

$$H(s) = \frac{U}{I}(s) = \cfrac{1}{sC + \cfrac{1}{R} + \cfrac{1}{sL}}$$

from which we get, by one multiplication step, the form (b25). The solution was in this case short because the circuit has only two nodes. Generally, one has to write a set of equations, by which the unknown voltages and currents can be eliminated.

The practicality of transfer functions is largely based on the fact that when system blocks are connected in cascade, their transfer functions may be multiplied together. In time domain analysis, there is no corresponding advantage. Based on equation (b20) (1st row), it is readily seen that each successive system only causes a new multiplication operation to the output phasor.

Figure B13a depicts a cascade connection of two blocks, the transfer functions H_1 and H_2 forming a product H_1H_2. The order of the combined transfer function will be, in the general case, the sum of the orders of the parts. Signals X and Y are in this representation either rotating phasor quantities or frequency spectrums, that are discussed in appendix C.

The transfer function of the parallel connection of blocks will be, as indicated in Fig. B13b, the sum of the individual transfer functions. For the order, the same applies as in the cascade connection.

Figure B14a shows a system that incorporates a feedback via block H_2. A feedback may be termed positive if H_2Y is in phase with X, and negative if H_2Y and X are in opposite phase. In the general case, this

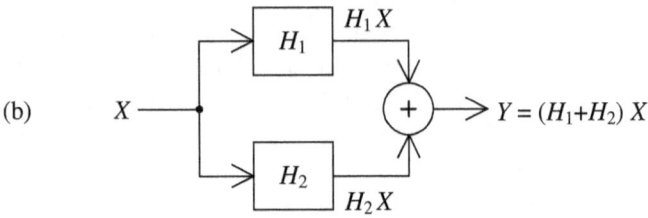

(a) $X \longrightarrow \boxed{H_1} \xrightarrow{H_1 X} \boxed{H_2} \longrightarrow Y = H_1 H_2 X$

(b) $X \longrightarrow$ ⟶ $\boxed{H_1}$ $\xrightarrow{H_1 X}$ ⊕ $\longrightarrow Y = (H_1 + H_2)\, X$ ⟶ $\boxed{H_2}$ $\;H_2 X$

Figure B13.   Series and parallel connection of system blocks. In the series (cascade) connection (a), the resulting transfer function is the product of the individual functions H_1 and H_2. The parallel connection (b) yields the sum of the individual transfer functions. The circle represents a summing block.

phase difference can, of course, assume any value. Often H_2 is a mere constant.

The transfer function, H, of this system can be found by examining the flow of signals in the loop and solving for Y/X:

$$Y = H_1(X + H_2 Y)$$

$$\Leftrightarrow\ Y(1 - H_1 H_2) = H_1 X$$

$$\Leftrightarrow H = \frac{Y}{X} = \frac{H_1}{1 - H_1 H_2} \qquad\qquad (\text{b}26)$$

In the diagram, a difference forming block is often used instead of the summing block. In that case, H_2 should be replaced by $-H_2$, that is, the minus sign in the denominator changes to a plus in the transfer function (b26).

The system of Fig. B14b has two feedback branches that lead to the same summing point. Reading the diagram, signals Z and Y can now be expressed as follows:

$$\begin{cases} Z = H_1(X + H_3 Z + H_4 Y) \\[2mm] Y = H_2 Z \Leftrightarrow Z = \dfrac{Y}{H_2} \end{cases}$$

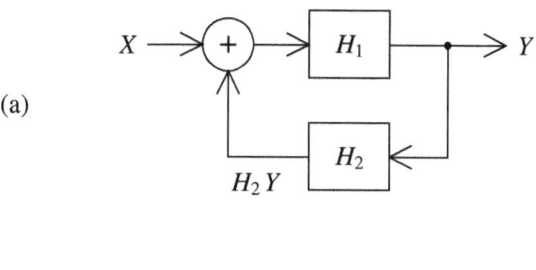

(a)

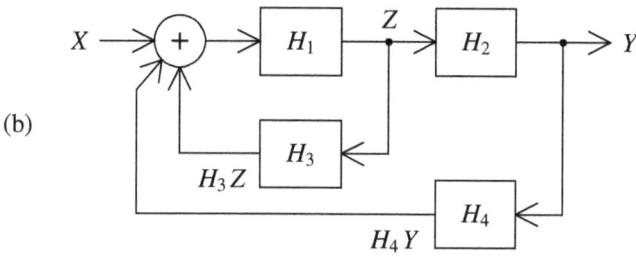

(b)

Figure B14. a) System with one feedback path. b) System with two feedback paths.

By substituting Z, obtained from the lower equation, to the upper equation and arranging terms, we obtain as the transfer function of this structure:

$$H = \frac{Y}{X} = \frac{H_1 H_2}{1 - H_1 H_3 - H_1 H_2 H_4} \tag{b27}$$

B6 Poles and Zeroes

The shape of the frequency response of a system is fully determined by the zeroes of the numerator and denominator of the transfer function. The numerator and denominator are always polynomials of s and can thus be factorized by finding the zeroes of the polynomial functions, i.e. the roots, whose number always equals the degree of the polynomial. Those values of s at which the denominator becomes zero, are called the *poles* of the system, and those values at which the numerator becomes zero, are called the *zeroes* of the system.

The example transfer function (b25) can be written, in terms of its poles (p_1 and p_2) and zero (z_1), in the factored form:

$$H(s) = \frac{1}{C} \frac{(s - z_1)}{(s - p_1)(s - p_2)} \tag{b28}$$

Immediately, it can be found that $z_1 = 0$, and thus the only zero of the system is located at the origin. Using the well-known solution formula of 2nd-degree equations, we have as the poles:

$$p_1, p_2 = \frac{-\dfrac{1}{RC} \pm \sqrt{\dfrac{1}{(RC)^2} - \dfrac{4}{LC}}}{2} = -\frac{1}{2RC} \pm \sqrt{\frac{1}{4R^2C^2} - \frac{1}{LC}}$$

The poles are real if $[1/(4R^2C^2) - 1/(LC)] \geq 0$ which is satisfied when $R^2 \leq L/(4C)$. Otherwise, the poles constitute a complex conjugate pair, that has a common real part $(-1/(2RC))$ but opposite imaginary parts.

Also generally, regardless of the degree of the polynomial, complex roots can only appear as conjugate pairs. Consequently, polynomials of odd degree have at least one real root.

Poles are usually marked in the complex plane (*s*-plane) by small crosses and zeroes by small circles. Figure B15 shows the pole-zero diagram of the example system in two different cases. In Fig. 'a', $R^2 < L/(4C)$, the poles being on the negative real axis. In Fig. b, $R^2 > L/(4C)$,

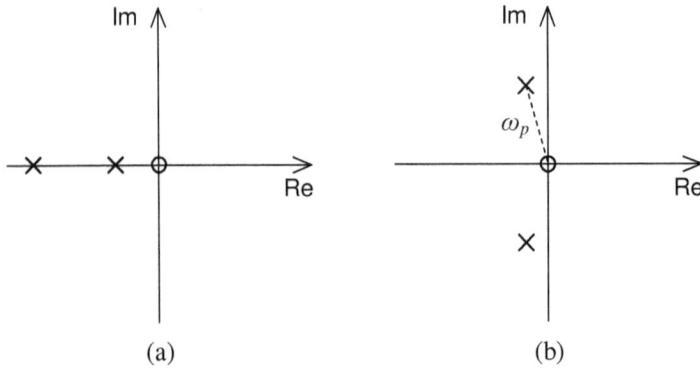

(a)   (b)

Figure B15.   Pole-zero diagram of the example system described by transfer function (b25), with two different values of R, C being 0.0005 and L being 0.01. In Fig. 'a', $R = 2$, system poles being on the real axis at -500 ± 224. In Fig. b, $R = 10$, the poles being symmetrically about the real axis, at points $-100 \pm j436$. The zero of the system lies at the origin. The dashed line in Fig. b indicates the distance of the pole pair from the origin, i.e. the characteristic frequency ω_p.

the poles forming a symmetrical pair. In literature, the real part of s is often designated by σ and the imaginary part by ω, the axis markings being hence σ and $j\omega$, respectively.

The pole-zero diagram, too, is a way to describe a linear system. If the poles and zeroes of the system are known, the transfer function is also known, except the gain factor, that in the example case, equation (b28), is $1/C$.

In frequency analysis, $s = j\omega$. A pole p introduces therefore, in the denominator of the frequency response function, a factor $(j\omega - p)$.

If p is real, we can note the following: When $\omega \ll |p|$, the factor is dominated by constant $-p$, and when $\omega \gg |p|$, the factor is dominated by $j\omega$, which is directly proportional to frequency. The boundary is reached at frequency $\omega = |p|$, which may called the characteristic frequency of the pole, or the pole frequency (designated ω_p). Factor $(j\omega - p)$ thus causes a turning point in the frequency response, at the angular frequency indicated by $|p|$.

A complex pole pair introduces a factor $(j\omega - p)(j\omega - p^*)$ in the denominator. When $\omega \ll |p|$, the factor is essentially constant, $pp^* = |p|^2$, and when $\omega \gg |p|$, the factor becomes essentially $(j\omega)^2 = -\omega^2$, which is proportional to the square of frequency. The asymptotes representing these two intersect when $\omega^2 = |p|^2$, that is, when $\omega = |p|$. The turning point in the frequency response, caused by a complex pole pair, is thus established at frequency $\omega_p = |p|$, as in the case of a real pole. A fully equivalent analysis naturally applies to the zeroes, that determine the frequency behavior of the numerator.

The pole frequency, ω_p, which graphically means the distance of the pole from the origin, is illustrated by a dashed line in Fig. B15b. If the system includes poles or zeroes that are relatively far from each other in frequency, it may be difficult to represent them in the same pole-zero diagram.

The essential behavior of the frequency response may be determined directly from the factored transfer function (like (b28)) by drawing, in the pole-zero diagram, the vectors representing each factor. The term $j\omega$ is, then, a point on the positive imaginary axis, at a height directly proportional to frequency. (In mathematical representations, negative frequencies are also possible, but in practice, frequency is normally deemed to be a positive quantity.)

Figure B16 illustrates the developing of the frequency response in the example case of Fig. B15b, the bode plot being shown before in Fig. B11. A vector is drawn from every pole and zero to the point $j\omega$. The magnitude behavior can be analyzed by examining the absolute values,

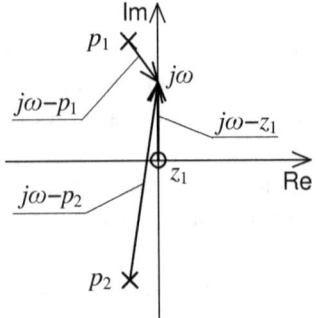

Figure B16. Visual determination of the frequency response from the pole-zero diagram; in the case of Fig. B15b. The factors that constitute $H(\omega)$ can be represented as vectors pointing to $j\omega$. As frequency increases, point $j\omega$ moves up on the positive imaginary axis, and the lengths and direction angles of the vectors change accordingly.

or lengths, of the vectors. So, from equation (b28):

$$|H(\omega)| = \frac{1}{C} \frac{|j\omega - z_1|}{|j\omega - p_1||j\omega - p_2|}$$

From Fig. B16 it can be reasoned that at low frequencies the denominator of $|H(\omega)|$ remains approximately constant, whereas the numerator increases directly proportional to frequency, as was already noticed in the context of equation (b23).

As frequency approaches the pole frequency, $|j\omega-p_1|$ falls off quite rapidly, causing the peak seen in Fig. B11, in the vicinity of the pole frequency. The characteristic frequency of complex poles and zeroes can also be called the *resonant frequency*. In this case, the issue is about the parallel resonance of the inductor and capacitor.

As frequency increases further, all the vectors increase in length, eventually turning in the same direction, whereupon the slope of $|H(\omega)|$ depends only on the difference of the counts of the poles and zeroes.

Generalizing, it can be stated:

As frequency increases, every zero introduces, at the point of its characteristic frequency, an unit increase (20 dB per decade) in the slope of $|H(\omega)|$; and every pole introduces, at the point of its characteristic frequency, an unit decrease in the slope of $|H(\omega)|$.

The phase shift is obtained by adding together the direction angles of the numerator factors and subtracting, from the sum, the direction angles of the denominator factors. Possible minus sign in the gain factor yet brings an additional 180 degree shift.

In the case at hand, we thus have:

$$\theta(\omega) = \angle(\, j\omega - z_1) - \angle(\, j\omega - p_1) - \angle(\, j\omega - p_2)$$

where "$\angle$" denotes the direction angle, or argument, defined relative to the positive real axis. From Fig. B16 we can detect that at low frequencies $\angle(\, j\omega - p_1)$ and $\angle(\, j\omega - p_2)$ cancel each other, leaving left $\angle(\, j\omega - z_1)$, which is constant, 90°. At the pole frequency, $\angle(\, j\omega - p_1) + \angle(\, j\omega - p_2)$ reaches 90°, whereupon $\theta(\omega)$ changes its sign, as seen in Fig. B11. At high frequencies, all vectors turn upwards, whereupon the phase shift approaches a multiple of 90°, here −90°, determined by the order difference between the numerator and denominator.

Poles have a crucial significance in determining the stability of the system. Stability means that the impulse response has to decay toward zero, as time increases to infinity. The impulse response always consists of terms that include, as a factor, an exponential function (like in section B3) whose exponent is the product of time and the real part of the pole. This exponent has to be negative, in order that the function would be decaying. Consequently, a system is stable when, and only when, the real parts of all poles are negative. In other words, all poles must be located left from the imaginary axis. By contrast, the zeroes may lie anywhere, but usually they too have negative real parts.

Stability cannot be directly decided from a polynomial form transfer function (such as (b25)), but a system is unstable at least if one or more of the coefficients of the denominator polynomial are negative or zero.

B7 Minimum-Phase Systems

As mentioned above, for stability, all system poles must lie in the left half of the complex plane. Based on the previous section, every real pole thus causes a 90° fall in the phase response $\theta(\omega)$, in the region of the pole frequency. Every complex pole pair causes, in turn, a 180° fall in $\theta(\omega)$, in the region of the characteristic frequency.

If the zeroes are also located in the left half of the complex plane, they introduce completely similar but positive shifts in the phase response. In that case, there exists a regular relationship between the system's amplitude and phase responses, and the latter is possible to be solved from the former, and vice versa (Hilbert transform). This kind of system, that doesn't introduce any extra phase lag, is called a *minimum phase* system.

If the system has zeroes also in the right half-plane, these zeroes introduce descending terms in the phase response, similarly as the poles in the left half do. With increasing frequency, the phase response descends then more than the shape of the amplitude response implies, so the system is non-minimum-phase. This, as though an additional, phase shift is called *excess phase*.

Amplifier stages and individual filters used in sound reproduction are usually minimum-phase unless they are specifically intended for modifying phase characteristics. In the cascade connection (Fig. B13a), the minimum-phase property is preserved because the positions of the poles and zeroes do not change. However, the parallel connection of minimum-phase systems does not necessarily yield a minimum-phase system because in the summing of the transfer functions the zeroes are not preserved. For this reason, the acoustic frequency response of multi-way loudspeakers is usually non-minimum-phase even though the reproduction of individual drivers would be minimum-phase.

If the system has zeroes on the imaginary axis, they cause discontinuities in the phase response, and the minimum-phase nature cannot be properly decided. If these zeroes are located at the origin, that is, at frequency 0, like in high-pass and band-pass filters, the system can, however, be regarded as minimum-phase because frequencies approaching zero don't have practical significance here.

Deviation from the condition of minimum-phase also occurs because the acoustic source points of the divers that serve different frequency bands are at different distances from the listener. This kind of delay-induced phase shift between frequency bands cannot be modelled with poles and zeroes, but it breaks, however, the regular phase behavior.

Pure delays can be described by the frequency response

$$H(\omega) = e^{-j\omega T} \tag{b29}$$

where T is the length of the delay. $|H(\omega)| = 1$, and $\theta(\omega) = -\omega T$, that is, the phase lag is directly proportional to frequency, as is natural for a delay. Expression (b29) differs therefore totally from the usual rational transfer functions, that are only able to approximate linear phase.

An important application of non-minimum-phase systems are all-pass filters, whose amplitude response is constant while the phase response is yet frequency-dependent. Figure B17 shows the pole-zero diagrams of first- and second-order all-pass systems. An all-pass transfer function has an equal number of zeroes and poles, and the zeroes form a mirror image of the poles about the imaginary axis. From this symmetry, it follows that, in the transfer function, each factor of the denomi-

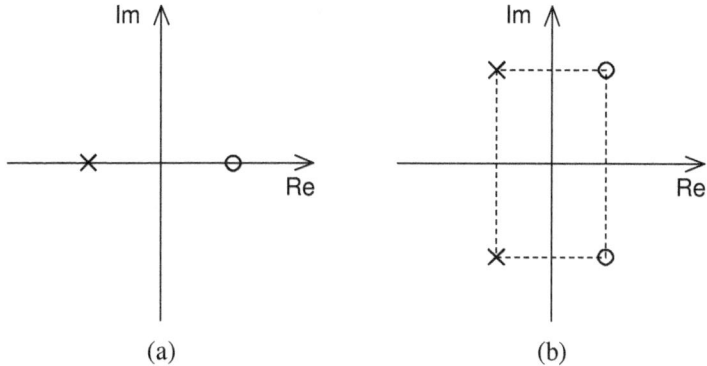

Figure B17. Pole-zero diagrams of all-pass filters. a) 1st-order all-pass. b) 2nd-order all-pass. Every pole is corresponded by a symmetrical zero in the right half of the complex plane.

nator polynomial is corresponded by a factor of equal absolute value in the numerator, and thus the amplitude response remains constant for all frequencies.

As noted above, every pole of the left half plane and every zero of the right half plane produces a 90° fall in the phase response, as frequency increases over the characteristic frequency. The total change in phase, from low to high frequencies, is therefore $N\cdot(-180°)$, where N is the order of the system. With all-pass filters, it is thus possible to realize only descending phase responses, never ascending, since the poles cannot be brought to the right half-plane without losing stability. Otherwise, it would be intriguing to null, with all-pass equalizers, the phase response of the entire reproduction chain; but unfortunately, this is possible to be accomplished only at single frequencies.

The condition of minimum-phase, in itself, is probably not hearable, but excessive phase errors between the frequency components within the signal should be minimized. The minimum-phase property is beneficial in circuit design and loudspeaker modelling since only the magnitude curve has to be fitted in place, and the phase curve finds its position automatically.

B8 Phase Linearity

When desiring to convey the waveform of signals as intact as possible, one must pay attention to the preservation of the phase relationships between different frequencies. Most ideal would be that the phase shift of the system would be negligibly small in the whole operating frequency band. However, when approaching the edges of the passband (the poles), the phase shift determined by the minimum-phase property is inevitable.

Wave shape modifications are minimized when the phase lag is directly proportional to frequency, since only then the mutual positions of different frequency components, on the time axis, remain unchanged. This kind of case is represented by phase response 'a' in Fig. B18. Unlike in Bode plots, ω-axis is here linearly divided since on a logarithmic frequency scale a linear phase response would produce a downward-bending curve. Response b is, in practice, fully identical to 'a'. Response b has only been drawn to start from 360°, which establishes a corresponding reference level as phase 0. Response c is also linear but not directly proportional to frequency, and hence, timing differences are introduced between different frequency components.

When the phase shift is 180° for all frequencies, a special case is established where the waveform of signals is preserved otherwise, except that the polarity is reversed. Response d represents a case in which the phase lag is directly proportional to frequency when measured with respect to this 180° reference level, so response d preserves the original waveform, with inversion. Depending on the case, this change of po-

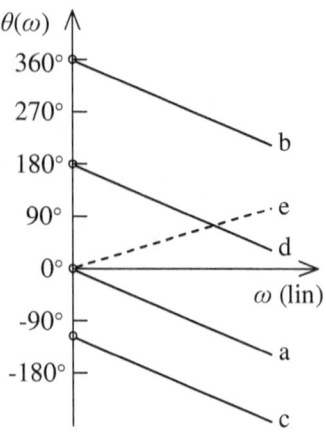

Figure B18. Linear phase responses. Responses a and b, that are in practice identical, do not modify signal waveform. Responses c and d are linear but modify the waveform. Monotonically rising response (e) is not possible.

larity may or may not be acceptable.

The graphs of phase responses are usually descending curves, and ascending responses may occur only in limited frequency ranges. Response e, which rises steadily at all frequencies, is thus not realizable.

A linear phase response directed at the origin is worth pursuing for the sake of the waveform, but linearity also influences the flatness of so-called *group delay*, that measures, in a certain way, the delaying of a group of waves, or a burst, in the system. Group delay is defined as

$$d_g = -\frac{d\theta}{d\omega} \tag{b30}$$

being thus the opposite of the derivative of phase. In case this derivative is constant (as in Fig. B18) and the amplitude response is flat, d_g indicates the delay of the signal's envelope curve. Otherwise, the quantity doesn't have any particular meaning.

Differing views exist about the ability of humans to hear phase errors. When a square or triangle wave, obtained from an ordinary wave generator, is led to a loudspeaker through an alterable all-pass filter, it is difficult to observe any difference in the sound between different phase settings. On the other hand, it seems that, at least when dealing with low fundamental frequencies and high-peaked transients, one can produce different tones only by manipulating the phasing ("phase organ") [2]. In any case, it is nowadays known that in headphone listening the phase between the fundamental and the 2nd harmonic is audible at frequencies below 1000 Hz.

Phase linearity should not be confused with phase coherence. For instance, a crossover filter can be realized so that the phases of the low- and high-range drivers are mutually coincident, but this is a totally different thing than phase linearity.

---

[1]   Robert A. Gabel & Richard A. Roberts, "Signals and Linear Systems", Wiley, 1987, p. 147-155.

[2]   John Borwick, "Loudspeaker and Headphone Handbook", Focal Press, 2001, p. 381.

APPENDIX C
Frequency Content of Signals

C1 Periodic Signals

In common discussion, it is often implied that sine wave testing of equipment does not provide sufficiently information about the sonic properties because actual music signals are anything but sinusoidal. It is, however, easy to disregard the fact that all existing periodic signals, or even an arbitrary portion of an aperiodic signal, can be constructed, however accurately, of sinusoidal components whose frequency, amplitude, and phase are appropriately chosen. If a system is completely linear and time-invariant and performs properly on sine wave (in both amplitude and phase), it cannot, then, due to the superposition principle, work incorrectly with any combination of sine waves either. A system containing nonlinearity, in turn, always generates harmonic and intermodulation distortion, whose definition involves the dividing of periodic signals into distinct frequency components.

Signal $x(t)$ is periodic if it repeats itself regularly so that $x(t+T) = x(t)$, where T is the length of the fundamental period. The fundamental frequency is, then, obtained directly from the fundamental period, that is, $\omega_0 = 2\pi/T$; and $x(t)$ can be expressed as a sum of sine and cosine functions as follows:

$$x(t) = \frac{a_0}{2} + \sum_{n=1}^{\infty} \left(a_n \cos(n\omega_0 t) + b_n \sin(n\omega_0 t) \right) \qquad (\text{c}1)$$

where

$$a_n = \frac{2}{T}\int_0^T x(t)\cos(n\omega_0 t)\,dt \ , \quad n = 0, 1, 2, \ldots \tag{c2}$$

and

$$b_n = \frac{2}{T}\int_0^T x(t)\sin(n\omega_0 t)\,dt \ , \quad n = 1, 2, 3, \ldots \tag{c3}$$

$a_0/2$ is the signal's DC component, to which is added the fundamental frequency and its multiples, or harmonics, weighted by the coefficients a_n and b_n. The representation is called the *Fourier series*. The coefficients a_n and b_n indicate how well the signal, $x(t)$, correlates with frequency $n\omega_0$.

A Fourier series can also be written using complex sinusoids [1] which makes the representation more compatible with frequency response functions, but the issue is not treated here.

We take as an example the half-wave rectified sine wave shown in Fig. C1a, having period T and amplitude 1. Thus:

$$x(t) = \begin{cases} \sin(\omega_0 t), & 0 \le t \le T/2, \quad \omega_0 = 2\pi/T \\ 0, & T/2 \le t \le T \end{cases}$$

The integrals (c2) and (c3), whose upper limit becomes now $T/2$, are solved in this case by making use of relations obtained from the addition and subtraction formulas of the sin and cos functions. We have:

$$\sin(\alpha)\cos(\beta) = [\sin(\alpha+\beta) + \sin(\alpha-\beta)]\,/2 \tag{c4}$$

$$\sin(\alpha)\sin(\beta) = [\cos(\alpha-\beta) - \cos(\alpha+\beta)]\,/2 \tag{c5}$$

(At $n = 1$, it must yet be noticed that $\alpha = \beta$, whereupon $\sin(\alpha-\beta) = 0$ and $\cos(\alpha-\beta) = 1$.) After routine calculations, we obtain as the result:

$$a_0 = \frac{2}{\pi}; \quad a_1 = 0; \quad a_2 = -\frac{2}{3\pi}; \quad a_3 = 0; \quad a_4 = -\frac{2}{15\pi}; \quad a_5 = 0;$$

$$a_6 = -\frac{2}{35\pi}; \quad \ldots \quad b_1 = \frac{1}{2}; \quad b_n = 0, \ n \ge 2.$$

The amplitudes of the sine terms are zero, except for the fundamental frequency, and the cosine terms are also zero at odd frequencies.

The Fourier series is thus

$$x(t) = \frac{1}{\pi} + \frac{1}{2}\sin(\omega_0 t) - \frac{2}{3\pi}\cos(2\omega_0 t) - \frac{2}{15\pi}\cos(4\omega_0 t) - \frac{2}{35\pi}\cos(6\omega_0 t)\ldots$$

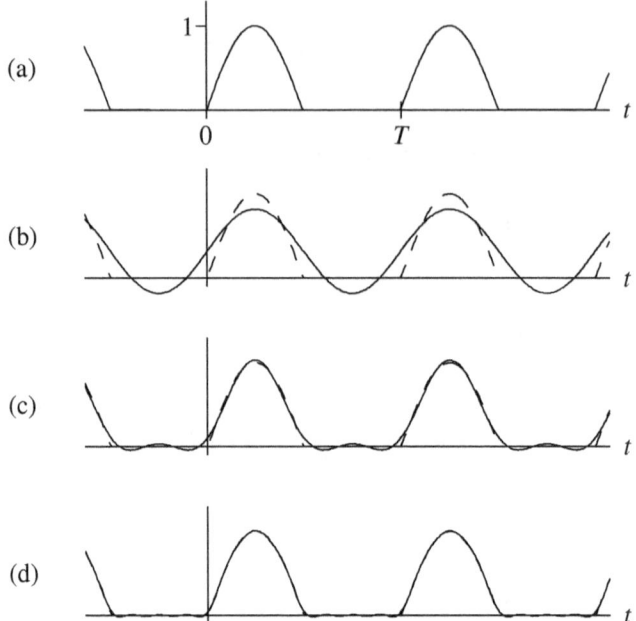

Figure C1. Composition of a periodic signal from its Fourier components. Figure 'a' shows the original half-wave rectified sine wave, marked by a dashed line also in the other Figures. In Fig. b, the signal is approximated using the DC component and the fundamental frequency. In Fig. c, the 2nd harmonic multiple has been added; and in Fig. d, the harmonics up to the 6th have been included.

In Fig. C1b, only the DC component and the fundamental frequency have been taken from the Fourier series. The more harmonic multiples are included, the better the waveform matches the original signal. In Fig. d, after 6th harmonic, the deviations have diminished to the order the line width.

According to Fig. C1, it seems that a waveform converges well even at rather low values of n. Complete conformity is, anyway, never achieved with any finite orders because, for example, the accurate representation of the corner points of the signal (like here at 0, $T/2$, T, and so on) would call for infinite frequencies. This means also that an actual signal can never contain accurate corners, since an infinitely wide frequency band cannot exist.

Waveforms like in Fig. C1a appear in practice frequently; for instance, in the currents of the output transistors in class A and AB amplifier stages. In design, it should be consequently minded that these transistors should be capable of producing at least tenfold frequencies compared to the highest frequency to be reproduced without distortion.

In the general case, both a_n and b_n ($n \geq 1$) differ from zero. Because $\sin(n\omega_0 t)$ and $\cos(n\omega_0 t)$ are always in 90° phase shift with each other, the total amplitude (A_n) at frequency $n\omega_0$ is obtained using Pythagoras' theorem: $A_n^2 = a_n^2 + b_n^2$. Figure C2 illustrates the example signal's amplitudes on a frequency scale, i.e. the amplitude spectrum. The quadratic mean, i.e. the RMS value, could also be used in the diagram, as is customary in spectrum analyzers.

The evenness or oddness of the harmonics has certain implications on the symmetry properties of the signal. If even harmonics are added to a sine wave, the positive and negative half-waves become mutually different. Adding only odd harmonics, instead, preserves the similarity of the half-waves.

C2 One-Time Signals

The spectrum obtained for a signal that is periodic, or regarded as such, only contains energy at discrete, equal-spaced frequencies. When the signal is transient-like or otherwise includes aperiodic components, its frequency distribution is characterized by a continuous function, that can be found using the *Fourier transform*. An ordinary music signal generally consists of both periodic components, or notes, and one-time bursts, called transients.

In the Fourier transform, a time function $x(t)$ that satisfies certain requirements is converted to a corresponding frequency domain function $X(\omega)$ in the following manner:

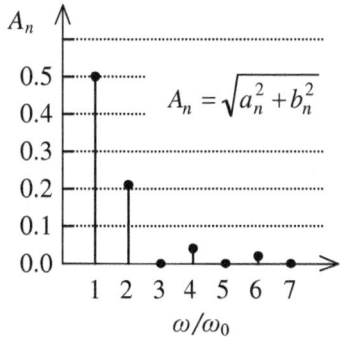

Figure C2. Amplitude spectrum of the periodic example signal of Fig. C1. Line length indicates the total amplitude A_n at each harmonic frequency. The representation is also called the line spectrum. (The DC component is not drawn because relating it to the sinusoids is not unambiguous.)

$$X(\omega) = \int_{-\infty}^{+\infty} x(t)e^{-j\omega t} dt \qquad (c6)$$

From the frequency domain representation, we get back to the time domain by the inverse Fourier transform:

$$x(t) = \frac{1}{2\pi} \int_{-\infty}^{+\infty} X(\omega)e^{j\omega t} d\omega \qquad (c7)$$

Here, $x(t)$ is expressed as a continuous sum, or integral, of complex sinusoids weighted by a specific function $X(\omega)$. Equation (c7) thus resembles, by its principle, the Fourier series (c1), where real sine waves are summed at discrete frequencies. (If notation $2\pi f$ is used, instead of ω, the coefficient $1/2\pi$ is canceled because then $d\omega = 2\pi df$.)

Signal $x(t)$ must be limited in time so that

$$\int_{-\infty}^{+\infty} |x(t)| dt < \infty \qquad (c8)$$

in order that formula (c6) would yield a finite result. The transform is thus not directly applicable to continuous-nature signals.

What, then, takes place in the Fourier transform? A complex sinusoid $\exp(-j\omega t)$ is weighted by the function $x(t)$ and integrated over all time. As a result, we obtain a complex number $X(\omega)$, that indicates how $x(t)$ correlates with the frequency in question. If $x(t)$ has nothing in common with frequency ω, the net result of the integration will be zero. The more $x(t)$ coincides with the referenced frequency, the greater the net result accumulated from the time integral.

A complex integral can also be conceived to consist of area slices, in the manner of real integrals. A complex area slice has, besides its magnitude, also a specified direction, that is perpendicular to the integration axis. In integral (c6), this direction angle is $-\omega t$ ($x(t)$ is real), so the area slices as though turn around the time axis, spiral-like. These slices are finally summed like vectors.

Transform (c6) is somewhat simplified if $x(t)$ is symmetric about instant 0. By writing $\exp(-j\omega t)$ in the trigonometric form and taking into account the symmetry properties of sin and cos, it can be seen that:

$$X(\omega) = \begin{cases} 2\int_{0}^{\infty} x(t)\cos(\omega t)dt, & x(-t) = x(t) \\ \\ -j2\int_{0}^{\infty} x(t)\sin(\omega t)dt, & x(-t) = -x(t) \end{cases} \qquad (c9)$$

When $x(-t) = x(t)$ (even symmetry), $X(\omega)$ is real; and when $x(-t) = -x(t)$ (odd symmetry), $X(\omega)$ is purely imaginary.

We will consider as an example the rectangular pulse shown in Fig. C3a, having height A and duration τ. Due to symmetry, we may apply formula (c9), yielding:

$$X(\omega) = 2\int_0^{\tau/2} A\cos(\omega t)\,dt$$

$$= 2A\left|_0^{\tau/2}\ \frac{\sin(\omega t)}{\omega}\right.$$

$$= \frac{2A}{\omega}\sin(\frac{\tau\omega}{2}), \quad \omega \neq 0 \tag{c10}$$

Since $\sin(\tau\omega/2) \to \tau\omega/2$ when $\omega \to 0$, we obtain as the result the amplitude spectrum, $|X(\omega)|$, shown in Fig. C3b. Most part of the pulse's frequency content is concentrated in the region $|\omega| < 2\pi/\tau$, but the spectrum continues, asymptotically decaying, to infinite frequencies.

Negative frequencies are also included in the Fourier transform, but in practice, $|X(\omega)|$ is equal at positive and negative frequencies since, when $x(t)$ is real, $X(\omega)$ and $X(-\omega)$ are complex conjugates of each other. However, when determining the energy or power of the signal, both po-

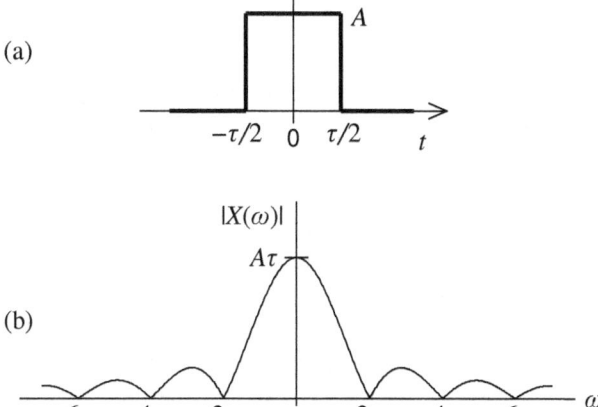

Figure C3. Single rectangular pulse (a) and its amplitude spectrum (b). (When necessary, the angle of $X(\omega)$ may be displayed in a separate phase spectrum.)

sitive and negative frequencies must be counted.

Expression (c10) is undefined at frequency 0, but it can readily be seen that $X(0) = A\tau$ (because $\cos(0) = 1$), and hence $X(\omega)$ is continuous also at frequency 0. Further, it follows from definition (c6) that $X(0)$ always equals the net area of $x(t)$ (because $\exp(0) = 1$).

$x(t)$ and its Fourier transform $X(\omega)$ are two alternate ways to describe the same one-time signal. Both representations contain the same information, but in a different form. Transform pairs derived for common functions can be found in general Fourier transform tables.

From Fig. C3 it can be reasoned that by increasing the duration of the pulse, τ, the frequency distribution becomes narrower, and vice versa. The same also applies generally for other than pulse-like signals as well. Scaling a signal on the time axis thus always causes an inverse scaling on the frequency axis.

The basis of frequency domain analysis is the simple relation that exists between the input and output signals of a linear system:

$$Y(\omega) = H(\omega) X(\omega) \tag{c11}$$

The Fourier spectrum of the output, $Y(\omega)$, is therefore obtained from the input spectrum $X(\omega)$ by multiplying it by the frequency response $H(\omega)$. Relation (c11) is analogous to (b20), that dealt with rotating phasors.

C3 Continuous-Nature Signals

It is reasonable to talk about the spectrum of continuous-nature signals only when the statistical properties of the signal stay sufficiently stable during the time the signal is to be examined. In other words, the signal should exhibit certain stationarity. In a non-stationary case (e.g. a fluctuating music signal), the spectrum only tells the average frequency behavior within the sample period.

Continuous signals are usually characterized by a *power spectral density function*, that, as the name implies, describes the distribution of signal power in the frequency domain. In these contexts, "power" does not necessarily mean the physical power, measured in watts, but the mathematical average of the squared signal, that is:

$$P_x = \frac{1}{T} \int_0^T x^2(t)\, dt \tag{c12}$$

where T is the length of the observation period.

Power spectral density is obtained as the Fourier transform of the signal's *autocorrelation function* (designated $R_x(\tau)$), which, in turn, is defined as the expected value of the product $x(t)x(t-\tau)$ as a function of time difference τ:

$$R_x(\tau) = \frac{1}{T}\int_0^T x(t)x(t-\tau)dt \ , \quad T \gg \tau \tag{c13}$$

The shape of autocorrelation functions is typically as shown in Fig. C4. $R_x(\tau)$ is always symmetric so that $R_x(-\tau) = R_x(\tau)$, and reaches its maximum at $\tau = 0$. This maximum value is P_x, as can be detected by comparing equations (c12) and (c13).

$R_x(\tau)$ approaches zero when $|\tau|$ increases, if $x(t)$ does not include DC or periodic components whose duration exceeds $|\tau|$. Then, $R_x(\tau)$ satisfies condition (c8), and the power spectral density (designated $G_x(f)$) is obtained by Fourier-transforming $R_x(\tau)$ according to formula (c9) as follows:

$$G_x(f) = 4\int_0^\infty R_x(\tau)\cos(2\pi f\tau)d\tau \ , \quad f \geq 0 \tag{c14}$$

The extraneous multiplication by 2 stems from the fact that frequency is here interpreted to be a positive quantity, whereupon the power spectral density contained at the negative frequencies, as produced by the Fourier transform, must be summed to the positive side. (In measurements, negative frequencies cannot be distinguished from positive ones, but e.g. in the inverse transform (c7), both are needed.)

When taken from the autocorrelation, $G_x(f)$ is always real-valued, and its unit will be the unit of $x(t)$ squared and divided by the unit of frequency. If $x(t)$ is, for example, a voltage, the unit of $G_x(f)$ is V^2/Hz.

The power spectral density corresponding to the autocorrelation of Fig. C4 might be like that shown in Fig. C5. Using the inverse transform, it can be shown that, with linear scales, the area left under the

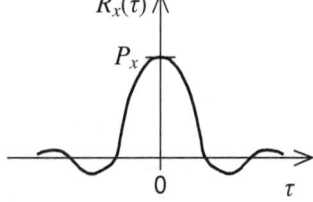

$R_x(\tau)$

P_x

0   τ

Figure C4. Example of a typical autocorrelation function. When time difference τ is zero, R_x reaches its highest value, which equals the mean square, or power (P_x), of the signal.

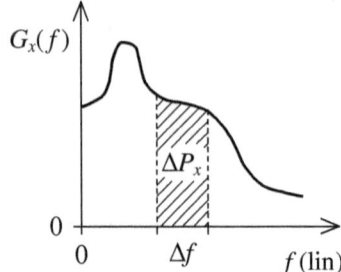

Figure C5. Example of a power spectral density curve. Power ΔP_x contained within range Δf equals the corresponding area under the curve.

$G_x(f)$ curve equals the total power of the signal, P_x. Consequently, the power ΔP_x contained within a frequency interval Δf can be evaluated as the product of Δf and its associated average spectral density, as illustrated by the shaded area in Fig. C5. The area interpretation of power applies directly only when f is the frequency variable. When using ω, the result must yet be divided by 2π.

It follows from the above that if $G_x(f)$ is constant, as in the case of so-called white noise, the power of the signal is directly proportional to the absolute bandwidth, in which case the power, when expressed on a logarithmic frequency scale, strongly concentrates at high frequencies. Hence, for example, frequency range 2-20 kHz would contain 10 times more power than range 200-2000 Hz and even 100 times more than range 20-200 Hz. Taking into account the actual power handling capacity of high frequency driver units, which is of the order of 5 watts, it would be quite problematic to play a program that has even spectrum. Luckily however, the spectrum of music signals resembles more so-called pink noise, where $G_x(f)$ is inversely proportional to frequency and where the power is equal in every frequency band of same relative width.

For one-time signals, the quantity corresponding to power spectral density is the *energy spectral density*, that is directly obtained by taking the square of the absolute value of the signal's Fourier transform, that is, $|X(f)|^2$ (Rayleigh's energy theorem). This, too, has to be multiplied by 2 when using only positive frequencies.

In a corresponding manner, the power spectral density of continuous-nature signals can also be estimated without the autocorrelation, by taking, from a signal sample, the energy spectral density and dividing it by the duration of the sample. Due to randomness of phase, the $G_x(f)$ estimate obtained in this way is, however, very inaccurate; and, in order to get a satisfactory result, several results have to be averaged. In addition, the cutting of samples always causes some spectral leakage, which can be minimized by using a suitably shaped time window.

In practice, the transforming of samples is generally carried out by so-called discrete Fourier transform, whose computationally efficient implementation is known as the "fast Fourier transform" (FFT).

Power spectral density can also be measured analogically on a fairly simple principle: The signal is led to a steep-edged narrow-band filter having bandwidth Δf and center frequency f_0. After the filter follows a squaring circuit, whose output signal is yet averaged by a low-pass filter. When the mean square thus obtained is divided by Δf, we have, as the result, the power spectral density at frequency f_0.

Power spectral density and energy spectral density are, by nature, quadratic quantities, that are modified in a linear system according to the square of the amplitude response. Therefore:

$$G_y(f) = |H(f)|^2 \, G_x(f) \qquad\qquad (c15)$$

where $G_x(f)$ and $G_y(f)$ are either power or energy spectral density functions, correspondingly.

Due to the above-described correspondence of power and area, it is evident that when summing together signals whose frequency bands do not overlap, the total power will be the sum of the individual powers.

The same also holds more generally:

In the summing of uncorrelated signals, the resulting power will be the sum of the individual powers.

Signals $x_1(t)$ and $x_2(t)$ are uncorrelated when the average of the product $x_1(t)x_2(t)$ is zero, and the above statement can be easily proved using definition (c12).

Signals originating from different sources are, in general, always uncorrelated. Thus, for instance, noise contributions caused by different noise sources are combined by adding together the mean square values. Sine waves having different frequencies are also uncorrelated. Instead, for example, the right and left channel signals of a stereo recording usually correlate strongly, at least at low frequencies.

If a signal includes continuous sinusoidal components, they appear in the power spectral density graph, in principle, as impulses, that have infinite height and zero width, and whose area corresponds to the power of the sinusoid ($=A^2/2$, where A is the amplitude). In practice, these impulses show up as peaks whose shape depends on the resolution used.

C4 The Relationship between Time and Frequency Domains

As already turned out in section C2, the spread of the signal in the frequency domain is inversely proportional to that in the time domain. Concerning Fourier transform, there are also many other primal regularities, that describe how certain operations in the time domain affect the spectrum, and vice versa; but these properties are not enumerated here.

Instead, we will consider the relationship between the frequency and time behaviors of a linear system. In appendix B, it was shown how the impulse response can be derived from the differential equation defining the system, as also the frequency response function. This suggests that there also exists a definite relationship between the impulse and frequency responses.

The commutative law applies to the convolution operation (b18), and hence the equation may be written as:

$$y(t) = \int_{-\infty}^{+\infty} h(\tau)x(t - \tau)d\tau$$

By using the complex sinusoid $\exp(j\omega t)$ as the input, we have:

$$y(t) = \int_{-\infty}^{+\infty} h(\tau)e^{j\omega(t-\tau)}d\tau$$

$$= e^{j\omega t}\int_{-\infty}^{+\infty} h(\tau)e^{-j\omega\tau}d\tau$$

By comparing this with the formerly obtained relation (b20), we observe that

$$H(\omega) = \int_{-\infty}^{+\infty} h(\tau)e^{-j\omega\tau}d\tau \qquad (c16)$$

Thus, the impulse response $h(t)$ and frequency response $H(\omega)$ constitute a Fourier transform pair, that is, the frequency response is in fact the spectrum of the impulse response. The result shows how directly the time and frequency properties of a system are linked to each other.

The frequency response, determining the sine wave behavior, and the impulse response, determining the transient behavior, both carry the same information, presented in different form; and one does not contain anything that is not contained

in the other also.

This being the case, if the system can be regarded as linear, all flaws possibly showing up in the transient behavior are always linked to some feature observed in the frequency response. ("Frequency response" also includes the phase.) For example, the characteristic vibration observable in the impulse response shown in Fig. B7 is related to the resonance peak seen in the amplitude response of Fig. B11.

By comparing the Fourier transform (c6) with its inverse transform (c7), one can see that they are very similar in form, as t and ω are interchanged. This dualism property makes it possible to easily transform functions that are the result of an earlier Fourier transform. It can namely be shown that if (arrows denoting the Fourier transform)

$$x(t) \to X(\omega)$$

then

$$X(t) \to 2\pi\, x(-\omega) \tag{c17}$$

Function X is real when function x is evenly symmetric, and in this case, even the minus sign can be omitted.

By applying rule (c17) to the transform (c10), we can derive a useful relationship between the frequency and impulse responses of an ideal filter. First, we obtain the transform pair:

$$X(t) = \frac{2A}{t}\sin(\frac{\tau}{2}t) \;\to\; 2\pi \cdot x(\omega) = \begin{cases} 2\pi A, & |\omega| \le \tau/2 \\ 0, & |\omega| > \tau/2 \end{cases}$$

which means that the spectrum of the time domain function $X(t)$, corresponding to (c10), is a rectangular function, corresponding to Fig. C3a, in the frequency domain. By dividing both sides by $2\pi A$ and designating $\tau/2 = B$ (= bandwidth), we can write:

$$\frac{1}{\pi t}\sin(Bt) \;\to\; \begin{cases} 1, & |\omega| \le B \\ 0, & |\omega| > B \end{cases} \tag{c18}$$

By interpreting the right side as a frequency response, the left side represents the corresponding impulse response. This is, however, non-zero also when t is negative, which means that the impulse response should begin already long before the impulse arrives. This is, of course, not possible, so the frequency response in question cannot be realized with any real-time (analogue) system. A similar conclusion also applies to other frequency responses that become zero above some limit frequency.

However, the realizability of the response improves conclusively if the system is allowed to contain enough group delay. By appending to the system, described by the transform pair (c18), a pure delay, T, we obtain a system for which:

$$h(t) = \frac{\sin(B(t-T))}{\pi(t-T)} \rightarrow H(\omega) = \begin{cases} e^{-j\omega T}, & |\omega| \le B \\ 0, & |\omega| > B \end{cases} \qquad (c19)$$

where $H(\omega)$ is obtained from equation (b29). Transform pair (c19) represents an ideal phase-linear low-pass filter, whose frequency response is depicted in Fig. C6a.

Figure C6b displays the filter's impulse response $h(t)$, which is identical in form with the spectrum derived earlier for the rectangular pulse, the peak occurring at instant T. Part of the response is still before instant 0, so the filter cannot be realized precisely. However, by increasing the delay T, we can get arbitrarily close to the abrupt model of Fig. 'a'.

In a corresponding manner, it can be shown that the impulse respon-

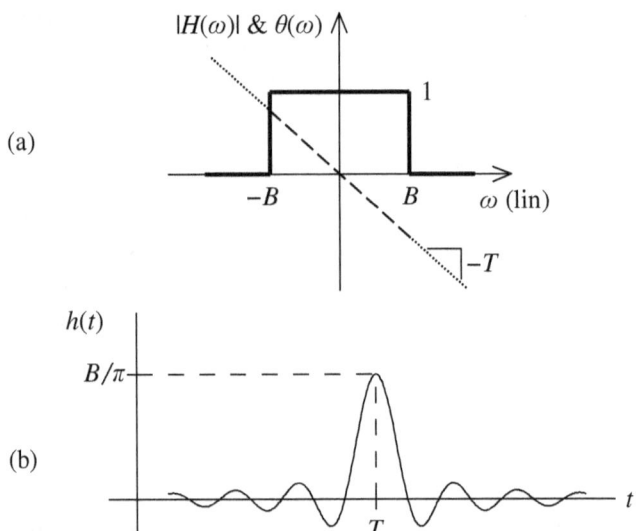

(a)

(b)

Figure C6. a) Amplitude and phase responses of an ideal low-pass filter. The slope of the phase $\theta(\omega)$ is $-T$. The frequency response is here two-sided, like Fourier transform results as a rule. b) Impulse response corresponding to the frequency response of Fig. 'a'. $h(t)$ undulates also at negative values of t, but only the portion after zero is realizable in real-time (causally).

se of an ideal band-pass filter is also like that in Fig. C6b, when the ratio of the upper and lower cut-off frequencies is high.

The employing of digital techniques does not alter these basic principles either, except that, by making use of memory, the delay T can be easily increased unendingly.

The operation of an ordinary CD player involves both digital and analogue filtering. When examining the impulse responses measured from CD players in various tests, one can observe that, at least in high quality gear, they are often similar to that in Fig. C6b. If the impulse response is, instead, asymmetric, there is reason to doubt the phase linearity of the apparatus.

The undulation seen in the Figure is a fully pertinent phenomenon, and there is no need to reduce it in any way. It does not introduce, in the passband, any "ringing" or other harms that would not show up in the frequency response. It can also be seen that the waves before and after the peak are in opposite phase and thus do not signify any extraneous resonation.

Generally, the steeper the filtering slope (also at the low end), the broader the corresponding impulse response. It is, however, erroneous to think that longer decay or undulations in the impulse response would themselves bring about any inappropriate behavior, for we never listen to impulses as such but what we really hear is the convolution of the impulse response with the applied signal, as given by equation (b18); and when this convolution process is performed, details of the impulse response are smoothed out.

C5 About Test Signals

Oftentimes, one can see circuits or equipment, intended for sound reproduction, being excited even with somewhat elaborate test signals, keeping as an ideal the perfect passing of the signal, without changes in shape or without any extraneous spectral components. Such a requirement is, however, many times unrealistic if the frequency content of the test signal has not been conformed to the usable frequency range of the unit under test. A continuous sine wave is the only test signal that is composed of only one frequency. All other signals always involve several distinct frequencies or energy continuously distributed over bands of frequency. We will thus consider the frequency distributions of some elementary test signals.

The spectrum of a unit impulse (see Fig. B6a) is obtained by applying the Fourier transform (c6):

$$\delta(t) \;\rightarrow\; \int_{-\infty}^{+\infty} \delta(t)e^{-j\omega t}\,dt \;=\; 1 \qquad\qquad (c20)$$

since the area of $\delta(t)$ is 1 and $\exp(0) = 1$. Therefore, the spectrum of an ideal impulse is a mere constant, that is, the frequency distribution of an infinitely brief signal extends flat to infinite frequencies, as is expectable. The result is consistent with the fact that the frequency response of a system is obtained by finding the spectrum of the impulse response, as seen from equation (c11) by substituting $X(\omega) = 1$.

By taking advantage of this basic property, one can measure loudspeaker frequency response also in domestic room conditions, without having to resort to the rare anechoic chambers or outdoor measurement. The length of the measured impulse response is, however, limited by the first reflection arriving at the microphone from room surfaces.

The unit step function (Fig. B6b) is special in the sense that it cannot be distinctly interpreted to be a one-time signal, nor truly continuous either. However, by leaving frequency 0 out of consideration, the spectrum is easily found, by making use of a rule which states that the Fourier transform of an integral function is the same as the transform of the integrand, divided by the factor $j\omega$.

Because the unit step (designated $\xi(t)$) is the integral function of the unit impulse, we obtain a transform pair:

$$\xi(t) \;\rightarrow\; \frac{1}{j\omega}, \qquad \omega \neq 0 \qquad\qquad (c21)$$

The Fourier spectrum of $\xi(t)$ is thus inversely proportional to frequency, and hence, as a whole, low frequencies are accentuated in a step signal. (This should not be confused with pink noise, where the *power* spectral density is inversely proportional to frequency.) It must be noted, however, that high frequencies concentrate here near instant 0, where the abrupt signal change occurs. Low frequencies, in turn, are needed to keep the signal value unchanged, as time increases.

The step response is the integral function of the impulse response (equation (b16)), so the step response may also exhibit corresponding undulations like e.g. in Fig. C6b, without anything being wrong in the system's phase behavior. However, wide back-and-forth swings in the step response usually signify the lack of phase linearity.

Square wave is a periodic test signal, that is often used when desiring to examine waveform modifications or the reacting of a system to steep changes; but only seldom it is realized how much more demanding it is to reproduce a square wave, compared with any audio signal occurring in reality.

A symmetric square wave, having period T, may be defined as

$$x(t) = \begin{cases} 1, & 0 < t \le T/2 \\ -1, & T/2 < t \le T \end{cases}$$

$x(t)$ can now be expressed by a Fourier series (c1), the coefficients a_n and b_n being easily solvable, resulting to:

$$a_n = 0 \; ; \qquad b_n = \begin{cases} 4/(\pi n), & n \text{ odd} \\ 0, & n \text{ even} \end{cases}$$

The Fourier series thus consists of sine terms representing odd harmonics, the amplitudes being inversely proportional to frequency. On the other hand, as frequency increases, every frequency decade comprises a 10-fold amount of harmonics compared to the previous, lower, decade; and this increases the weight of the upper decade by a factor of $\sqrt{10}$ (in terms of RMS values). The result is that the RMS value corresponding to each decade is as much as about $1/3$ of that of the previous decade. Hence, for example, in a square wave of 1 kHz, the RMS of the sum of harmonics contained in the range 100 kHz - 1MHz is still about 10% of the RMS of the sum of harmonics contained below 10 kHz; the latter approximately representing the RMS of the whole signal. Taking yet into account that the distortion of all amplifier circuits rapidly increases as frequency rises sufficiently, it is worth recognizing that:

When using square wave as a test signal or its component, it should be minded that a considerable part of the signal's frequency content may fall in a region that the device is not even intended to reproduce properly, and that these overtones can, when becoming distorted by various reasons, induce modulation products also in the assigned operation band.

We will yet consider the sine wave burst shown in Fig. C7a, truncated at both ends. The signal is oddly symmetric, and we may apply formula (c9):

$$X(\omega) = -j2 \int_0^{\tau/2} A \sin(\omega_0 t) \sin(\omega t) \, dt \qquad (\omega_0 = \frac{2\pi}{T})$$

from which we obtain, using relation (c5):

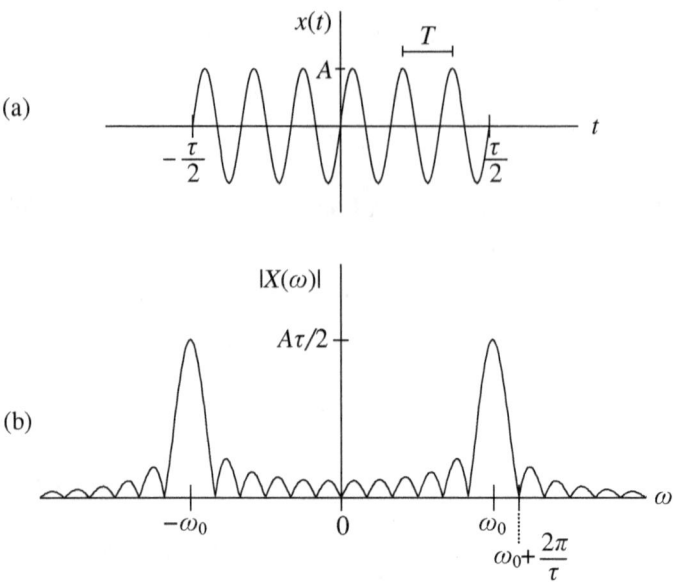

Figure C7. a) A burst of sine wave, having duration τ, amplitude A, and period T. b) Amplitude spectrum of the signal of Fig. 'a' ($\omega_0 = 2\pi/T$).

$$X(\omega) = -jA\left[\int_0^{\tau/2}\cos((\omega_0-\omega)t)\,dt - \int_0^{\tau/2}\cos((\omega_0+\omega)t)\,dt\right]$$

$$= -jA\left[\left|_0^{\tau/2}\frac{\sin((\omega_0-\omega)t)}{\omega_0-\omega} - \left|_0^{\tau/2}\frac{\sin((\omega_0+\omega)t)}{\omega_0+\omega}\right.\right]$$

$$= -jA\left[\frac{1}{\omega_0-\omega}\sin(\frac{\tau}{2}(\omega_0-\omega)) - \frac{1}{\omega_0+\omega}\sin(\frac{\tau}{2}(\omega_0+\omega))\right]$$

$$= jA\left[\frac{1}{\omega+\omega_0}\sin(\frac{\tau}{2}(\omega+\omega_0)) - \frac{1}{\omega-\omega_0}\sin(\frac{\tau}{2}(\omega-\omega_0))\right]$$

The amplitude spectrum $|X(\omega)|$ is now as in Fig. C7b. It consists of two halves, that partly overlap. The smaller the duration, τ, of the burst, the wider the dispersion of the frequency content on both sides of the actual sine frequency ω_0. The truncation of a signal thus introduces, in principle, infinite frequencies into the spectrum; and hence it is impossible for any band-limited system to reproduce this kind of break points unchanged.

The spectrum obtained resembles very much that of the rectangular pulse (expression (c10)). The primal peak has only moved to $\pm\omega_0$, and its height has reduced to half. In fact, this is a case of amplitude modulation in which the level of a sine wave has been modulated by the rectangular signal of Fig. C3a. The amplitude spectrum of the signal then shifts, in frequency, by the value of the carrier frequency ω_0. In terms of the sine wave, modulation brings about the spreading of a singular point frequency into a band whose width corresponds to the modulating signal.

---

[1] A. Bruce Carlson, "Communication Systems", McGraw-Hill, 1981, p. 27.

Appendix D - Solving HDEs

(Referred to in section B3)

When determining the impulse response from the differential equation of the system, in the manner shown in section B3, one has to solve a homogenous constant-coefficient differential equation, that generally is of the form:

$$a_n \frac{d^n y(t)}{dt^n} + a_{n-1} \frac{d^{n-1} y(t)}{dt^{n-1}} + \ldots + a_1 \frac{dy(t)}{dt} + y(t) = 0 \qquad (d1)$$

The solution $y(t)$ consists of n linearly independent solution functions, $y_i(t)$ $(i = 1, 2, \ldots, n)$, as follows:

$$y(t) = c_1 y_1(t) + c_2 y_2(t) + \ldots + c_n y_n(t) \qquad (d2)$$

where the coefficients $c_1 \ldots c_n$ are determined by the initial conditions set by the particular problem.

The functions $y_i(t)$ are determined according to the roots of so-called characteristic equation, that pertains to equation (d1):

$$f(r) = a_n r^n + a_{n-1} r^{n-1} + \ldots + a_1 r + 1 = 0 \qquad (d3)$$

We will limit ourselves here to the very general case in which these roots are all distinct. Then:

- Every real pole r introduces a solution function e^{rt}.

- Every complex pair of roots $r = a \pm jb$ introduces two solution functions: $e^{at}\cos(bt)$ and $e^{at}\sin(bt)$.

Thus, for example, if the roots of the characteristic equation are -10, -50, and $-20 \pm j100$, the solution of the homogenous equation (d1) will be:

$$y(t) = c_1 e^{-10t} + c_2 e^{-50t} + c_3 e^{-20t}\cos(100t) + c_4 e^{-20t}\sin(100t)$$

Appendix E - Decibel Scale

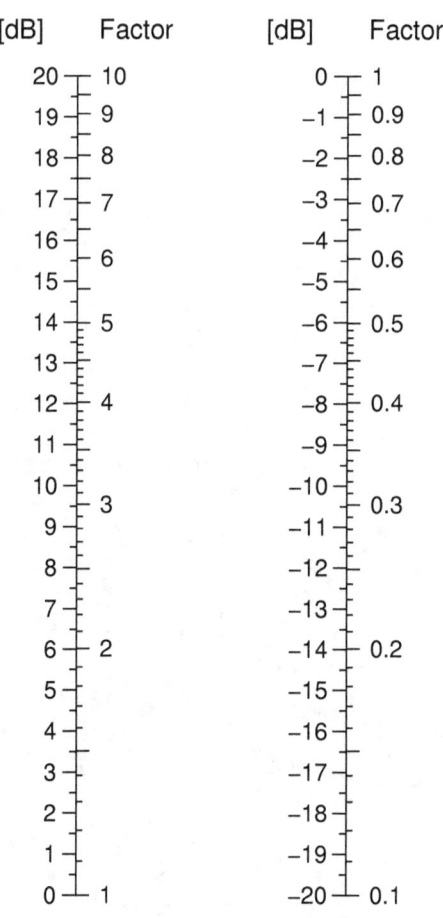

[dB]	Factor	[dB]	Factor
20	10	0	1
19	9	−1	0.9
18	8	−2	0.8
17	7	−3	0.7
16		−4	
	6		0.6
15		−5	
14	5	−6	0.5
13		−7	
12	4	−8	0.4
11		−9	
10		−10	
	3		0.3
9		−11	
8		−12	
7		−13	
6	2	−14	0.2
5		−15	
4		−16	
3		−17	
2		−18	
1		−19	
0	1	−20	0.1

Appendix F - Amplifier Artwork

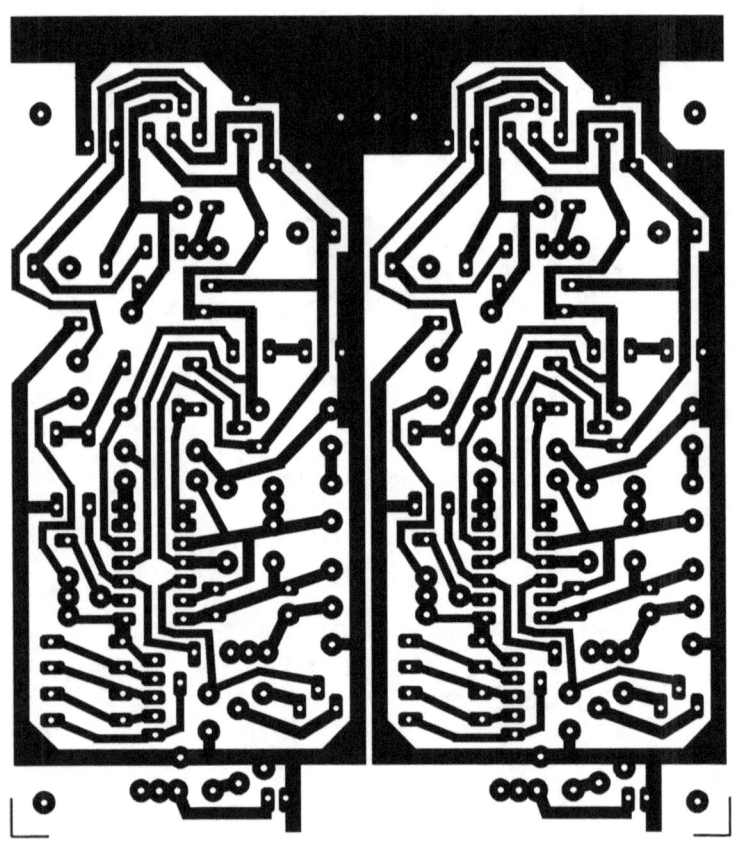

Appendix G - Power Supply Artwork

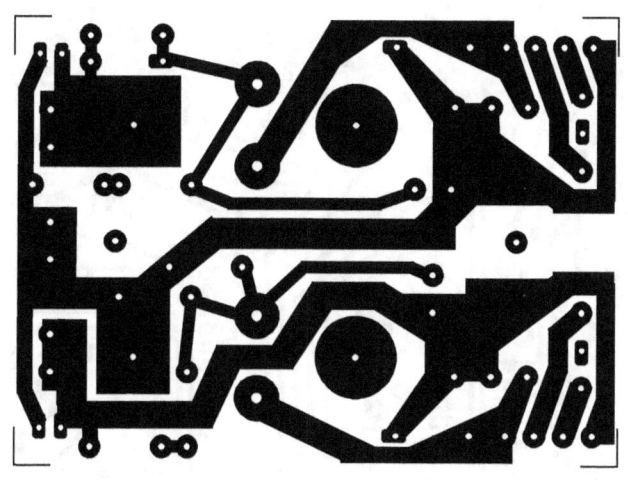

Appendix H - EMF Extractor Artwork

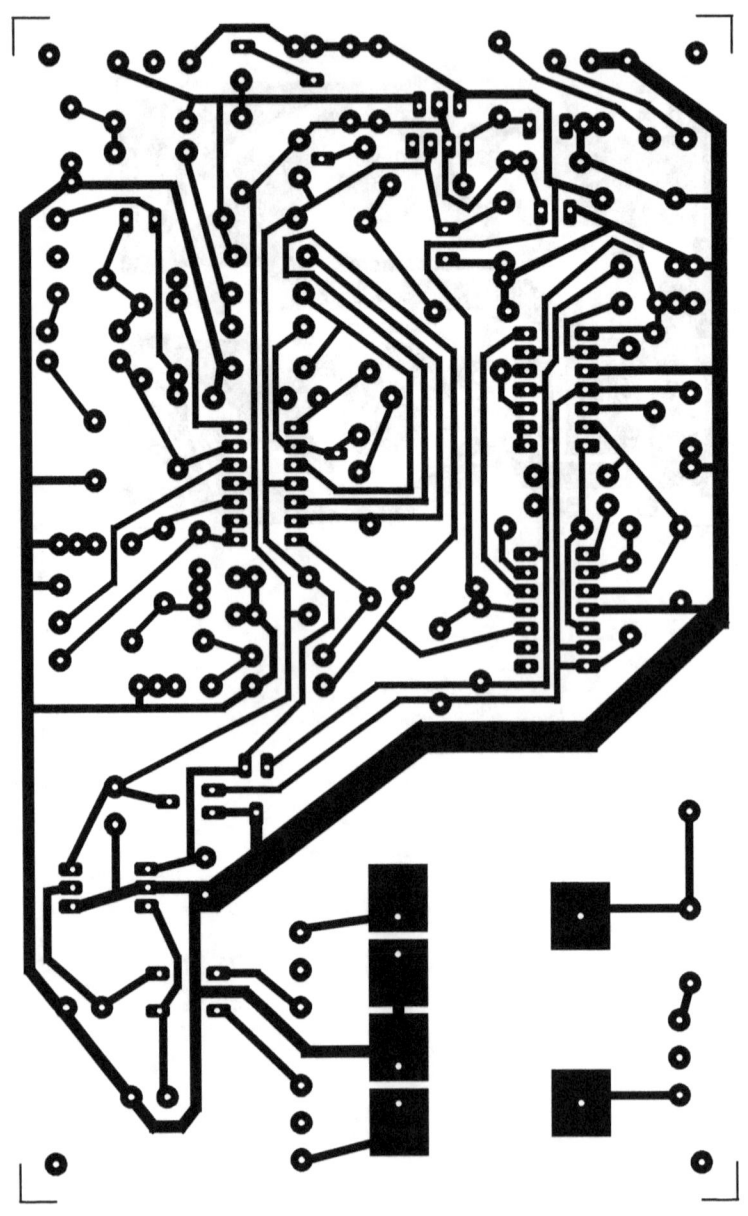